FUZZY SETS AND INTERACTIVE MULTIOBJECTIVE OPTIMIZATION

CONTROL OF COMPLEX SYSTEMS: Methods and Technology
M. Drouin, H. Abou-Kandil, and M. Mariton

CONTROL OF NONLINEAR MECHANICAL SYSTEMS
Janislaw M. Skowronski

FUZZY SETS AND INTERACTIVE MULTIOBJECTIVE OPTIMIZATION
Masatoshi Sakawa

KNOWLEDGE-BASED SYSTEM DIAGNOSIS, SUPERVISION, AND CONTROL
Edited by Spyros G. Tzafestas

PARALLEL PROCESSING TECHNIQUES FOR SIMULATION
Edited by M. G. Singh, A. Y. Allidina, and B. K. Daniels

ROBUSTNESS IN IDENTIFICATION AND CONTROL
Edited by M. Milanese, R. Tempo, and A. Vicino

PREFACE

The main characteristics of the real-world decision-making problems facing humans today are multidimensional and have multiple objectives including economic, environmental, social, and technical ones. Hence, it seems natural that the consideration of many objectives in the actual decision-making process requires multiobjective approaches rather than single-objective. One of the major systems-analytic multiobjective approaches to decision-making under constraints is multiobjective optimization as a generalization of traditional single-objective optimization. Although multiobjective optimization problems differ from single-objective optimization problems only in the plurality of objective functions, it is significant to realize that multiple objectives are often noncommensurable and conflict with each other in multiobjective optimization problems. With this observation, in multiobjective optimization, the notion of Pareto optimality or efficiency has been introduced instead of the optimality concept for single-objective optimization. However, decisions with Pareto optimality or efficiency are not uniquely determined; the final decision must be selected from among the set of Pareto optimal or efficient solutions. Therefore, the question is, how does one find the preferred point as a compromise or satisficing solution with rational procedure? This is the starting point of multiobjective optimization. To be more specific, the aim is to determine how one derives a compromise or satisficing solution of a decision maker (DM), which well represents the subjective judgments, from a Pareto optimal or an efficient solution set. An enormous number of articles together with several significant monographs and books have been published since the First International Conference on Multiple Criteria Decision Making was held at the University of South Carolina in 1972, and nowadays, multiobjective optimization is considered to make a major contribution to decision-making under certainty or some probabilistic settings.

However, recalling the imprecision or fuzziness inherent in human judgments, two types of inaccuracies of human judgments should be incorporated in multiobjective optimization problems. One is the experts' ambiguous understanding of the nature of the parameters in the problem-formulation process, and the other is the fuzzy goals of the DM for each of the objective functions. For handling and tackling such kinds of imprecision or vagueness in human beings, it is not hard to imagine that the conventional multiobjective optimization approaches, such as a deterministic or even a probabilistic approach, cannot be applied. The motivation for multiobjective optimization under imprecision or fuzziness comes from this observation. For this reason, multiobjective optimization under imprecision or fuzziness seems to be particularly promising and applicable for dealing with human-centered decision-making problems in most practical situations. Fortunately, due to the theory of fuzzy sets initiated by Zadeh, if we look at recent developments in multiobjective optimization under uncertainty and imprecision, we can see remarkable advances in the field of so-called fuzzy multiobjective optimization.

Although a number of books in the field of multiobjective optimization have already been published in recent years, they focus mainly on multiobjective optimization under certainty or some probabilistic settings. In spite of its urgent necessity, there seems to be no book which is designed to incorporate both types of fuzziness of human judgments into multiobjective optimization in a unified way.

In this book, the author is concerned with not only presenting a unified presentation of some of the most important methods for both multiobjective optimization and fuzzy multiobjective optimization, but also introducing the latest advances in the new field of interactive multiobjective optimization under fuzziness together with a wide range of actual applications on the basis of the author's continuing research works. Special stress is placed on interactive decision-making aspects of fuzzy multiobjective optimization for human-centered systems in most realistic situations under fuzziness.

In addition to presenting the several interactive methods as well as basic notions, this book contains the interactive computer programs for almost all of the linear programming-based interactive methods introduced in the book with numerical examples. These programs are written in C language and compiled by MS-C (Microsoft C compiler) Ver. 6.0 for IBM PCs. The programs are provided on floppy diskettes for implementation on IBM PCs.

The intended readers of this book are senior undergraduate students, graduate students, and specialists in systems analysis. This includes scientists and executives who are mainly interested in decision-making problems with several conflicting objectives under fuzziness, such as public decision-making, administrative planning, and managerial decision-making, together with practitioners for social engineering in any field. In order to master all the material discussed in this book, the readers would probably be required to have some background in linear algebra and mathematical programming. However, by skipping the mathematical details together with using the interactive computer programs, much can be learned about interactive decision-making methods for human-centered systems in most realistic settings without prior mathematical sophistication.

This book was written while the author was an honorary visiting professor at the University of Manchester Institute of Science and Technology (UMIST), Computation Department, sponsored by the Japan Society for Promotion of Science (JSPS) from March to December 1991 on leave from Hiroshima University. The author would like to express his sincere appreciation to Professor Madan G. Singh of UMIST, whose arrangements and warm encouragement during his stay in Manchester made it possible for him to write this book. Special thanks should be extended to Professor Yoshikazu Sawaragi, chairman of the Japan Institute of Systems Research and emeritus professor of Kyoto University, Department of Applied Mathematics and Physics, for his invariant stimulus and encouragement ever since the author's student days at Kyoto University. The author also wishes to thank all of his undergraduate and graduate students at Kobe University, Iwate University, and Hiroshima University. Special thanks go to his former graduate students Hitoshi Yano, Ichiro Nishizaki, Toru Yumine, Yuji Nango of Kobe University, and Junya Takahashi of Iwate University for their invaluable assistance through discussions and computer works. The author would further like to thank Mr. Colin Lloyd, a teacher at Ducie High School, Manchester, who kindly read the manuscript and made corrections. Finally, the author would like to thank Dr. Ken Derham and John Matzka, Plenum Publishing Corporation, London and New York, for their assistance in the publication of this book.

Masatoshi Sakawa

Manchester

CONTENTS

CHAPTER 1

INTRODUCTION

1.1 Introduction and historical remarks

Since the First International Conference on Multiple Criteria Decision Making, held at the University of South Carolina in 1972 (Cochrane and Zeleny (eds.) 1973), it has been increasingly recognized that most of the real-world decision-making problems usually involve multiple, noncommensurable, and conflicting objectives which should be considered simultaneously. One of the major systems-analytic multiobjective approaches to decision-making under constraints is multiobjective optimization as a generalization of traditional single-objective optimization. For such multiobjective optimization problems, it is significant to realize that multiple objectives are often noncommensurable and cannot be combined into a single objective. Moreover, the objectives usually conflict with each other and any improvement of one objective can be achieved only at the expense of another. With this observation, in multiobjective optimization, the notion of Pareto optimality or efficiency has been introduced instead of the optimality concept for single-objective optimization. However, decisions with Pareto optimality or efficiency are not uniquely determined; the final decision must be selected from among the set of Pareto optimal or efficient solutions. Consequently, the aim in solving multiobjective optimization problems is to derive a compromise or satisficing† solution of a decision maker (DM) which is also Pareto optimal based on subjective value judgments (see, for example, Chankong and Haimes 1983a; Cohon 1978; Goicoechea, Hansen and Duckstein 1982; Grauer, Lewandowski, and Wierzbicki (eds.) 1982; Grauer and Wierzbicki (eds.) 1984; Haimes, Hall, and Freedman 1975; Hwang and Masud 1979; Nijkamp 1979; Seo and Sakawa 1988; Steuer 1986; Zeleny 1982).

Two approaches for the determination of a compromise or satisficing solution for a DM in multiobjective optimization problems have been developed. They are

(1) goal programming approaches,

(2) interactive programming approaches.

The goal programming approaches, which assume that the DM can specify the goals of the objective functions, first appeared in a 1961 text by Charnes and

† "Satisficing" is a term or concept defined by J.G. March and H.A. Simon (1958). An alternative is satisficing if: (1) there exists a set of criteria that describes minimally satisfactory alternatives, and (2) the alternative in question meets or exceeds all these criteria.

Cooper (1961) to deal with multiobjective linear programming (MOLP) problems. Subsequent works on goal programming approaches have been numerous including Charnes and Cooper (1977), Ignizio (1976, 1982, 1983), and Lee (1972).

The interactive programming approaches, which assume that the DM is able to give some preference information on a local level to a particular solution, were first initiated by Geoffrion et al. (1972) and further developed by many researchers such as Chankong and Haimes (1983a), Choo and Atkins (1980), Korhonen and Laakso (1985), Musselman and Talavage (1980), Oppenheimer (1978), Sakawa (1981, 1982), Sakawa and Mori (1983a, 1983b), Sakawa and Seo (1980, 1982, 1983), Sakawa and Yano (1984), Steuer (1977), Steuer and Choo (1983),Wierzbicki (1979, 1980), and Zionts and Wallenius (1976, 1983).

The interactive goal programming method proposed by Dyer (1972) is a first attempt to provide a link between goal programming and interactive approaches. Since then, several goal programming-based interactive methods which combine the attractive features from both goal programming and interactive approaches have been proposed (e.g. Masud and Hwang 1981; Monarchi, Kisiel, and Duckstein 1973; Weistroffer 1982, 1983, 1984).

However, considering the imprecise nature of the DM's judgments in multiobjective optimization problems, fuzzy programming approaches (e.g. Carlsson 1982; Dubois 1987; Inuiguchi, Ichihashi, and Tanaka 1990; Kacprzyk and Orlovski (eds.) 1987; Kickert 1978; Luhandlula 1989; Slowinski 1986; Slowinski and Teghem (eds.) 1990; Zimmermann 1978, 1983, 1987, 1991; Zimmermann, Gaines, and Zadeh (eds.) 1984; Verdegay and Delgado (eds.) 1990) seem to be very applicable and promising for solving multiobjective optimization problems.

The concept of fuzzy mathematical programming on a general level was first proposed by Tanaka et al. (1974) in the framework of the fuzzy decision of Bellman and Zadeh (1970). Since then, fuzzy mathematical programming has been developed not only on a general level but also on a more practical level. Specifically, a relatively practical introduction of fuzzy set theory (Zadeh 1965) into conventional multiobjective linear programming problems was first presented by Zimmermann (1978) and further studied by Leberling (1981) and Hannan (1981). Following the fuzzy decision or the minimum operator proposed by Bellman and Zadeh (1970) together with linear, hyperbolic, or piecewise linear membership functions respectively, they proved that there exist equivalent linear programming problems. However, suppose that the interaction with the DM establishes that the first membership function should be linear, the second hyperbolic, the third piecewise linear and so forth. In such a situation, the resulting problem becomes a nonlinear programming problem and cannot be solved by a linear programming technique.

In order to overcome such difficulties, Sakawa (1983) has proposed a new method using a combination of the bisection method and the linear programming method together with five types of membership functions: linear, exponential, hyperbolic, hyperbolic inverse, and piecewise linear functions. This method was further extended for solving multiobjective linear fractional (Sakawa and Yumine

1983) and nonlinear programming problems (Sakawa 1984b). A brief survey of major approaches to so-called fuzzy mathematical programming proposed before 1984 can be found in the paper of Slowinski (1986) together with the proposed method for solving a water supply planning problem under fuzziness. More comprehensive surveys of the major fuzzy programming approaches proposed through the mid-1980s can also be found in Dubois (1987) and Kacprzyk and Orlovski (1987).

In these fuzzy approaches, however, it has been implicitly assumed that the fuzzy decision or the minimum operator of Bellman and Zadeh (1970) is the proper representation of the DM's fuzzy preferences. Therefore, these approaches are preferable only when the DM feels that the fuzzy decision or the minimum operator is appropriate when combining the fuzzy goals and/or constraints. However, such situations seem to occur rarely, and consequently, it becomes evident that an interaction with the DM is necessary.

Under these circumstances, assuming that the DM has a fuzzy goal for each of the objective functions in multiobjective linear, linear fractional, and nonlinear programming problems, several interactive fuzzy decision-making methods have been proposed by incorporating the desirable features of both the goal programming methods and the interactive approaches into the fuzzy approaches (Sakawa 1984c, 1986; Sakawa and Yano 1984a, 1984b, 1985b, 1986a, 1986b, 1986f, 1988b; Sakawa, Yano, and Yumine, 1987; Sakawa and Yumine 1983; Sakawa, Yumine, and Yano 1984, 1987).

However, when formulating a multiobjective programming problem which closely describes and represents the real decision situation, various factors of the real system should be reflected in the description of the objective functions and constraints. Naturally, these objective functions and constraints involve many parameters whose possible values may be assigned by the experts. In the traditional approaches, such parameters are fixed at some values in an experimental and/or subjective manner through the experts' understanding of the nature of the parameters.

In most practical situations, however, it is natural to consider that the possible values of these parameters are often only ambiguously known to the experts. In this case, it may be more appropriate to interpret the experts' understanding of the parameters as fuzzy numerical data which can be represented by means of fuzzy subsets of the real line known as fuzzy numbers (Dubois and Prade 1978, 1980b). The resulting multiobjective programming problem involving fuzzy parameters would be viewed as the more realistic version of the conventional one.

From this point of view, Orlovski (1983, 1984) first formulated general multiobjective nonlinear programming problems with fuzzy parameters. He presented two approaches to the formulated problems by making systematic use of the extension principle of Zadeh (1965, 1975) and demonstrated that there exist, in some sense, equivalent nonfuzzy formulations.

Along a similar line of fuzzy linear programming proposed thus far, without the explicit interpretation of fuzzy parameters as was pointed out by Orlovski

(1983, 1984), Tanaka and Asai (1984a, 1984b) formulated the multiobjective linear programming problems with fuzzy parameters. Following the fuzzy decision or the minimum operator proposed by Bellman and Zadeh (1970) together with triangular membership functions for fuzzy parameters, they considered two types of fuzzy multiobjective linear programming problems; one was to decide the nonfuzzy solution and the other was to decide the fuzzy solution.

Recently, in order to deal with the multiobjective linear, linear fractional, and nonlinear programming problems with fuzzy parameters characterized by fuzzy numbers, Sakawa and Yano (1985c, 1986c, 1986d, 1986e, 1986g, 1986h, 1987a, 1987b, 1987c, 1987d, 1988a, 1989a, 1989b, 1990b, 1990g) introduced the concept of α-multiobjective programming and (M-) α-Pareto optimality based on the α-level sets of the fuzzy numbers. They then presented several interactive decision-making methods, both without and with the fuzzy goals of the DM, to derive the satisficing solution of the DM efficiently from among an (M-) α-Pareto optimal solution set for multiobjective linear, linear fractional, and nonlinear programming problems as a generalization of their previous results. Recent excellent survey papers of Luhandlula (1989) and Inuiguchi et al. (1990) are devoted to reviewing and classifying the numerous major papers in the area of so-called fuzzy mathematical programming.

Finally, it is appropriate to mention some areas of application for multiobjective and/or fuzzy multiobjective programming approaches. Although most of the early practical applications have been accomplished in the areas of water resources planning (e.g. Haimes 1977, Cohon 1978, Slowinski 1986), regional planning (e.g. Rietveld 1980; Czyzak 1990), and environmental planning (e.g. Sommer and Pollatschek 1978; Nijkamp 1979; Sakawa and Seo 1980, 1982, 1983; Sakawa and Yano 1985d), many other real-world problems are inherently multiobjective. As we look at recent engineering, industrial, and management applications of the multiobjective and/or fuzzy multiobjective approaches, we can see continuing advances. They can be found, for example, in the areas of optimal design of shallow arches (e.g. Stadler 1983a, 1983b), electronic circuit design (e.g. Lightner 1979), media selection in advertising (Wiedey and Zimmermann 1978), operation of a packaging system in automated warehouses (e.g. Nakayama et al. 1980; Sakawa 1983; Sakawa, Yano, and Yumine 1987), management of the erection of a cablestayed bridge (Ishido, Nakayama, Furukawa, Inoue, and Tanikawa 1986) and, pass scheduling for hot tandem mills (Sakawa, Narazaki, Konishi, Nose, and Morita 1986; Sakawa, Narazaki, Nose, and Konishi 1987).

1.2 Organization of the book

In multiobjective optimization problems, multiple objectives are often noncommensurable and conflict with each other. Consequently, the aim is to find a compromise or satisficing solution for a decision maker (DM) from a Pareto optimal solution set on the basis of subjective value judgments.

However, considering the imprecise or fuzzy nature of human judgments, a fuzzy set approach seems to be very applicable and promising for multiobjective optimization problems under fuzziness. Two types of fuzziness of human judgments should be incorporated in multiobjective optimization problems. One is the experts' ambiguous understanding of the nature of the parameters in the problem-formulation process; the other is the fuzzy goals of the DM for each of the objective functions.

Organization of each chapter is briefly summarized as follows.

Chapter 2 reviews the fundamentals of the basic fuzzy set theory initiated by Zadeh, which will be used throughout the remainder of this book. Starting with several basic definitions involving fuzzy sets, Zadeh's extension principle is presented. It provides a general method for extending nonfuzzy mathematical concepts to the fuzzy framework. With the extension principle, operations on fuzzy sets, especially fuzzy numbers, are systematically developed. Bellman and Zadeh's approach to decision-making in a fuzzy environment, called fuzzy decision, is then examined in detail.

In Chapter 3, linear programming is briefly reviewed and the fuzzy linear programming approach proposed by Zimmermann is presented. Fundamental notions and methods of multiobjective linear programming are reviewed and fuzzy multiobjective linear programming is also introduced. Finally, interactive fuzzy multiobjective linear programming is explained in detail.

Chapter 4 can be viewed as the nonlinear version of Chapter 3 and is mainly concerned with interactive fuzzy multiobjective nonlinear programming as well as fuzzy nonlinear programming and fuzzy multiobjective nonlinear programming.

In Chapter 5, in contrast to the multiobjective linear programming problems discussed thus far, by considering the experts' imprecise or fuzzy understanding of the nature of the parameters in the problem-formulation process, the multiobjective linear programming problems involving fuzzy parameters are formulated. Through the introduction of extended Pareto optimality concepts, linear programming-based interactive decision-making methods, both without and with the fuzzy goals of the DM, for deriving a satisficing solution of the DM from among the extended Pareto optimal solution set are presented together with detailed numerical examples.

Chapter 6 is devoted to a nonlinear generalization along the same lines as Chapter 5. Through the use of nonlinear programming, considerable effort is devoted to the development of some refined interactive decision-making methods for multiobjective nonlinear programming problems with fuzzy parameters.

Chapter 7 presents interactive computer programs, which have been developed by the author's group based on the methods introduced in Chapters 3, 4, 5, and 6, in order to facilitate the interaction processes. Moreover, it demonstrates the feasibility and efficiency of both the methods and the corresponding computer programs. Interaction processes for several numerical examples for multiobjective linear and nonlinear programming problems, both without and with fuzzy parameters, are shown under the hypothetical decision maker together with the corresponding computer outputs.

Chapter 8 is concerned with some application aspects of interactive fuzzy optimization. As examples of Japanese case studies, interactive fuzzy optimization methods presented in this book are applied to the operation of a packaging system in automated warehouses, pass scheduling for hot tandem mills, and environmental planning.

Finally, Chapter 9 outlines related topics including multiobjective possibilistic programming, multiobjective programming with fuzzy decision variables and fuzzy parameters, fuzzy multiobjective 0-1 programming, and fuzzy regression analysis.

The Appendix presents generalized scalarizing methods for multiobjective optimization problems, called the hyperplane methods, by putting the special emphasis not only on generating Pareto optimal solutions but also on obtaining trade-off information. The results presented in the Appendix are the theoretical basis for the trade-off information used in Chapters 3, 4, 5, and 6.

CHAPTER 2

FUNDAMENTALS OF FUZZY SET THEORY

In 1965, L.A. Zadeh published his famous paper "Fuzzy sets" in Information and Control providing a new mathematical tool which enables us to describe and handle vague or ambiguous notions such as "a set of all real numbers which are much greater than 1," "a set of beautiful women," or "the set of tall men." Since then, fuzzy set theory has been rapidly developed by Zadeh himself and numerous researchers, and an increasing number of successful real applications of this theory in a wide variety of unexpected fields has been appearing. The main idea of fuzzy set theory is quite intuitive and natural: Instead of determining the exact boundaries as in an ordinary set, a fuzzy set allows no sharply defined boundaries because of a generalization of a characteristic function to a membership function.

This chapter reviews the fundamentals of basic fuzzy set theory initiated by Zadeh, which will be used throughout the remainder of this book. Starting with several basic definitions involving fuzzy sets, the extension principle of Zadeh is presented. It provides a general method for extending nonfuzzy mathematical concepts to the fuzzy framework. With the extension principle, operations on fuzzy sets, especially fuzzy numbers, are systematically developed. Bellman and Zadeh's approach to decision-making in a fuzzy environment, in which the fuzzy objective functions as well as the fuzzy constraints are characterized by their membership functions, is then discussed.

2.1 Fuzzy sets

In general, a fuzzy set initiated by Zadeh (1965) is defined as follows:

Definition 2.1 (Fuzzy sets)

Let X denote a universal set. Then a fuzzy subset $\tilde{A}$ of X is defined by its membership function

$$\mu_{\tilde{A}} : X \to [0,1], \tag{2.1}$$

which assigns to each element $x \in X$ a real number $\mu_{\tilde{A}}(x)$ in the interval $[0,1]$, where the value of $\mu_{\tilde{A}}(x)$ at x represents the grade of membership of x in $\tilde{A}$. Thus, the nearer the value of $\mu_{\tilde{A}}(x)$ is unity, the higher the grade of membership of x in $\tilde{A}$.

A fuzzy subset $\tilde{A}$ can be characterized as a set of ordered pairs of element x and grade $\mu_{\tilde{A}}(x)$ and is often written

$$\tilde{A} = \{(x, \mu_{\tilde{A}}(x)) \mid x \in X\}. \tag{2.2}$$

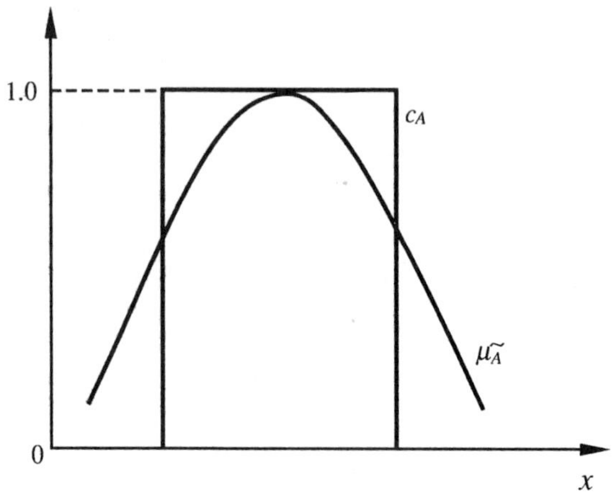

Figure 2.1 Membership function and characteristic function

When the membership function $\mu_{\tilde{A}}(x)$ contains only the two points 0 and 1, then $\mu_{\tilde{A}}(x)$ is identical to the characteristic function $c_A : X \to \{0, 1\}$, and hence, $\tilde{A}$ is no longer a fuzzy subset, but an ordinary set A.

As is well known, an ordinary set A is expressed as

$$A = \{x \in X \mid c_A(x) = 1\}, \tag{2.3}$$

through its characteristic function

$$c_A(x) = \begin{cases} 1 & x \in A \\ 0 & x \notin A. \end{cases} \tag{2.4}$$

Figure 2.1 illustrates the membership function $\mu_{\tilde{A}}(x)$ of a fuzzy subset $\tilde{A}$ together with the characteristic function $c_A(x)$ of an ordinary set A.

Observe that the membership function is an obvious extension of the idea of a characteristic function of an ordinary set because it takes on values between 0 and 1, not only 0 and 1.

As can be easily understood from the definition, a fuzzy subset is always defined as a subset of a universal set X. For the sake of convenience, a fuzzy subset is usually called a fuzzy set by omitting the term "sub." To distinguish an ordinary set from a fuzzy set, an ordinary set is called a nonfuzzy set or a crisp set. A fuzzy set is often denoted by $\tilde{A}$, $\tilde{B}$, $\tilde{C}$, ... , but it is sometimes written as $A, B, C, \ldots$ for simplicity in the notation.

Example 2.1 (the young)

Suppose that ages are denoted by a numerical-valued variable which ranges over the interval $X = [0, \infty)$. Then the set of ages less than or equal to 20 is obviously a crisp set. However, the set of "young ages" has no sharply defined

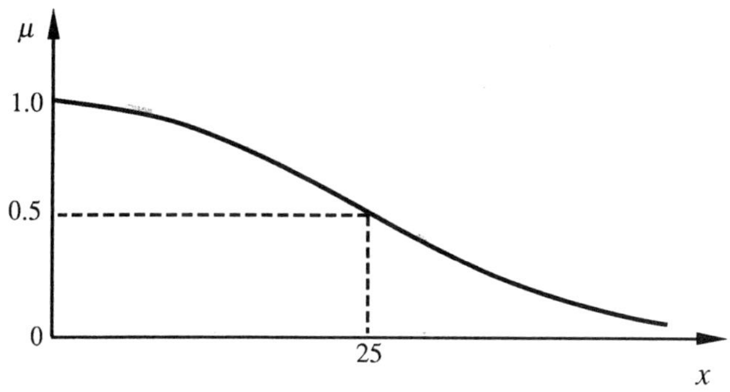

Figure 2.2 Membership function for young ages

boundaries and can be interpreted as a fuzzy set A of X. It may be subjectively characterized, for example, by the following membership function:

$$\mu_A(x) = (1 + (0.04x)^2)^{-1}. \tag{2.5}$$

This membership function $\mu_A(x)$ is illustrated in Figure 2.2. In this case, the degree to which a numerical age, say $x = 25$, is compatible with the concept young is 0.5.

Example 2.2 (real numbers considerably larger than 10)

In the real line $R^1 = (-\infty, \infty)$, the set of real numbers larger than 10 is obviously a crisp set. However, the set of "real numbers considerably larger than 10" has no sharply defined boundaries and can be interpreted as a fuzzy set B of the real line. It may be subjectively expressed, for example, by the following membership function:

$$\mu_B(x) = \begin{cases} 0 & ; x \leq 10 \\ 1 - (1 + (0.1(x - 10))^2)^{-1} & ; x > 10. \end{cases} \tag{2.6}$$

This membership function $\mu_B(x)$ is illustrated in Figure 2.3.

When X is a finite set whose elements are $x_1, x_2, \ldots, x_n$, i.e.,

$$X = \{x_1, x_2, \ldots, x_n\}, \tag{2.7}$$

a fuzzy set A on X is expressed as

$$A = \{(x_1, \mu_A(x_1)), (x_2, \mu_A(x_2)), \ldots, (x_n, \mu_A(x_n))\}. \tag{2.8}$$

In this expression, for simplicity, the pairs with $\mu_A(x) = 0$ are usually omitted.

According to the notation proposed by Zadeh (1965), this fuzzy set A on X is often written as

$$A = \mu_A(x_1)/x_1 + \mu_A(x_2)/x_2 + \cdots + \mu_A(x_n)/x_n \tag{2.9}$$

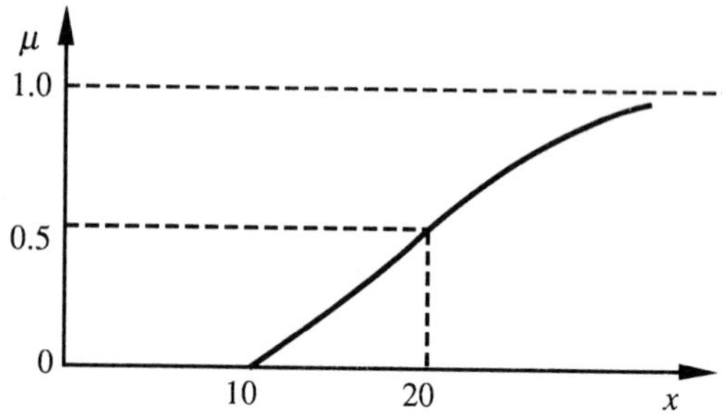

Figure 2.3 Membership function for real numbers considerably larger than 10

or more simply

$$A = \sum_{i=1}^{n} \mu_A(x_i)/x_i. \tag{2.9$'$}$$

This expression means that the grade of x_1 is $\mu_A(x_1)$, the grade of x_2 is $\mu_A(x_2), \ldots,$ the grade of x_n is $\mu_A(x_n)$, and the operations "+" and "$\sum$" do not refer to the ordinary addition but the set-theoretic "or."

When X is infinite, in addition to the notation of (2.2), a fuzzy set A is frequently written as

$$A = \int_X \mu_A(x)/x. \tag{2.10}$$

Here, the integral $\int$ can be viewed as a natural extension of $\sum$ in (2.9)$'$.

Example 2.3 (integers approximately equal to 5)
Let $X = \{1, 2, 3, 4, 5, 6, 7, 8, 9\}$ and consider the fuzzy set A expressed as integers approximately equal to 5. Such a fuzzy set A may be subjectively defined by

$$A = \{(3, 0.4), (4, 0.8), (5, 1), (6, 0.8), (7, 0.4)\}$$

or

$$A = 0.4/3 + 0.8/4 + 1/5 + 0.8/6 + 0.4/7.$$

The following basic notions are defined for fuzzy sets.

(1) Support: The support of a fuzzy set A on X, denoted by supp(A), is the set of points in X at which $\mu_A(x)$ is positive, i.e.,

$$\text{supp}\,(A) = \{x \in X \mid \mu_A(x) > 0\}. \tag{2.11}$$

(2) Height: The height of a fuzzy set A on X, denoted by hgt(A), is the least upper bound of $\mu_A(x)$, i.e.,

$$\text{hgt}\,(A) = \sup_{x \in X} \mu_A(x). \tag{2.12}$$

(3) Normal: A fuzzy set A on X is said to be normal if its height is unity, i.e., if there is $x \in X$ such that $\mu_A(x) = 1$. If it is not normal, a fuzzy set is said to be subnormal.

(4) Empty: A fuzzy set A on X is empty, denoted by $\emptyset$, if and only if $\mu_A(x) = 0$ for all $x \in X$. Obviously, the universal set X can be viewed as a fuzzy set whose membership function is $\mu_X(x) = 1$ for all $x \in X$.

Observe that a nonempty subnormal fuzzy set A can be normalized by dividing $\mu_A(x)$ by its hgt (A).

Several set-theoretic operations involving fuzzy sets originally proposed by Zadeh (1965) are as follows:

(1) Equality: The fuzzy sets A and B on X are equal, denoted by $A = B$, if and only if their membership functions are equal everywhere on X:

$$A = B \Leftrightarrow \mu_A(x) = \mu_B(x) \text{ for all } x \in X. \tag{2.13}$$

(2) Containment: The fuzzy set A is contained in B (or a subset of B), denoted by $A \subseteq B$, if and only if their membership function is less or equal to that of B everywhere on X:

$$A \subseteq B \Leftrightarrow \mu_A(x) \leq \mu_B(x) \text{ for all } x \in X. \tag{2.14}$$

(3) Complementation: The complement of a fuzzy set A on X, denoted by $\bar{A}$, is defined by

$$\mu_{\bar{A}}(x) = 1 - \mu_A(x) \text{ for all } x \in X. \tag{2.15}$$

(4) Intersection: The intersection of two fuzzy sets A and B on X, denoted by $A \cap B$, is defined by

$$\mu_{A \cap B}(x) = \min\{\mu_A(x), \mu_B(x)\} \text{ for all } x \in X. \tag{2.16}$$

(5) Union: The Union of two fuzzy sets A and B on X, denoted by $A \cup B$, is defined by

$$\mu_{A \cup B}(x) = \max\{\mu_A(x), \mu_B(x)\} \text{ for all } x \in X. \tag{2.17}$$

Observe that the intersection $A \cap B$ is the largest fuzzy set which is contained in both A and B and the union $A \cup B$ is the smallest fuzzy set containing both A and B since any fuzzy set C such that $C \subseteq A$, $C \subseteq B$ satisfies $C \subseteq A \cap B$ and any fuzzy set D such that $D \supseteq A$, $D \supseteq B$ satisfies $D \supseteq A \cup B$.

The intersection and the union of two fuzzy sets A and B, and the complement of a fuzzy set A are illustrated in Figure 2.4, from which it can be easily

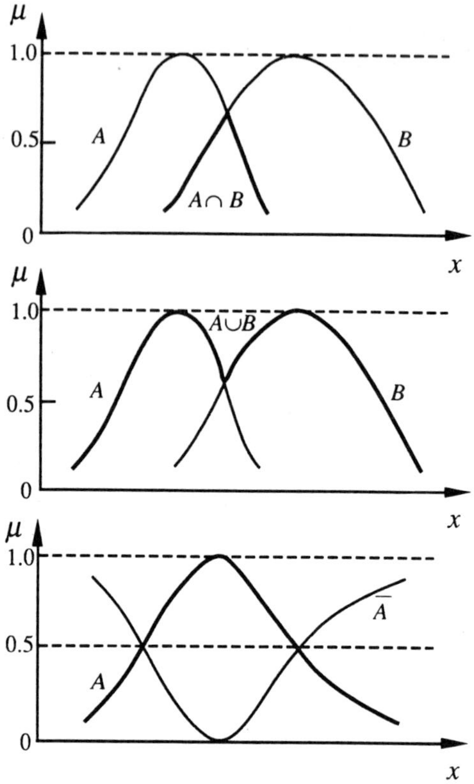

Figure 2.4 Set-theoretic operations for fuzzy sets

understood that these set-theoretic operations for fuzzy sets can be viewed as a natural extension of those for ordinary sets.

In fuzzy set theory, the minimum or the maximum of a and b is frequently expressed as

$$\min(a, b) = a \wedge b, \qquad \max(a, b) = a \vee b. \tag{2.18}$$

By the above notation, the intersection or the union of two fuzzy sets A and B can be written as

$$A \cap B \Leftrightarrow \mu_{A\cap B}(x) = \mu_A(x) \wedge \mu_B(x), \qquad (2.16)'$$

$$A \cup B \Leftrightarrow \mu_{A\cup B}(x) = \mu_A(x) \vee \mu_B(x). \qquad (2.17)'$$

Furthermore, using the notation of (2.10), the intersection and the union of two fuzzy sets A and B, and the complement of a fuzzy set A are represented as follows:

$$A \cap B = \int_X \mu_A(x)/x \cap \int_X \mu_B(x)/x$$

$$= \int_X (\mu_A(x) \wedge \mu_B(x))/x, \tag{2.19}$$

$$A \cup B = \int_X (\mu_A(x) \vee \mu_B(x))/x, \tag{2.20}$$

$$\bar{A} = \int_X (1 - \mu_A(x))/x. \tag{2.21}$$

Based on the definitions of the set-theoretic operations for fuzzy sets, Zadeh (1965) pointed out that it is possible to extend many of the basic identities which hold for ordinary sets to fuzzy sets. As examples, the following properties for union, intersection, and complementation hold for fuzzy sets similar to ordinary sets.

(1) **Commutativity laws**:

$$A \cup B = B \cup A, \quad A \cap B = B \cap A$$

(2) **Associativity laws**:

$$A \cup (B \cup C) = (A \cup B) \cup C, \quad A \cap (B \cap C) = (A \cap B) \cap C$$

(3) **Distributivity laws**:

$$A \cup (B \cap C) = (A \cup B) \cap (A \cup C), \quad A \cap (B \cup C) = (A \cap B) \cup (A \cap C)$$

(4) **De Morgan's laws**:

$$\overline{(A \cup B)} = \overline{A} \cap \overline{B}, \quad \overline{(A \cap B)} = \overline{A} \cup \overline{B}$$

(5) **Involution**:

$$\overline{\overline{A}} = A$$

For example, the proof of the first equation of De Morgan's laws follows directly by

$$\mu_{\overline{(A \cup B)}}(x) = 1 - \max[\mu_A(x), \mu_B(x)] = \min[1 - \mu_A(x), 1 - \mu_B(x)] = \mu_{\overline{A} \cap \overline{B}}(x).$$

It should be noted here that the only law which is no longer valid for fuzzy sets is the excluded-middle law:

$$\overline{A} \cup A \neq X, \ \overline{A} \cap A \neq \emptyset.$$

This fact can be easily understood by considering the example $\mu_A(x) = 0.5$ for all $x \in X$.

In addition to the set-theoretic operations, the following algebraic operations for fuzzy sets are also useful.

(1) Algebraic product: The algebraic product of two fuzzy sets A and B, denoted by AB, is defined by

$$AB \Leftrightarrow \mu_{AB}(x) = \mu_A(x)\mu_B(x). \tag{2.22}$$

(2) Algebraic sum: The algebraic sum of two fuzzy sets A and B, denoted by $A + B$, is defined by

$$A + B \Leftrightarrow \mu_{A+B}(x) = \mu_A(x) + \mu_B(x) - \mu_A(x)\mu_B(x). \tag{2.23}$$

(3) Bounded product: The bounded product of two fuzzy sets A and B, denoted by $A \odot B$, is defined by

$$\begin{aligned} A \odot B \Leftrightarrow \mu_{A \odot B} &= \max(0, \mu_A(x) + \mu_B(x) - 1) \\ &= 0 \vee (\mu_A(x) + \mu_B(x) - 1). \end{aligned} \tag{2.24}$$

(4) Bounded sum: The bounded sum of two fuzzy sets A and B, denoted by $A \oplus B$, is defined by

$$\begin{aligned} A \oplus B \Leftrightarrow \mu_{A \oplus B}(x) &= \min(1, \mu_A(x) + \mu_B(x)) \\ &= 1 \wedge (\mu_A(x) + \mu_B(x)). \end{aligned} \tag{2.25}$$

(5) Bounded difference: The bounded difference of two fuzzy sets A and B, denoted by $A \ominus B$, is defined by

$$\begin{aligned} A \ominus B \Leftrightarrow \mu_{A \ominus B}(x) &= \max(0, \mu_A(x) - \mu_B(x)) \\ &= 0 \vee (\mu_A(x) - \mu_B(x)). \end{aligned} \tag{2.26}$$

Example 2.4

Let $X = \{1, 2, 3, 4, 5, 6, 7, 8, 9, 10\}$ and consider the following two fuzzy sets A and B

$$A = 0.8/3 + 1/5 + 0.6/6, \qquad B = 0.7/3 + 1/4 + 0.5/6.$$

Then

$$\begin{aligned} A \cap B &= 0.7/3 + 0.5/6 \\ A \cup B &= 0.8/3 + 1/4 + 1/5 + 0.6/6 \\ \bar{A} &= 1/1 + 1/2 + 0.2/3 + 1/4 + 0.4/6 + 1/7 + 1/8 + 1/9 + 1/10 \\ AB &= 0.56/3 + 0.3/6 \\ A + B &= 0.94/3 + 1/4 + 1/5 + 0.8/6 \\ A \odot B &= 0.5/3 + 0.1/6 \\ A \oplus B &= 1/3 + 1/4 + 1/5 + 1/6 \\ A \ominus B &= 0.1/3 + 1/5 + 0.1/6. \end{aligned}$$

The concept of α-level sets serves as an important transfer between ordinary sets and fuzzy sets. It also plays an important role in the construction of a fuzzy set by a series of ordinary sets.

Definition 2.2 **(α-level set)**

The α-level set of a fuzzy set A is defined as an ordinary set A_α for which the degree of its membership function exceeds the level α:

$$A_\alpha = \{x \mid \mu_A(x) \geq \alpha\}, \ \alpha \in [0,1]. \tag{2.27}$$

Observe that the α-level set A_α can be defined by the characteristic function

$$c_{A_\alpha} = \begin{cases} 1 & \text{if } \mu_A(x) \geq \alpha \\ 0 & \text{if } \mu_A(x) < \alpha \end{cases} \tag{2.28}$$

since it is an ordinary set. Actually, an α-level set is an ordinary set whose elements belong to the corresponding fuzzy set to a certain degree α.

It is clear that the following evident property holds for the α-level sets:

$$\alpha_1 \leq \alpha_2 \Leftrightarrow A_{\alpha_1} \supseteq A_{\alpha_2}. \tag{2.29}$$

This relationship is illustrated in Figure 2.5.

From the definition of the α-level sets, it can be easily understood that the following basic properties hold:

$$(A \cup B)_\alpha = A_\alpha \cup B_\alpha, \tag{2.30}$$

$$(A \cap B)_\alpha = A_\alpha \cap B_\alpha. \tag{2.31}$$

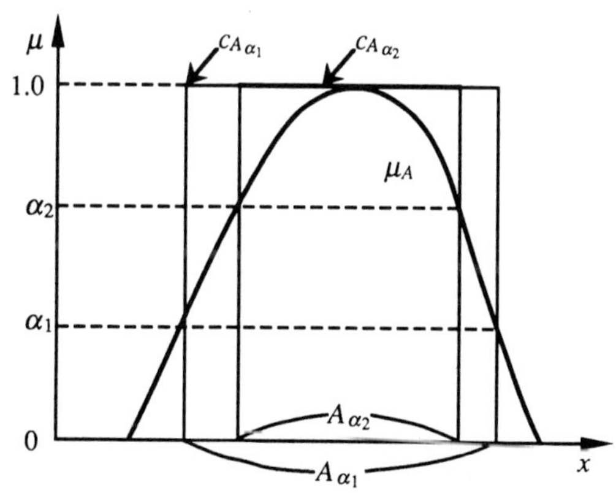

Figure 2.5 Examples of α-level sets

For example, the first property can be shown as follows:

$$\begin{aligned}(A \cup B)_\alpha &= \{x \mid \max(\mu_A(x), \mu_B(x)) \geq \alpha\} \\ &= \{x \mid \mu_A(x) \geq \alpha \text{ or } \mu_B(x) \geq \alpha\} \\ &= \{x \mid \mu_A(x) \geq \alpha\} \cup \{x \mid \mu_B(x) \geq \alpha\} \\ &= A_\alpha \cup B_\alpha.\end{aligned}$$

Using the concept of α-level sets, the relationships between ordinary sets and fuzzy sets can be characterized by the following theorem.

Theorem 2.1 (Decomposition theorem)

A fuzzy set A can be represented by

$$A = \bigcup_{\alpha \in [0,1]} \alpha A_\alpha \tag{2.32}$$

where αA_α denotes the algebraic product of a scalar α with the α-level set A_α, i.e., its membership function (characteristic function) is given by

$$\mu_{\alpha A_\alpha}(x) = \alpha \mu_{A_\alpha}(x) = \alpha c_{A_\alpha}(x), \quad \forall x \in X. \tag{2.33}$$

Actually, in view of

$$\begin{aligned}\mu_{\underset{\alpha\in[0,1]}{\cup} \alpha A_\alpha}(x) &= \sup_{\alpha \in [0,1]} \mu_{\alpha A_\alpha}(x) \\ &= \sup_{\alpha \in [0,1]} \alpha \cdot c_{A_\alpha}(x) \\ &= \sup_{\alpha \leq \mu_A(x)} \alpha = \mu_A(x),\end{aligned}$$

the decomposition theorem follows directly.

Theorem 2.1 states that a fuzzy set A can be decomposed into a series of α-level sets A_α by which A can be reconstructed. Thus any fuzzy set can be viewed as a family of ordinary sets.

Figure 2.6 depicts $\mu_A(x)$ together with $\mu_{\alpha_i A_{\alpha_i}}(x), i = 1, 2, 3, 4$, for $0 < \alpha_1 < \alpha_2 < \alpha_3 < \alpha_4$. As shown in Figure 2.6, it may be intuitively understood that the supremum of $\mu_{\alpha A_\alpha}(x)$ for all $0 \leq \alpha \leq 1$ tends to $\mu_A(x)$.

Example 2.5

Let

$$A = 0.1/2 + 0.3/1 + 0.5/7 + 0.9/6 + 1/9.$$

Then the fuzzy set A can be represented as

$$\begin{aligned}A = 0.1/2 + 0.1/1 + 0.1/7 + 0.1/6 + 0.1/9 \\ +0.3/1 + 0.3/7 + 0.3/6 + 0.3/9 \\ +0.5/7 + 0.5/6 + 0.5/9 \\ +0.9/6 + 0.9/9 \\ +1/9\end{aligned}$$

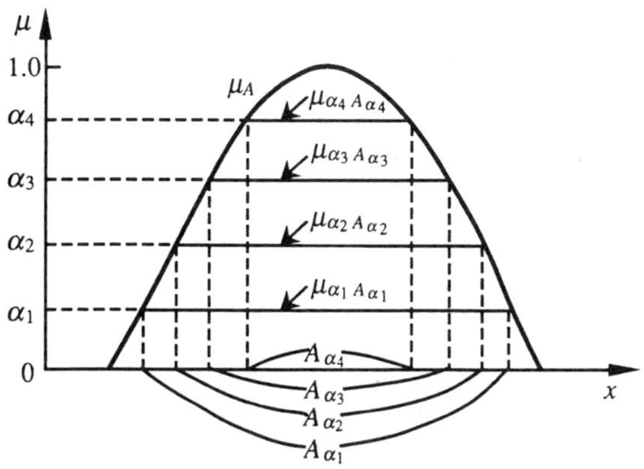

Figure 2.6 Illustration of decomposition theorem

or

$$
\begin{aligned}
A \;=\; & 0.1(1/2+1/1+1/7+1/6+1/9) \\
\cup\; & 0.3(1/1+1/7+1/6+1/9) \\
\cup\; & 0.5(1/7+1/6+1/9) \\
\cup\; & 0.9(1/6+1/9) \\
\cup\; & 1(1/9).
\end{aligned}
$$

By omitting 1/ for simplicity, each level set is given by

$$
\begin{aligned}
A_{0.1} &= 2+1+7+6+9 \\
A_{0.3} &= 1+7+6+9 \\
A_{0.5} &= 7+6+9 \\
A_{0.9} &= 6+9 \\
A_{1} &= 9.
\end{aligned}
$$

As an alternative representation of the decomposition theorem, observe that the membership function $\mu_A(x)$ of a fuzzy set A can be represented as

$$\mu_A(x) = \sup_{\alpha \in [0,1]} \min\left[\alpha, c_{A_\alpha}(x)\right] \tag{2.34}$$

through the characteristic function $c_{A_\alpha}(x)$ of the α-level sets.

2.2 Extension principle

As discussed in the previous section, both the α-level sets and the decomposition theorem are shown to be fundamental in relating fuzzy sets to ordinary sets and vice versa. The extension principle introduced by Zadeh (1965, 1973) is to provide a general method for extending nonfuzzy mathematical concepts to the fuzzy framework.

Consider a mapping f from a set X to a set Y written

$$f : X \to Y \tag{2.35}$$

where the set X is the domain of f and Y is the range of f.

Such a mapping f may be viewed as a rule which assigns some element $y = f(x)$ in Y to each element x in X. When both X and Y are sets of real numbers, the mapping f is a function.

Naturally, for a subset A of X, the image of A through f, denoted by

$$f(A) = \{y \mid y = f(x), x \in A\}, \tag{2.36}$$

is a subset of Y. Also, for a subset B of Y, the inverse image of B through f, denoted by

$$f^{-1}(B) = \{x \mid f(x) = y, y \in B\}, \tag{2.37}$$

is a subset of X, where f^{-1} is an inverse mapping of f.

Now, let B be a fuzzy set in Y whose membership function is $\mu_B(y)$. Then it seems to be quite natural to consider that the inverse mapping f^{-1} induces a fuzzy set A in X with its membership function defined by

$$\mu_A(x) = \mu_B(f(x)), \qquad \forall y \in Y \tag{2.38}$$

for all $x \in X$ which are mapped by f into y.

Conversely, when A is no longer an ordinary set but a fuzzy set in X, our main question is, what is the membership function for the fuzzy set B in Y which is induced through the mapping $f : X \to Y$?

If f is one-to-one, it is reasonable to define

$$\mu_{f(A)}(y) = \mu_A(x). \tag{2.39}$$

However, if f is not one-to-one, then an ambiguity arises when two or more distinct points in X, say x_1 and x_2, are mapped into the same point y in Y. To resolve this ambiguity, Zadeh (1965) proposed to assign the larger of the two grades of membership to y. More generally, the membership function for B is defined by

$$\mu_B(y) = \sup_{x \in f^{-1}(y)} \mu_A(x) \tag{2.40}$$

where $f^{-1}(y)$ is the set of points in X which are mapped into y by f.

Such an elementary form of the so-called extension principle was already suggested by the original paper of Zadeh (1965), and since then some modifications have been proposed by Zadeh himself (1973), Gaines (1976), Jain (1976, 1977), and others. The extension principle originally suggested by Zadeh is formally defined as follows.

Definition 2.3 (Extension principle)

Let $f : X \to Y$ be a mapping from a set X to a set Y. Then the extension principle allows us to define the fuzzy set B in Y induced by the fuzzy set A in X through f as follows:

$$B = \{(y, \mu_B(y)) \mid y = f(x),\ x \in X\} \tag{2.41}$$

with

$$\mu_B(y) \triangleq \mu_{f(A)}(y) = \begin{cases} \sup\limits_{y=f(x)} \mu_A(x) & f^{-1}(y) \neq \phi \\ 0 & f^{-1}(y) = \phi \end{cases} \tag{2.42}$$

where $f^{-1}(y)$ is the inverse image of y.

As mentioned before, in the extension principle, observe that when f is one-to-one, (2.42) simply becomes $\mu_{f(A)}(y) = \mu_A(x)$ if $f^{-1}(y) \neq \phi$, but when f is not one-to-one, $\mu_B(y)$ is defined as the supremum of $\mu_A(x)$ for all x such that $y = f(x)$.

Also, it must be observed here that the inverse image $f^{-1}(B)$ of a fuzzy set B in Y through f is a fuzzy set A in X with its membership function defined by

$$\mu_A(x) = \mu_{f^{-1}(B)}(x) = \mu_B(f(x)), \qquad \forall x \in f^{-1}(y). \tag{2.43}$$

It should be noted here that both the image $f(A)$ of a fuzzy set A and the inverse image $f^{-1}(B)$ of a fuzzy set B through the mapping f have the same properties similar to nonfuzzy sets.

Figure 2.7 illustrates the extension principle.

Example 2.6

Let

$$\begin{aligned} A &= 0.3/(-2) + 0.5/(-1) + 0.8/0 + 1/1 + 0.4/2, \\ f(x) &= x^2. \end{aligned}$$

Then, from the extension principle, it follows that

$$B = f(A) = 0.8/0 + 1/1 + 0.4/4.$$

This relationship is illustrated in Figure 2.8.

To generalize the extension principle to the Cartesian space, we introduce the concept of the Cartesian product of fuzzy sets.

Definition 2.4 (Cartesian product)

Let $A_1, \ldots, A_n$ be fuzzy sets in $X_1, \ldots, X_n$ with the corresponding membership functions $\mu_{A_1}(x_1), \ldots, \mu_{A_n}(x_n)$, respectively. Then the Cartesian product of the fuzzy sets $A_1, \ldots, A_n$, denoted by $A_1 \times \cdots \times A_n$, is defined as a fuzzy set in $X_1 \times \cdots \times X_n$ whose membership function is expressed by

$$\mu_{A_1 \times \cdots \times A_n}(x_1, \ldots, x_n) = \min(\mu_{A_1}(x_1), \ldots, \mu_{A_n}(x_n)). \tag{2.44}$$

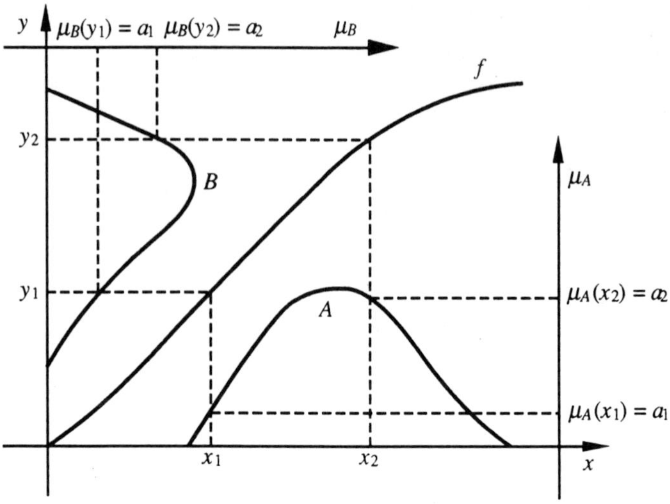

Figure 2.7 Illustration of extension principle

Example 2.7

Let $X_1 = X_2 = \{3, 5, 7\}$ and consider the following two fuzzy sets A_1 in X_1 and A_2 in X_2

$$A_1 = 0.5/3 + 1/5 + 0.6/7, \qquad A_2 = 1/3 + 0.6/5.$$

Then

$$A_1 \times A_2 = 0.5/(3,3) + 1/(5,3) + 0.6/(7,3) + 0.5/(3,5) + 0.6/(5,5) + 0.6/(7,5).$$

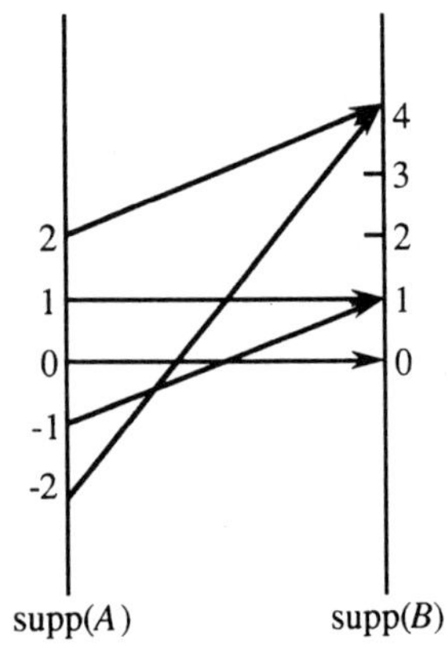

Figure 2.8 Extension principle for Example 2.6

Observe that, when $A_1, \ldots, A_n$ are nonfuzzy, (2.44) reduces to the conventional definition of the Cartesian product of nonfuzzy sets.

From the definition of the Cartesian product of fuzzy sets, it can be easily understood that the extension principle is generalized as follows.

Definition 2.5 (Extension principle in Cartesian space)

Let

$$f : X_1 \times \cdots \times X_n \to Y$$

be a mapping from $X_1 \times \cdots \times X_n$ *to a set* Y *such that* $y = f(x_1 \ldots x_n)$. *Then the extension principle allows us to define the fuzzy set* B *in* Y *induced by the fuzzy set* $A_1 \times \cdots \times A_n$ *in* $X_1 \times \cdots \times X_n$ *through* f *as follows:*

$$B = \{(y, \mu_B(y)) \mid y = f(x_1, \ldots, x_n),\ (x_1, \ldots, x_n) \in X_1 \times \cdots \times X_n\} \tag{2.45}$$

with

$$\mu_B(y) = \begin{cases} \sup\limits_{\substack{(x_1,\ldots,x_n) \in X_1 \times \cdots \times X_n \\ y = f(x)}} \mu_{A_1 \times \cdots \times A_n}(x_1, \ldots, x_n) & f^{-1}(y) \neq \phi \\ 0 & f^{-1}(y) = \phi \end{cases} \tag{2.46}$$

where $f^{-1}(y)$ *is the inverse image of* y.

Other extension principles than (2.42) or (2.46), such as replacing "sup" by the algebraic product, etc., are also possible. But since this classical extension principle proposed by Zadeh using "sup" is widely used, we do not consider other types of extension principles.

Using the concept of the α-level set, H.T. Nguyen (1978) showed that the extension principle (2.46) can be equivalently represented as follows.

Theorem 2.2 (Nguyen)

If there exist $x_1, \ldots, x_n$ *such that* $\mu_B(y) = \mu_{A_1 \times \cdots \times A_n}(x_1, \ldots, x_n)$ *for any* $y \in Y$, *i.e., if the supremum of (2.46) is attained for some* $x_1, \ldots, x_n$, *then it holds that*

$$[f(A_1, \ldots, A_n)]_\alpha = f(A_{1_\alpha}, \ldots, A_{n_\alpha}). \tag{2.47}$$

Although the proof of this theorem for the general case is omitted here, the proof for the special case in which all of the fuzzy sets $A_1, \ldots, A_n$ are fuzzy numbers can be found in the next section.

2.3 Fuzzy numbers

Before introducing the definition of fuzzy numbers such as "approximately m," "near zero," etc., consider an extension of ordinary convex sets to fuzzy sets. Naturally, a convex fuzzy set presented by Zadeh (1965) is defined through its membership function in a real n-dimensional Euclidean space R^n.

Definition 2.6 (Convex fuzzy set)

A fuzzy set A in $X = R^n$ is said to be a convex fuzzy set if and only if its α-level sets are convex. An alternative and more direct definition of a convex fuzzy set is as follows:

A fuzzy set A is convex if and only if

$$\mu_A(\lambda x_1 + (1-\lambda)x_2) \geq \min\left(\mu_A(x_1), \mu_A(x_2)\right) \tag{2.48}$$

for all $x_1, x_2 \in X$ and $\lambda \in [0,1]$.

Observe that this definition does not imply that $\mu_A(x)$ is a convex function of x.

The proof of the equivalence between the above definitions follows.

Let A be convex in the sense of the first definition and assume that there exist $x_1', x_2' \in X$ and $\lambda' \in [0,1]$ such that

$$\mu_A(\lambda' x_1' + (1-\lambda')x_2') < \min\left[\mu_A(x_1'), \mu_A(x_2')\right].$$

Let $\alpha' = \min\left[\mu_A(x_1'), \mu_A(x_2')\right]$. Then $x_1', x_2' \in A_{\alpha'}$ and $\lambda' x_1' + (1-\lambda')x_2' \notin A_{\alpha'}$, which means $A_{\alpha'}$ is not convex and leads to a contradiction.

Conversely, assume that a fuzzy set A is convex in the sense of the second definition and take $x, y \in A_\alpha$ for any $\alpha \in [0,1]$. Then $\mu_A(x) \geq \alpha$ and $\mu_A(y) \geq \alpha$. Hence, (2.48) implies $\mu_A(\lambda x + (1-\lambda)y) \geq \alpha$ for any $\lambda \in [0,1]$. Thus, we have $\lambda x + (1-\lambda)y \in A_\alpha$.

Thus, the proof of the assertion is established. □

From the definition of convexity, it is easy to show that if two fuzzy sets A and B are convex, their intersection $A \cap B$ is also convex.

Figure 2.9 illustrates a convex fuzzy set and a nonconvex fuzzy set.

Among fuzzy sets, numbers such as "approximately m" or "about n" can be defined as fuzzy sets of the real line R^1. Such fuzzy numbers are formally defined as follows (Dubois and Prade 1978, 1980b; Zimmermann 1991).

Definition 2.7 (Fuzzy numbers)

A fuzzy number is a convex normalized fuzzy set of the real line R^1 whose membership function is piecewise continuous.

Frequently, a fuzzy number M is called positive (negative), denoted by $M > 0$ ($M < 0$), if its membership function $\mu_M(x)$ satisfies $\mu_M(x) = 0$, $\forall x < 0$ ($\forall x > 0$).

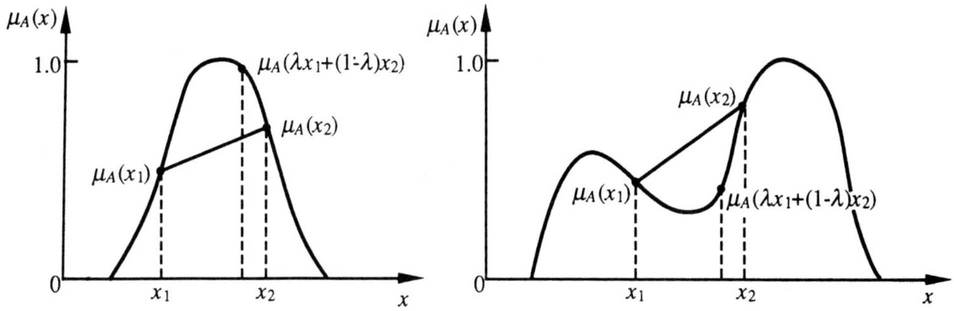

Figure 2.9 Convex and nonconvex fuzzy sets

Example 2.8

As examples of membership functions for a fuzzy number M, such as approximately m, a triangular membership function

$$\mu_M(x) = \max(0, 1 - \mid x - m \mid / a) \qquad a > 0 \tag{2.49}$$

or a bell-shaped membership function

$$\mu_M(x) = e^{-b(x-m)^2} \qquad b \geq 1 \tag{2.50}$$

is widely used. Such membership functions are illustrated in Figure 2.10.

From the definition of a fuzzy number M, it is significant to note that the α-level set M_α of a fuzzy number M can be represented by the closed interval which depends on the value of α as is shown in Figure 2.11. Namely,

$$M_\alpha = \{x \in R^1 \mid \mu_M(x) \geq \alpha\} = [m_\alpha^L, m_\alpha^R] \tag{2.51}$$

where m_α^L or m_α^R represents the left or right extreme point of the α-level set M_α, respectively.

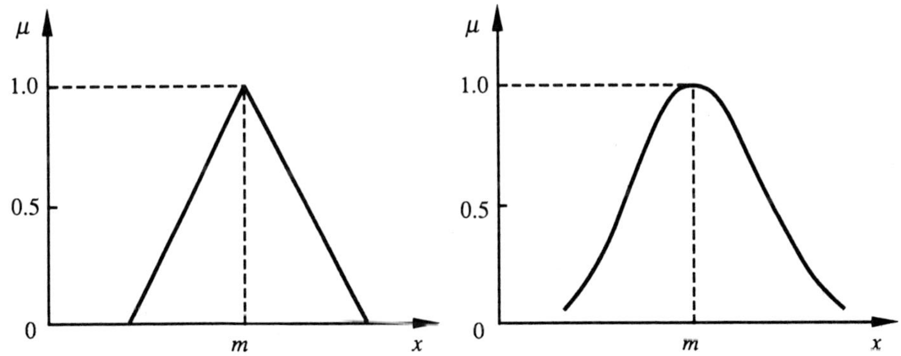

Figure 2.10 Examples of fuzzy numbers

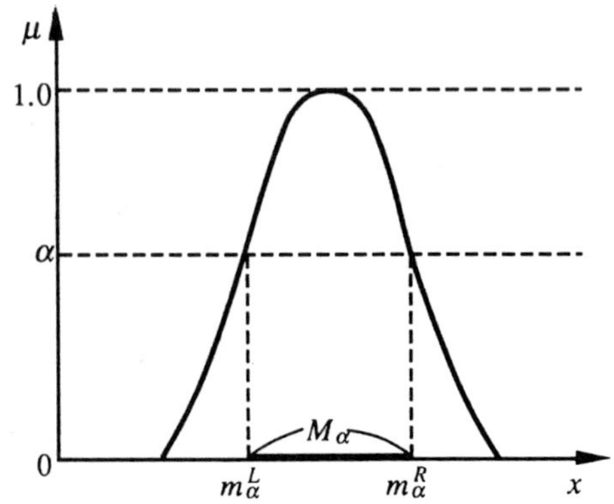

Figure 2.11 α-level set of fuzzy number M

One of the major applications of the extension principle in fuzzy set theory is the extension of the algebraic operations "$+$", "$-$", "$\times$", and "$/$" in ordinary numbers to fuzzy numbers. Such an extension can be made via the extension principle.

Let M and N be two fuzzy numbers with the membership functions $\mu_M(x)$ and $\mu_N(x)$, respectively. Then according to the extension principle of Zadeh, the binary operation "$*$" in R^1 can be extended to the binary operation "$\circledast$" of fuzzy numbers M and N as follows:

$$\mu_{M \circledast N}(z) = \sup_{z=x*y} \min\left(\mu_M(x), \mu_N(y)\right). \tag{2.52}$$

As the binary operation "$*$" of two numbers in R^1, if we take "$+$", "$-$", "$\times$", and "$/$", it is possible to define the extended addition $\oplus$, subtraction $\ominus$, multiplication $\otimes$, and division $\oslash$ of two fuzzy numbers M and N as follows.

(1) Addition: $M \oplus N$

$$\begin{aligned} \mu_{M\oplus N}(z) &= \sup_{z=x+y} \min\left(\mu_M(x), \mu_N(y)\right) \\ &= \sup_{x\in R^1} \min\left(\mu_M(x), \mu_N(z-x)\right) \end{aligned} \tag{2.53}$$

(2) Subtraction: $M \ominus N$

$$\begin{aligned} \mu_{M\ominus N}(z) &= \sup_{z=x-y} \min\left(\mu_M(x), \mu_N(y)\right) \\ &= \sup_{x\in R^1} \min\left(\mu_M(x), \mu_N(x-z)\right) \end{aligned} \tag{2.54}$$

(3) Multiplication: $M \otimes N$

$$\mu_{M\otimes N}(z) = \sup_{z=x\times y} \min(\mu_M(x), \mu_N(y))$$
$$= \begin{cases} \sup_{x\in R^1} \min(\mu_M(x), \mu_N(z/x)) & z \neq 0 \\ \max\left\{ \sup_{x\in R^1} \min(\mu_M(x), \mu_N(0)), \sup_{y\in R^1} \min(\mu_M(0), \mu_N(y)) \right\} & z = 0 \end{cases} \quad (2.55)$$

(4) Division: $M \oslash N$

$$\begin{aligned} \mu_{M\oslash N}(z) &= \sup_{z=x/y} \min(\mu_M(x), \mu_N(y)) \\ &= \sup_{x\in R^1} \min(\mu_M(x), \mu_N(x/z)) \\ &= \sup_{\substack{y\in \mathrm{supp}(N) \\ 0\notin \mathrm{supp}(N)}} \min(\mu_M(z\cdot y), \mu_N(y)) \end{aligned} \quad (2.56)$$

Example 2.9

Let M and N be two fuzzy numbers with their membership functions given by the following special forms

$$\mu_M(x) = e^{-(x-m)^2}, \qquad \mu_N(x) = e^{-(x-n)^2}.$$

Then the membership function $\mu_{M\oplus N}$ of their extended sum $M \oplus N$ can be explicitly obtained as follows:

$$\begin{aligned} \mu_{M\oplus N}(z) &= \sup_{z=x+y} \min(e^{-(x-m)^2}, e^{-(y-n)^2}) \\ &= \sup_{x\in R^1} \min(e^{-(x-m)^2}, e^{-(z-x-n)^2}) \\ &= e^{-(z-m-n)^2/4}. \end{aligned}$$

This means that the membership function of $M \oplus N$ increases its ambiguity compared with the membership function

$$\mu_{(M+N)}(z) = e^{-(z-m-n)^2}$$

of the fuzzy number $(M + N)$ representing the numbers approximately $(m + n)$.

In the same way, the membership function $\mu_{M\ominus N}$ of their extended difference can also be explicitly expressed as

$$\mu_{M\ominus N}(z) = e^{-(z-m+n)^2/4}.$$

Here, especially, if we choose $N = M$, it reduces to

$$\mu_{M\ominus M}(z) = e^{-z^2/4}$$

which implies that, in the extended subtraction of two fuzzy numbers, the difference of the same fuzzy numbers does not always become zero, i.e.,

$$M \ominus M \neq 0.$$

Obviously, the explicit representation of the membership function resulting from the extended multiplication or division of fuzzy numbers becomes much more difficult compared with the extended addition or subtraction of fuzzy numbers.

To get fast computation formulas for the operations of fuzzy numbers, Dubois and Prade (1978) introduced the concept of *L*-*R* fuzzy numbers.

Definition 2.8 (***L*-*R* fuzzy numbers**)

A fuzzy number M is said to be an L-R fuzzy number if

$$\mu_M(x) = \begin{cases} L\left(\dfrac{m-x}{\alpha}\right) & x \leq m, \quad \alpha > 0 \\ R\left(\dfrac{x-m}{\beta}\right) & x \geq m, \quad \beta > 0 \end{cases} \tag{2.57}$$

where m is the mean value of M and α and β are left and right spreads, respectively, and a function L(·) is a left shape function satisfying

(1) $L(x) = L(-x)$

(2) $L(0) = 1$

(3) $L(x)$ *is nonincreasing on* $[0, \infty)$.

Naturally, a right shape function R(·) is similarly defined as L(·).

Using its mean value, and left and right spreads, and shape functions, such an *L*-*R* fuzzy number M is symbolically written

$$M = (m, \alpha, \beta)_{LR}. \tag{2.58}$$

Figure 2.12 illustrates an example of an *L*-*R* fuzzy number.

Observe that, when its spreads α and β are zero, M reduces to $M = (m, 0, 0)_{LR}$ and is an ordinary number.

As examples of typical shape functions, the following functions are well-known:

$$\begin{aligned} L(x) &= \max(0, 1- \mid x \mid^p), \quad p > 0 \\ L(x) &= \exp(- \mid x \mid^p), \quad p > 0 \\ L(x) &= 1/(1- \mid x \mid^p), \quad p > 0 \end{aligned}$$

Concerning the basic operations of such *L*-*R* fuzzy numbers, Dubois and Prade (1978, 1980) showed the following exact formulas for ⊕ and ⊖ together with the approximate formulas for ⊗ and ⊘.

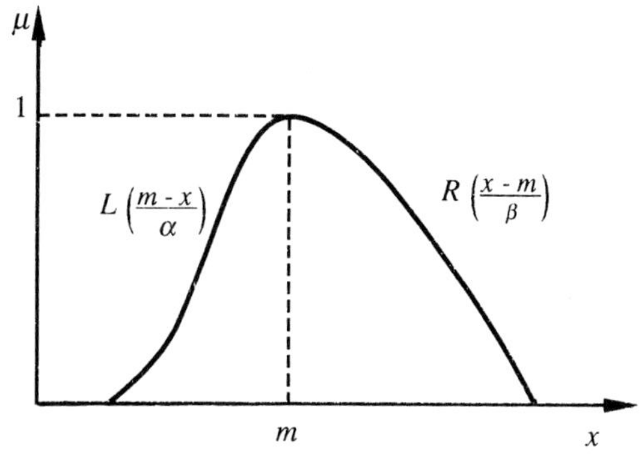

Figure 2.12 Illustration of L-R fuzzy number

For two L-R fuzzy numbers $M = (m, \alpha, \beta)_{LR}$ and $N = (n, \gamma, \delta)_{LR}$, the formula for the extended addition becomes the following.

(1) Addition: $M \oplus M$

$$(m, \alpha, \beta)_{LR} \oplus (n, \gamma, \delta)_{LR} = (m + n, \alpha + \gamma, \beta + \delta)_{LR} \tag{2.59}$$

This formula can be shown as follows.

For any fixed $\omega \in [0, 1]$, let

$$L((m - x)/\alpha) = \omega = L((n - y)/\gamma).$$

Then

$$x = m - \alpha L^{-1}(\omega), \qquad y = n - \gamma L^{-1}(\omega).$$

This implies

$$z = x + y = m + n - (\alpha + \gamma)L^{-1}(\omega).$$

Hence, it holds that

$$L\left(\frac{m + n - z}{\alpha + \gamma}\right) = \omega.$$

In the same way, for $R(\omega)$, it follows that

$$R\left(\frac{z - (m + n)}{\beta + \delta}\right) = \omega.$$

Thus, the formula for the extended addition is proved. □

For an L-R fuzzy number $M = (m, \alpha, \beta)_{LR}$ and an R-L fuzzy number $N = (n, \gamma, \delta)_{RL}$, the following formulas for the extended opposite and subtraction hold.

(2) Opposite: $-M$

$$-(m, \alpha, \beta)_{LR} = (-m, \beta, \alpha)_{RL} \tag{2.60}$$

(3) Subtraction: $M \ominus N$

$$(m, \alpha, \beta)_{LR} \ominus (n, \gamma, \delta)_{RL} = (m - n, \alpha + \delta, \beta + \gamma)_{LR} \tag{2.61}$$

The formula for the opposite of a fuzzy number is quite easy to understand, but observe that the shape functions are exchanged. Then in view of the formulas for the extended addition and opposite, the formula for the extended subtraction of two opposite types of fuzzy numbers follows immediately.

Example 2.10 (Kaufmann and Gupta 1985, pp. 64-65)
Let

$$M = (m, \alpha, \beta)_{LR} = (4, 2, 3)_{LR}, \qquad N = (n, \gamma, \delta)_{LR} = (8, 3, 5)_{LR}$$

with their shape functions

$$L(x) = \begin{cases} 0 & x < -1 \\ (1 + x)^{1/2} & -1 \le x \le 0 \end{cases}$$
$$R(x) = \begin{cases} 1 - x^2 & 0 \le x \le 1 \\ 0 & x > 1. \end{cases}$$

Then the membership functions of M and N are represented as

$$\mu_M(x) = \begin{cases} 0 & x < 2 \\ (1 + (x-4)/2)^{1/2} & 2 \le x < 4 \\ 1 & x = 4 \\ 1 - ((x-4)/3)^2 & 4 < x \le 7 \\ 0 & x > 7 \end{cases}$$

$$\mu_N(x) = \begin{cases} 0 & x < 5 \\ (1 + (x-8)/3)^{1/2} & 5 \le x < 8 \\ 1 & x = 8 \\ 1 - ((x-8)/5)^2 & 8 < x \le 13 \\ 0 & x > 13. \end{cases}$$

Using the formula for the extended addition of two L-R fuzzy numbers, we obtain

$$M \oplus N = (m + n, \alpha + \gamma, \beta + \delta)_{LR} = (12, 5, 8)_{LR}$$

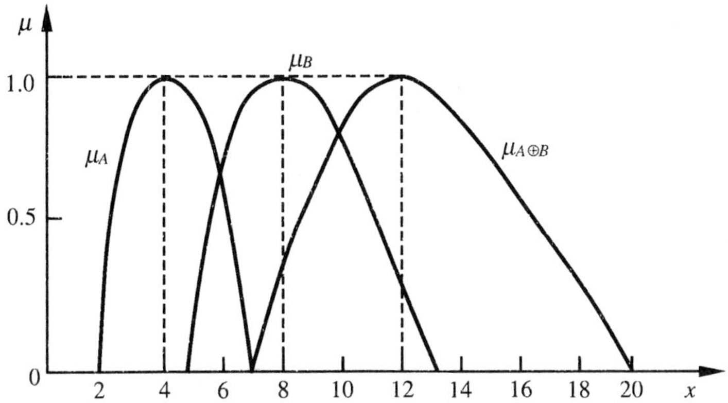

Figure 2.13 Addition of two *L*-*R* fuzzy numbers

with the corresponding membership function

$$\mu_{M\oplus N}(x) = \begin{cases} 0 & x < 7 \\ (1 + (x-12)/5)^{1/2} & 7 \le x < 12 \\ 1 & x = 12 \\ 1 - ((x-12)/8)^2 & 12 \le x \le 20 \\ 0 & x > 20. \end{cases}$$

Figure 2.13 illustrates the extended addition of these two *L*-*R* fuzzy numbers.

A proof similar to the one used for the extended addition of two fuzzy numbers shows that the following formula for the extended multiplication of two positive fuzzy numbers $M > 0$ and $N > 0$ holds:

$$z = x \cdot y = m \cdot n - (m\gamma + n\alpha)L^{-1}(\omega) + \alpha\gamma(L^{-1}(\omega))^2.$$

Here, if we assume that α and γ are small enough compared with m and n, we can neglect the term $\alpha\gamma(L^{-1}(\omega))^2$. By making a similar assumption for $R^{-1}(\omega)$, we can also obtain the approximate formula for the multiplication of two positive fuzzy numbers. When $M < 0, N > 0$ or $M < 0, N < 0$, it is easy to get the similar formulas for their multiplication. Thus, the approximate formulas for the extended multiplication of two *L*-*R* fuzzy numbers can be summarized as follows.

(4) Multiplication: $M \otimes N$

If $M > 0$ and $N > 0$, then

$$(m, \alpha, \beta)_{LR} \otimes (n, \gamma, \delta)_{LR} \cong (mn, m\gamma + n\alpha, m\delta + n\beta)_{LR}. \tag{2.62}$$

If $M < 0$ and $N > 0$, then

$$(m, \alpha, \beta)_{LR} \otimes (n, \gamma, \delta)_{LR} \cong (mn, n\alpha - m\delta, n\beta - m\gamma)_{RL}. \tag{2.63}$$

If $M < 0$ and $N < 0$, then

$$(m, \alpha, \beta)_{LR} \otimes (n, \gamma, \delta)_{LR} \cong (mn, -n\beta - m\delta, -n\alpha - m\gamma)_{RL}. \tag{2.64}$$

As a special case, observe that scalar multiplication for an L-R fuzzy number $M = (m, \alpha, \beta)_{LR}$ can be given by the following formula depending on the sign of λ.

(5) Scalar multiplication: $\lambda \otimes M$

$$\lambda \otimes (m, \alpha, \beta)_{LR} = \begin{cases} (\lambda m, \lambda\alpha, \lambda\beta)_{LR} & \lambda > 0 \\ (\lambda m, -\lambda\beta, -\lambda\alpha)_{RL} & \lambda < 0 \end{cases} \tag{2.65}$$

Consider the inverse of L-R fuzzy number $M = (m, \alpha, \beta)_{LR}$ denoted by M^{-1}. Then for the positive L-R fuzzy number $M = (m, \alpha, \beta)_{LR}$, it holds that

$$\mu_{M^{-1}}(x) = L\left(\frac{1 - mx}{\alpha x}\right) \qquad x \geq 1/m. \tag{2.66}$$

Although m^{-1} is neither an L-R nor an R-L fuzzy number, because $(1-mx)/\alpha x \cong ((1/m) - x)/(\alpha/m^2)$ in the neighborhood of $1/m$, the inverse of a fuzzy number M becomes an R-L fuzzy number and can be approximated by the following formula.

(6) Inverse: M^{-1}

$$(m, \alpha, \beta)_{LR}^{-1} \cong (m^{-1}, \beta m^{-2}, \alpha m^{-2})_{RL} \tag{2.67}$$

Observing that $M \oslash N = M \cdot N^{-1}$, the following approximate formula for the extended division of two positive L-R and R-L fuzzy numbers can be obtained.

(7) Division: $M \oslash N$

If $M > 0$ and $N > 0$, then

$$(m, \alpha, \beta)_{LR} \oslash (n, \gamma, \delta)_{RL} \cong \left(\frac{m}{n}, \frac{\delta m + \alpha n}{n^2}, \frac{\gamma m + \beta n}{n^2}\right)_{LR} \tag{2.68}$$

Similar approximate formulas hold when $M < 0, N > 0$ or $M < 0, N < 0$.

The concept of fuzzy numbers has been extended to fuzzy intervals which can be viewed as intervals whose boundaries are not sharp (Dubois and Prade 1978). Following the definition of fuzzy numbers introduced by Dubois and Prade (1978), for the remainder of this book, we consider the fuzzy numbers whose membership functions are not only continuous but also strictly increasing on the left and strictly decreasing on the right. To be more explicit, we deal with a fuzzy number M whose membership function $\mu_M(x)$ satisfies the following conditions:

(1) A continuous mapping from R^1 to the closed interval $[0, 1]$

(2) $\mu_M(x) = 0$ for all $x \in (-\infty, c]$

(3) Strictly increasing and continuous on $[c, a]$

(4) $\mu_M(x) = 1$ for all $x \in [a, b]$

(5) Strictly decreasing and continuous on $[b, d]$

(6) $\mu_M(x) = 0$ for all $x \in [d, +\infty)$

where a, b, c, and d are real numbers and $(a + b)/2$ is the mean value of M.

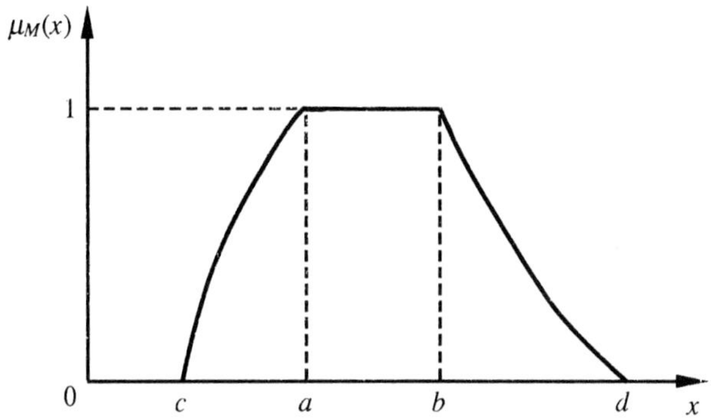

Figure 2.14 Fuzzy numbers

Observe that (1) if $a = c = b = d$, M is an ordinary number and (2) if $a = b$, M is a fuzzy number representing approximately a discussed thus far. Figure 2.14 illustrates the graph of the possible shape of a fuzzy number M.

Then as the special case of Theorem 2.2 (Nguyen's theorem) where all of the fuzzy sets are fuzzy numbers, it is significant to prove that the following relation holds.

Theorem 2.3

Let $y = h(z_1, \ldots, z_n)$ be a real-valued ordinary continuous function with n variables $z_1, \ldots, z_n$. Also let $Z_1, \ldots, Z_n$ be fuzzy numbers with their corresponding α-level sets denoted by $[z^L_{1\alpha}, z^R_{1\alpha}], \ldots, [z^L_{n\alpha}, z^R_{n\alpha}]$ as was shown in Figure 2.11. Then the α-level set of the image of $Z_1, \ldots, Z_n$ through h denoted by $h(Z_1, \ldots, Z_n)$ can be represented by

$$[h(Z_1, \ldots, Z_n)]_\alpha = \{h(z_1, \ldots, z_n) \in R^1 \mid z_i \in [z^L_{i\alpha}, z^R_{i\alpha}], i = 1, \ldots, n\}. \quad (2.69)$$

Proof

Let $\mu_{Z_i}(z_i)$ and $\mu_{h(Z_1,\ldots,Z_n)}(y)$ be membership functions of the fuzzy number Z_i and $h(Z_1, \ldots, Z_n)$, respectively. Then, via the extension principle, it holds that

$$\mu_{h(Z_1,\ldots,Z_n)}(y) = \max_{y=h(z_1,\ldots,z_n)} \min_{i=1,\ldots,n} \mu_{Z_i}(z_i). \tag{2.70}$$

Assume $z_i \in [z_{i\alpha}^L, z_{i\alpha}^R]$, $i = 1, \ldots, n$. Then, from (2.70), it holds that $\mu_{h(z_1,\ldots,z_n)}(y) \geq \alpha$ for any $y = h(z_1, \ldots, z_n)$, which implies $y \in [h(Z_1, \ldots, Z_n)]_\alpha$. Conversely, assume $y \in [h(Z_1, \ldots, Z_n)]_\alpha$. Then from (2.70), it holds that $\min_{i=1,\ldots,n} \mu_{Z_i}(z_i) \geq \alpha$ for some $(z_1, \ldots, z_n) \in R^n$ such that $y = h(z_1, \ldots, z_n)$. This implies $\mu_{Z_i}(z_i) \geq \alpha$, $i = 1, \ldots, n$, i.e., $z_i \in [z_{1\alpha}^L, z_{p\alpha}^R]$, $i = 1, \ldots, n$.

Thus, the proof of the theorem is completed. □

The following corollary, as an immediate result of Theorem 2.3, is sometimes very useful.

Corollary 2.1

Let M and N be two fuzzy numbers and let $f : R^1 \times R^1 \to R^1$ be a continuous function. Then for all $\alpha \in [0, 1]$, it holds that

$$[f(M, N)]_\alpha = f(M_\alpha, N_\alpha) \tag{2.71}$$

where M_α or N_α is the α-level set of the fuzzy number M or N, respectively, i.e.,

$$M_\alpha = [m_\alpha^L, m_\alpha^R], \ N_\alpha = [n_\alpha^L, n_\alpha^R]. \tag{2.72}$$

2.4 Fuzzy decision

Before discussing the original definitions of the fuzzy decision proposed by Bellman and Zadeh (1970), it is quite significant to go back to their first explanation indicating the necessity for incorporating fuzzy sets into decision-making processes. In their 1970 paper "Decision making in a fuzzy environment," Bellman and Zadeh explain:

> Much of the decision making in the real world takes place in an environment in which the goals, the constraints, and the consequences of possible actions are not known precisely. To deal quantitatively with imprecision, we usually employ the concepts and techniques of probability theory and, more particularly, the tools provided by decision theory, control theory, and information theory. In doing so, we are tacitly accepting the premise that imprecision - whatever its nature - can be equated with randomness. This, in our view, is a questionable assumption.
>
> Specially, our contention is that there is a need for differentiation between *randomness* and *fuzziness*, with the later being a major source of imprecision in many decision processes. By fuzziness, we mean a type of imprecision which is associated with *fuzzy sets*, that is, classes in which

> there is no sharp transition from membership to nonmembership. For example, the class of *green objects* is a fuzzy set. So are the classes of objects characterized by such commonly used adjectives as large, small, substantial, significant, important, serious, simple, accurate, approximate, etc. Actually, in sharp contrast to the notion of a class or a set in mathematics, most of the classes in the real world do not have crisp boundaries which separate those objects which belong to a class from those which do not. In this connection, it is important to note that, in the discourse between humans, fuzzy statements such as "John is *several* inches taller than Jim," "x is *much larger* than y," "Corporation X has a *bright future*," "the stock market has suffered a *sharp decline*," convey information despite the imprecision of the meaning of the italicized word. In fact, it may be argued that the main distinction between human intelligence and machine intelligence lies in the ability of humans - an ability which present-day computers do not possess - to manipulate fuzzy concepts and respond to fuzzy instructions.

With this observation, Bellman and Zadeh (1970) introduced three basic concepts: fuzzy goal, fuzzy constraint, and fuzzy decision and explored the application of these concepts to decision-making processes under fuzziness.

Let us now introduce the conceptual framework for decision-making in a fuzzy environment.

Let X be a given set of possible alternatives which contains the solution of a decision-making problem under consideration.

A fuzzy goal G is a fuzzy set on X characterized by its membership function

$$\mu_G : X \to [0,1]. \tag{2.73}$$

A fuzzy constraint C is a fuzzy set on X characterized by its membership function

$$\mu_C : X \to [0,1]. \tag{2.74}$$

Realizing that both the fuzzy goal and the fuzzy constraint are desired to be satisfied simultaneously, Bellman and Zadeh (1970) defined the fuzzy decision D resulting from the fuzzy goal G and fuzzy constraint C as the intersection of G and C.

To be more explicit, the fuzzy decision of Bellman and Zadeh is the fuzzy set D on X defined as

$$D = G \cap C \tag{2.75}$$

and is characterized by its membership function

$$\mu_D(x) = \min(\mu_G(x), \mu_C(x)). \tag{2.76}$$

The maximizing decision is then defined as

$$\underset{x \in X}{\text{maximize}}\, \mu_D(x) = \underset{x \in X}{\text{maximize}}\ \min(\mu_G(x), \mu_C(x)). \tag{2.77}$$

More generally, the fuzzy decision D resulting from k fuzzy goals $G_1, \ldots, G_k$ and m fuzzy constraints $C_1, \ldots, C_m$ is defined by

$$D = G_1 \cap \ldots \cap G_k \cap C_1 \cap \cdots \cap C_m \tag{2.78}$$

and the corresponding maximizing decision is defined as

$$\underset{x \in X}{\text{maximize}}\, \mu_D(x) = \underset{x \in X}{\text{maximize}}\ \min(\mu_{G_1}(x), \ldots, \mu_{G_k}(x), \mu_{C_1}(x), \ldots, \mu_{C_m}(x)). \tag{2.79}$$

It is significant to realize here that in the fuzzy decision defined by Bellman and Zadeh (1970), the fuzzy goals and the fuzzy constraints enter into the expression for D in exactly the same way. In other words, in the definition of the fuzzy decision, there is no longer a difference between the fuzzy goals and the fuzzy constraints.

However, depending on the situations, other aggregation patterns for the fuzzy goal G and the fuzzy constraint C may be worth considering. When fuzzy goals and fuzzy constraints have unequal importance, Bellman and Zadeh (1970) also suggested the convex fuzzy decision defined by

$$\mu_D^{co}(x) = \sum_{i=1}^{k} \alpha_i \mu_{G_i}(x) + \sum_{j=1}^{m} \beta_j \mu_{C_j}(x),$$

$$\sum_{i=1}^{k} \alpha_i + \sum_{j=1}^{m} \beta_j = 1, \quad \alpha_i, \beta_j \geq 0 \tag{2.80}$$

where the weighting coefficients reflect the relative importance among the fuzzy goals and constraints.

As an example of an alternative definition of a fuzzy decision, the product fuzzy decision defined by

$$\mu_D^{pr}(x) = \left(\prod_{i=1}^{k} \mu_{G_i}(x)\right) \cdot \left(\prod_{j=1}^{m} \mu_{C_j}(x)\right) \tag{2.81}$$

has been proposed.

For the convex fuzzy decision or the product fuzzy decision, similar to the maximizing decision for the fuzzy decision, the maximizing decision to select x^* such that

$$\mu_D^{co}(x^*) = \max_{x \in X} \left[\sum_{i=1}^{k} \alpha_i \mu_{G_i}(x)\right] \tag{2.82}$$

or

$$\mu_D^{pr}(x^*) = \max_{x \in X} \left[\left(\prod_{i=1}^{k} \mu_{G_i}(x)\right) \cdot \left(\prod_{j=1}^{m} \mu_{C_j}(x)\right)\right] \tag{2.83}$$

is also defined.

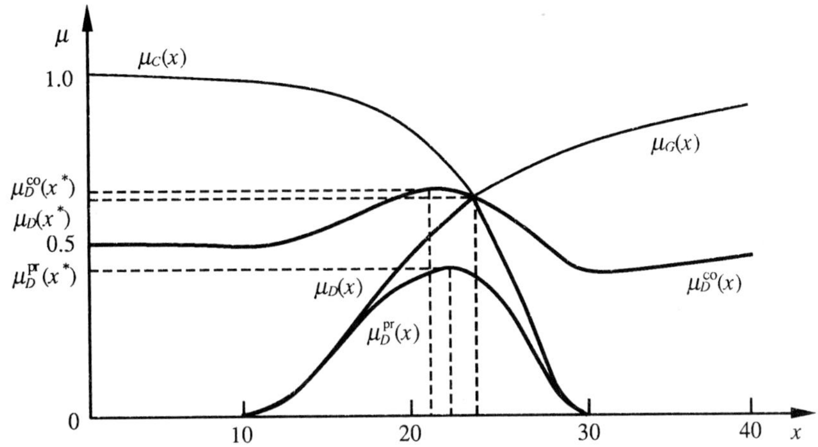

Figure 2.15 Fuzzy decision, convex fuzzy decision, and product fuzzy decision

It should be noted here that among these three types of fuzzy decisions $\mu_D^{co}(x)$, $\mu_D^{pr}(x)$, and $\mu_D(x)$, the following relation holds:

$$\mu_D^{pr}(x) \leq \mu_D(x) \leq \mu_D^{co}(x). \tag{2.84}$$

Example 2.11

Let $X = [0, \infty]$ be a set of alternatives. Suppose that we have a fuzzy goal G and a fuzzy constraint C expressed as "x should be much larger than 10" and "x should be substantially smaller than 30" where their membership functions are subjectively defined by

$$\mu_G(x) = \begin{cases} 0 & ; x \leq 10 \\ 1 - (1 + (0.1(x-10))^2)^{-1} & ; x > 10, \end{cases} \tag{2.85}$$

$$\mu_C(x) = \begin{cases} 0 & ; x \geq 30 \\ (1 + x(x-30)^{-2})^{-1} & ; x < 30. \end{cases} \tag{2.86}$$

The fuzzy decision, the convex fuzzy decision, and the product fuzzy decision for this situation are depicted in Figure 2.15.

CHAPTER 3

FUZZY LINEAR PROGRAMMING

In 1976, H.-J. Zimmermann first introduced fuzzy set theory into conventional linear programming problems. He considered linear programming problems with a fuzzy goal and fuzzy constraints. Following the fuzzy decision proposed by Bellman and Zadeh (1970) together with linear membership functions, he proved that there exists an equivalent linear programming problem. Since then, fuzzy linear programming has been developed in a number of directions with many successful applications. Nowadays, fuzzy programming is considered an important area of multiobjective optimization under fuzziness.

In this chapter, after an overview of the basic concepts of linear programming via a simple numerical example, fuzzy linear programming as proposed by Zimmermann is presented. Then, fundamental notions and methods of multiobjective linear programming are presented and fuzzy multiobjective linear programming is introduced. Finally, interactive fuzzy multiobjective linear programming is explained in detail.

3.1 Linear programming and fuzzy linear programming

First, consider the following simple production planning problem as an example of a problem that can be solved using linear programming.

Example 3.1 (Production planning problem)

Company C would like to maximize the total profit producing two products P_1 and P_2 utilizing three different materials M_1, M_2, and M_3. The company knows that to produce 1 ton of product P_1 requires 2 tons of material M_1, 8 tons of material M_2, and 3 tons of material M_3, while to produce 1 ton of product P_2 requires 6 tons of material M_1, 6 tons of material M_2 and 1 ton of material M_3. The total amounts of available materials are limited to 27 tons, 45 tons and 15 tons for M_1, M_2, and M_3, respectively. It also knows that product P_1 yields a profit of 1 million yen per ton, while P_2 yields 2 million yen (see Table 3.1). Given these limited materials, the company is trying to figure out how many units of products P_1 and P_2 should be produced to maximize the total profit.

Table 3.1 Production conditions and profit

	P_1	P_2	Amounts available
M_1 (ton)	2	6	27
M_2 (ton)	8	6	45
M_3 (ton)	3	1	15
Profit (million yen)	1	2	

Let x_1 and x_2 denote the number of tons produced of products P_1 and P_2, respectively. Then this production planning problem can be formulated as the following linear programming problem:

maximize the linear profit function

$$x_1 + 2x_2$$

subject to the linear constraints

$$\begin{aligned} 2x_1 + 6x_2 &\leq 27 \\ 8x_1 + 6x_2 &\leq 45 \\ 3x_1 + x_2 &\leq 15 \\ x_1 \geq 0,\ x_2 &\geq 0. \end{aligned}$$

For convenience in our subsequent discussion, let the opposite of the total profit be

$$z = -x_1 - 2x_2 \tag{3.1}$$

and convert the problem to the minimization problem to minimize z under the above constraints, i.e.,

$$\left.\begin{array}{lrl} \text{minimize} & z = & -x_1 - 2x_2 \\ \text{subject to} & & 2x_1 + 6x_2 \leq 27 \\ & & 8x_1 + 6x_2 \leq 45 \\ & & 3x_1 + x_2 \leq 15 \\ & & x_1 \geq 0,\ x_2 \geq 0. \end{array}\right\} \tag{3.2}$$

It is easy to see that, in the $x_1 - x_2$ plane, the linearly constrained set of points (x_1, x_2) satisfying all of the constraints of (3.2) becomes the boundary lines and interior points of the convex pentagon $ABCDE$ shown in Figure 3.1.

The set of points satisfying $-x_1 - 2x_2 = z$ for fixed z is a line. As z is varied, the line is moved parallel to itself. The optimal value of this example is the smallest value of z for which the corresponding line has at least one point in common with the linearly constrained set. As can been seen from Figure 3.1, this occurs at point D. Hence, the optimal solution to this problem is

$$x_1 = 3, \quad x_2 = 3.5, \quad z = -10. \tag{3.3}$$

It is significant to realize here that the optimal solution occurs at a vertex point of the linearly constrained set since the constrained set has a finite number of vertices and the contours of constant value of the objective function are linear. Note that vertices are usually called extreme points in linear programming.

The values of the objective function corresponding to the extreme points A, B, C, D, and E are 0, -5, -7.5, -10, and -9, respectively. Therefore, for

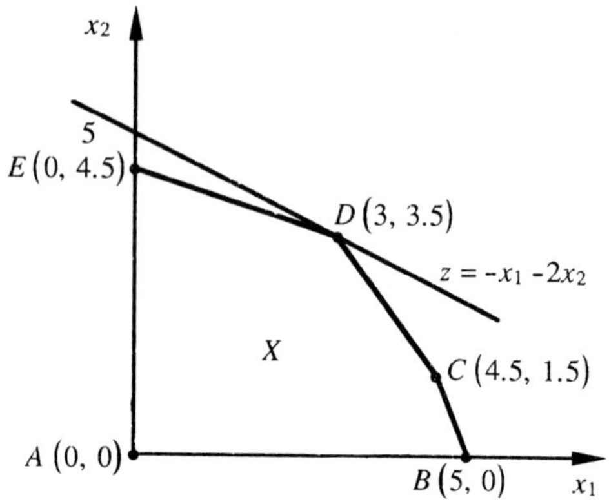

Figure 3.1 Feasible region for Example 3.1

example, starting from any extreme point, if we move $A \to B \to C \to D$ or $A \to E \to D$ such that the values of the objective function are decreasing, it seems to be possible to reach the extreme point which gives the minimum value of z.

Obviously, for more than two or three variables, such a graphical solution cannot be applied and it becomes necessary to characterize extreme points algebraically. The simplex method of linear programming originated by G. B. Dantzig (1947) is well-known and widely used as a powerful computational procedure for solving linear programming problems. The simplex method consists of two phases. Phase 1 finds an initial extreme point of the constrained set or gives the information that none exists due to the inconsistency of the constraints. In Phase 2, starting from an initial extreme point, it determines whether it is optimal or not. If not, it finds an adjacent extreme point at which the value of z is less than or equal to the previous value. The process is repeated until it finds an optimal solution or gives the information that the problem has an unbounded optimal solution. The details of linear programming can be found in standard texts including Dantzig (1963), Hadley (1962), and Luenberger (1973, 1984).

The preceding simple production planning problem is generalized to the general linear programming problem written in the following form‡:

minimize the linear objective function

$$z = c_1x_1 + c_2x_2 + \cdots + c_nx_n \tag{3.4}$$

‡ For convenience, we start from inequality constraints, but similar discussions can be made for equality constraints.

subject to the m linear inequality constraints

$$\left.\begin{array}{c} a_{11}x_1 + a_{12}x_2 + \cdots + a_{1n}x_n \leq b_1 \\ a_{21}x_1 + a_{22}x_2 + \cdots + a_{2n}x_n \leq b_2 \\ \cdots\cdots\cdots\cdots\cdots\cdots \\ a_{m1}x_1 + a_{m2}x_2 + \cdots + a_{mn}x_n \leq b_m \end{array}\right\} \tag{3.5}$$

and nonnegativity conditions for all variables

$$x_j \geq 0, \quad j = 1, 2, ..., n \tag{3.6}$$

where the a_{ij}, b_i, and c_j are given constants.

By introducing an n-dimensional row vector $c = (c_1, \ldots, c_n)$, an n-dimensional column vector $x = (x_1, \ldots, x_n)^T$, an m-dimensional column vector $b = (b_1, \ldots, b_m)^T$, and an $m \times n$ matrix $A = [a_{ij}]$, this problem can be expressed in a more compact vector-matrix form

$$\left.\begin{array}{rl} \text{minimize} & z = cx \\ \text{subject to} & Ax \leq b \\ & x \geq 0. \end{array}\right\} \tag{3.7}$$

In contrast to conventional linear programming problem, Zimmermann (1976) proposed to soften the rigid requirements of the decision maker (DM) to strictly minimize the objective function and to strictly satisfy the constraints. Namely, by considering the imprecision or fuzziness of the DM's judgment, he softened the usual linear programming problem into the following fuzzy version:

$$\left.\begin{array}{r} cx \precsim z_0 \\ Ax \precsim b \\ x \geq 0 \end{array}\right\} \tag{3.8}$$

where the symbol " $\precsim$ " denotes a relaxed or fuzzy version of the ordinary inequality " $\leq$ ". To be more explicit, these fuzzy inequalities representing the DM's fuzzy goal and fuzzy constraints mean that "the objective function cx should be essentially smaller than or equal to an aspiration level z_0 of the DM" and "the constraints Ax should be essentially smaller than or equal to b," respectively.

In the same spirit as the fuzzy decision of Bellman and Zadeh (1970), considering the fuzzy goal $cx \precsim z_0$ and fuzzy constraints $Ax \precsim b$ as equally important, Zimmermann (1976) expressed the problem as follows:

$$\left.\begin{array}{r} Bx \precsim b' \\ x \geq 0 \end{array}\right\} \tag{3.9}$$

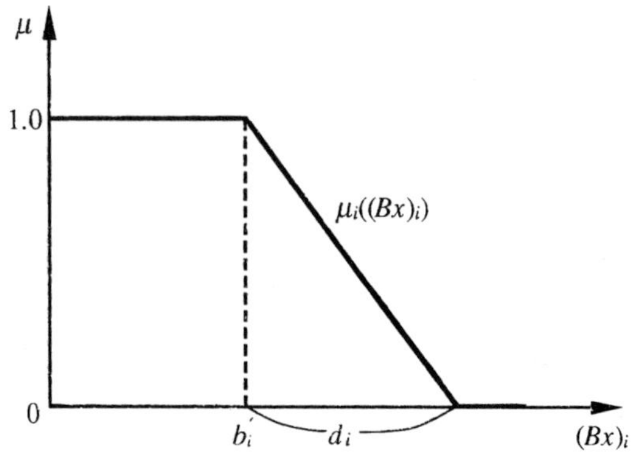

Figure 3.2 Linear membership function

where

$$B = \begin{bmatrix} c \\ A \end{bmatrix}, \quad b' = \begin{bmatrix} z_0 \\ b \end{bmatrix}. \tag{3.10}$$

For treating the ith fuzzy inequality $(Bx)_i \preceq b_i'$, $i = 0, ..., m$, of the DM's fuzzy inequalities $Bx \preceq b'$, he proposed the following linear membership function:

$$\mu_i((Bx)_i) = \begin{cases} 1 & ; \quad (Bx)_i \leq b_i' \\ 1 - \dfrac{(Bx)_i - b_i'}{d_i} & ; \quad b_i' \leq (Bx)_i \leq b_i' + d_i \\ 0 & ; \quad (Bx)_i \geq b_i' + d_i \end{cases} \tag{3.11}$$

where each d_i is a subjectively chosen constant expressing the limit of the admissible violation of the ith inequality. It is assumed that the ith membership function should be 1 if the ith constraint is well satisfied, 0 if the ith constraint is violated beyond its limit d_i, and linear from 0 to 1. Such a linear membership function is illustrated in Figure 3.2.

Following the fuzzy decision of Bellman and Zadeh (1970) together with the linear membership functions, the problem of finding the maximum decision is to choose x^* such that

$$\mu_D(x^*) = \max_{x \geq 0} \min_{i=0,...,m} \{\mu_i((Bx)_i)\}. \tag{3.12}$$

In other words, the problem is to find the $x^* \geq 0$ which maximizes the minimum membership function values.

Substituting

$$b_i'' = b_i'/d_i, \quad (B'x)_i = (Bx)_i/d_i, \tag{3.13}$$

the problem is rewritten as

$$\mu_D(x^*) = \max_{x \geq 0} \min_{i=0,...,m} \{1 + b_i'' - (B'x)_i\}. \tag{3.14}$$

By introducing the auxiliary variable λ, this problem can be transformed into the following equivalent conventional linear programming problem:

$$\left.\begin{array}{ll} \text{maximize} & \lambda \\ \text{subject to} & \lambda \leq 1 + b''_i - (B'x)_i, \ i = 0, ..., m \\ & x \geq 0. \end{array}\right\} \quad (3.15)$$

The fuzzy decision of Bellman and Zadeh is sometimes called the minimum operator, since it is expressed as $\min_{i=0,\ldots,m}\{\mu_i((Bx)_i)\}$.

Example 3.2 (Production planning problem)

Recall the production planning problem of Example 3.1:

$$\left.\begin{array}{lll} \text{minimize} & z = & -x_1 - 2x_2 \\ \text{subject to} & & 2x_1 + 6x_2 \leq 27 \\ & & 8x_1 + 6x_2 \leq 45 \\ & & 3x_1 + \ x_2 \leq 15 \\ & & x_1 \geq 0, \ x_2 \geq 0. \end{array}\right\} \quad (3.16)$$

The optimal solution to this problem is

$$x_1 = 3, \quad x_2 = 3.5, \qquad \min z = -10 \quad (3.17)$$

as discussed in Example 3.1.

Instead of giving rigid numerical values for the right-hand side constants as in the conventional linear programming problem, assume that the DM has the fuzzy goal and the fuzzy constraints shown in Table 3.2.

Assuming the linear membership functions from $\mu = 0$ to $\mu = 1$ for these fuzzy equalities, the flexible formulation of the original problem in the form of a fuzzy linear programming problem can be transformed into the following

Table 3.2 Nonfuzzy and fuzzy constraints

	Nonfuzzy	Fuzzy	
		$\mu = 0$	$\mu = 1$
Objective function	−10	−9.5	−10.5
First constraint	27	30	27
Second constraint	45	50	45
Third constraint	15	18	15

equivalent conventional linear programming problem:

$$\left.\begin{array}{ll} \text{minimize} & \lambda \\ \text{subject to} & \frac{2}{3}x_1 + \ \ 2x_2 + \lambda \leq 10 \\ & 1.6x_1 + 1.2x_2 + \lambda \leq 10 \\ & 1.5x_1 + 0.5x_2 + \lambda \leq 8.5 \\ & x_1 + \ \ 2x_2 - \lambda \geq 9.5 \\ & x_1 \geq 0,\ x_2 \geq 0. \end{array}\right\} \tag{3.18}$$

Solving this problem by the simplex method of linear programming yields the optimal solution

$$x_1 = 3.0789, \quad x_2 = 3.5921, \qquad \lambda = 0.76316. \tag{3.19}$$

In Table 3.3, the optimal solutions corresponding to the fuzzy linear programming problem as well as the nonfuzzy linear programming problem are shown.

In this fuzzy linear programming problem, observe that all of the original rigid constraints are softened by admitting the admissible violations. For example, the first constraint is softened to be "substantially less than or equal to 27" with linear function from $\mu_1(30) = 0$ to $\mu_1(27) = 1$. Moreover, instead of the minimization of the objective function, the admissible region roughly from -10.5 to -9.5 is considered the satisfaction level for the objective function. As a result, in this example, it may be natural that the DM has about 2.6% additional total profit at the expense of some additional material as shown in Table 3.3. Thus, in fuzzy linear programming, the DM is no longer forced to formulate the problem in precise and rigid form as in linear programming, which seems to be one of the major advantages of fuzzy linear programming.

In 1978, G. Sommer and M. A. Pollastschek proposed the adoption of the add operator for aggregating the DM's fuzzy goal and fuzzy constraints instead of the minimum operator. Their add operator can be viewed as a special case of the convex fuzzy decision by setting all α_i and β_j equal to 1.

Using the linear membership functions in (3.11) for representing the DM's fuzzy goal and fuzzy constraints and adopting the add operator instead of the

Table 3.3 Solutions of nonfuzzy and fuzzy LP

Nonfuzzy			Fuzzy		
x_1	=	3	x_1	=	3.0789
x_2	=	3.5	x_2	=	3.5921
z	=	-10	z	=	-10.2631
Constraints			Constraints		
1	:	27	1	:	27.71
2	:	45	2	:	46.18
3	:	12.5	3	:	12.83

minimum operator, the fuzzy version of the original linear programming problem becomes

$$\left.\begin{array}{ll}\text{maximize} & \sum_{i=0}^{m} \mu_i((Bx)_i) \\ \text{subject to} & x \geq 0.\end{array}\right\} \tag{3.20}$$

Obviously, this is a linear programming problem and can be easily solved by the simplex method of linear programming.

Sommer and Pollastschek (1978) also showed that constraints such as $Ax \geq b$ and $Ax = b$ can be softened similar to $Ax \leq b$ and applied their method successfully to the air pollution regulation problem.

3.2 Multiobjective linear programming

3.2.1 Problem formulation and solution concepts

Recall the production planning problem discussed in Example 3.1.

Example 3.1 (Production planning problem)

Company C would like to maximize the total profit producing two products P_1 and P_2 utilizing three different materials M_1, M_2, and M_3. The company knows that to produce 1 ton of product P_1 requires 2 tons of material M_1, 8 tons of material M_2, and 3 tons of material M_3, while to produce 1 ton of product P_2 requires 6 tons of material M_1, 6 tons of material M_2, and 1 ton of material M_3. The total amounts of available materials are limited to 27 tons, 45 tons, and 15 tons for M_1, M_2, and M_3, respectively. It also knows that product P_1 yields a profit of 1 million yen per ton, while P_2 yields 2 million yen (see Table 3.1). Given these limited materials, the company is trying to figure out how many units of products P_1 and P_2 should be produced to maximize the total profit.

Example 3.3 (Production planning with environmental quality)

Unfortunately, however, in this problem, it is pointed out that product P_1 yields 3 units of pollution per ton and product P_2 yields 2 units of pollution per ton. Thus, the manager should not only maximize the total profit but minimize the amount of pollution.

For simplicity, assume that the amount of pollution is a linear function of two variables x_1 and x_2 such as

$$3x_1 + 2x_2$$

where x_1 and x_2 denote the number of tons produced of products P_1 and P_2, respectively.

Then the production planning problem considering environmental quality can be formulated as the following two-objective linear programming problem:

$$\left.\begin{array}{lrl} \text{minimize} & z_1 = & -x_1 - 2x_2 \\ \text{minimize} & z_2 = & 3x_1 + 2x_2 \\ \text{subject to} & & 2x_1 + 6x_2 \leq 27 \\ & & 8x_1 + 6x_2 \leq 45 \\ & & 3x_1 + \ \ x_2 \leq 15 \\ & & x_1 \geq 0,\ x_2 \geq 0. \end{array}\right\} \tag{3.21}$$

The problem to optimize such multiple conflicting linear objective functions simultaneously under the given linear constraints is called the multiobjective linear programming (MOLP) problem and can be generalized as follows:

minimize k linear objective functions

$$\left.\begin{array}{c} z_1(x) = c_1 x \\ z_2(x) = c_2 x \\ \cdots\cdots \\ z_k(x) = c_k x \end{array}\right\} \tag{3.22}$$

subject to the linear inequality constraints

$$Ax \leq b \tag{3.23}$$

and the nonnegativity conditions

$$x \geq 0 \tag{3.24}$$

where

$$c_i = (c_{i1}, \ldots, c_{in}), \quad i = 1, \ldots, k, \tag{3.25}$$

$$x = (x_1, \ldots, x_n)^T, \tag{3.26}$$

$$A = \begin{bmatrix} a_{11}, \ldots, a_{1n} \\ \cdots\cdots\cdots \\ a_{m1}, \ldots, a_{mn} \end{bmatrix}, \tag{3.27}$$

$$b = (b_1, \ldots, b_n)^T. \tag{3.28}$$

Such a multiobjective linear programming problem is sometimes expressed as the following vector minimization problem

$$\left.\begin{array}{rl} \text{minimize} & z(x) \triangleq Cx \\ \text{subject to} & x \in X \triangleq \{x \in R^n \mid Ax \leq b,\ x \geq 0\} \end{array}\right\} \tag{3.29}$$

where $z(x) = (z_1(x), \ldots, z_k(x))^T = (c_1 x, \ldots, c_k x)^T$ is a k-dimensional vector and $C = (c_1, \ldots, c_k)^T$ is a $k \times n$ matrix.

If we directly apply the notion of optimality for single-objective linear programming to this multiobjective linear programming, we arrive at the following notion of a complete optimal solution.

Definition 3.1 (Complete optimal solution)

x^* is said to be a complete optimal solution, if and only if there exists $x^* \in X$ such that $z_i(x^*) \leq z_i(x), i = 1, \cdots, k$, for all $x \in X$.

However, in general, such a complete optimal solution that simultaneously minimizes all of the multiple objective functions does not always exist when the objective functions conflict with each other. Thus, instead of a complete optimal solution, a new solution concept, called Pareto optimality, is introduced in multiobjective linear programming.

Definition 3.2 (Pareto optimal solution)

x^* is said to be a Pareto optimal solution if and only if there does not exist another $x \in X$ such that $z_i(x) \leq z_i(x^*)$ for all i and $z_j(x) \neq z_j(x^*)$ for at least one j.

As can be seen from the definition, a Pareto optimal solution consists of an infinite number of points. A Pareto optimal solution is sometimes called a noninferior solution since it is not inferior to other feasible solutions.

In addition to Pareto optimality, the following weak Pareto optimality is defined as a slightly weaker solution concept than Pareto optimality.

Definition 3.3 (Weak Pareto optimal solution)

x^* is said to be a weak Pareto optimal solution if and only if there does not exist another $x \in X$ such that $z_i(x) < z_i(x^*), i = 1, \cdots, k$.

For notational convenience, let X^{CO}, X^P, or X^{WP} denote complete optimal, Pareto optimal, or weak Pareto optimal solution sets, respectively. Then from their definitions, it can be easily understood that the following relation holds:

$$X^{CO} \subseteq X^P \subseteq X^{WP}. \tag{3.30}$$

To understand the notion of Pareto optimal solutions of multiobjective linear programming problems, consider the two objective linear programming problem in Example 3.3. The feasible region X for this problem in the $x_1 - x_2$ plane becomes the boundary lines and interior points of the convex pentagon $ABCDE$ in Figure 3.3. Among the five extreme points A, B, C, D, and E, observe that z_1 is minimized at the extreme point $D(3, 3.5)$ while z_2 is minimized at the extreme point $A(0, 0)$. These two extreme points A and D are obviously Pareto optimal solutions since they cannot improve respective objective functions z_1 or z_2 any more. In addition to the extreme points A and D, the extreme point E and all of the points of the segments AE and ED are Pareto optimal solutions since they can be improved only at the expense of either z_1 or z_2. However, all of the

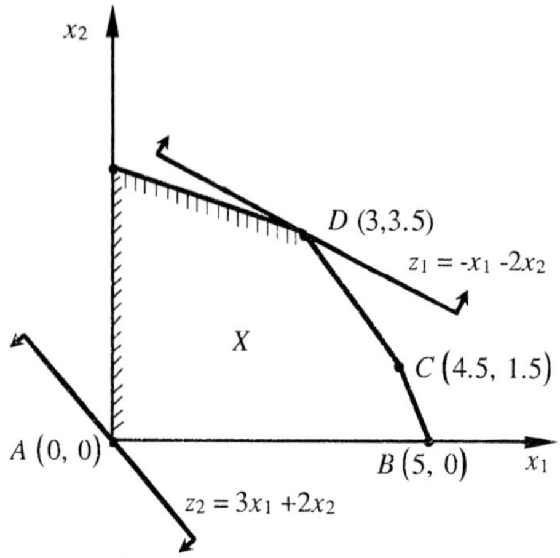

Figure 3.3 Feasible region for Example 3.3 in $x_1 - x_2$ plane

remaining feasible points are not Pareto optimal since there always exist other feasible points which improve at least one of the objective functions.

This situation can be more easily understood by observing the feasible region

$$Z = \{(z_1, z_2) \mid x \in X\} \tag{3.31}$$

in the $z_1 - z_2$ plane as depicted in Figure 3.4.

It is not hard to check that if the objective functions of this problem are changed to

$$z_1 = -x_1 - 4x_2, \quad z_2 = 2x_1 - x_2, \tag{3.32}$$

there exists a complete optimal solution $(z_1, z_2) = (-18, -4.5)$.

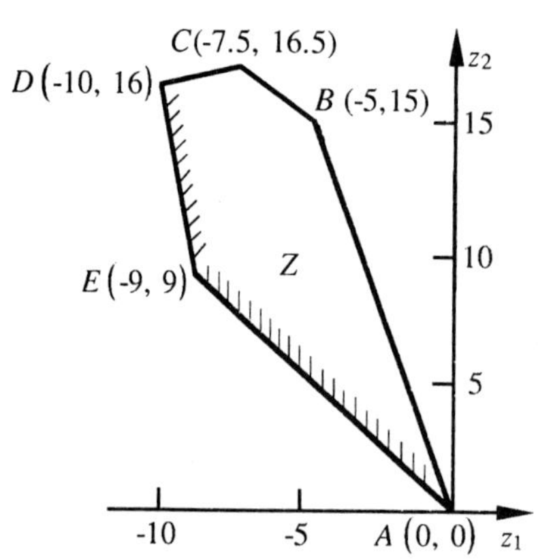

Figure 3.4 Feasible region for Example 3.3 in $z_1 - z_2$ plane

Moreover, if they are changed to

$$z_1 = -(2/3)x_1 - 2x_2, \quad z_2 = 3x_1 + 2x_2, \tag{3.33}$$

there exist weak Pareto optimal solutions.

3.2.2 Scalarization methods

Several computational methods have been proposed for characterizing Pareto optimal solutions depending on the different methods to scalarize the multiobjective linear programming (MOLP) problems. Among the many possible ways of scalarizing the MOLP, the weighting method, the constraint method, and the weighted minimax method have been studied as a means of characterizing Pareto optimal solutions of the MOLP.

(1) The weighting method

The weighting method for obtaining a Pareto optimal solution is to solve the weighting problem formulated by taking the weighted sum of all of the objective functions of the original MOLP (Kuhn and Tucker 1951, Zadeh 1963). Thus, the weighting problem is defined by

$$\left.\begin{array}{ll} \text{minimize} & wz(x) \triangleq \sum_{i=1}^{k} w_i z_i(x) \\ \text{subject to} & x \in X, \end{array}\right\} \tag{3.34}$$

where $w = (w_1, \ldots, w_k)$ is the vector of weighting coefficients assigned to the objective functions, and assumed to be

$$w = (w_1, \ldots, w_k) \geq 0. \tag{3.35}$$

The relationships between the optimal solution x^* of the weighting coefficient problem and the Pareto optimality concept of the MOLP can be characterized by the following theorems.

Theorem 3.1

If $x^* \in X$ is an optimal solution of the weighting problem for some $w > 0$, then x^* is a Pareto optimal solution of the MOLP.

Proof

If an optimal solution x^* of the weighting problem is not a Pareto optimal solution of the MOLP, then there exists $x \in X$ such that $z_j(x) < z_j(x^*)$ for some j and $z_i(x) \leq z_i(x^*)$, $i = 1, \ldots, k; i \neq j$. Noting $w = (w_1, \ldots, w_k) > 0$, this implies $\sum_i w_i z_i(x) < \sum_i w_i z_i(x^*)$. However, this contradicts the assumption that x^* is an optimal solution of the weighting problem for some $w > 0$. □

It should be noted here that the condition of Theorem 3.1 can be replaced by a unique optimal solution of the weighting problem for $w \geq 0$.

Theorem 3.2

If $x^* \in X$ is a Pareto optimal solution of the MOLP, then x^* is an optimal solution of the weighting problem for some $w = (w_1, \ldots, w_k) \geq 0$.

Proof

First we prove that x^* is an optimal solution of the linear programming problem

$$\left.\begin{array}{rl} \text{minimize} & 1^T Cx \\ \text{subject to} & Cx \leq Cx^* \\ & Ax \leq b, \; x \geq 0 \end{array}\right\} \tag{3.36}$$

where 1 is a k-dimensional column vector whose elements are 1.

If x^* is not an optimal solution of this problem, then there exists $x \in X$ such that $Cx \leq Cx^*$ and $1^T Cx < 1^T Cx^*$. This means that there exists $x \in X$ such that

$$1^T Cx = \sum_{i=1}^{k} c_i x < \sum_{i=1}^{k} c_i x^* = 1^T Cx^* \quad \text{and} \quad c_i x \leq c_i x^*, \; i = 1, \ldots, k$$

or equivalently

$$c_j x < c_j x^* \text{ for some } j \quad \text{and} \quad c_i x \leq c_i x^*, \; i = 1, \ldots, k; i \neq j$$

which contradicts the fact that x^* is a Pareto optimal solution. Hence, x^* is an optimal solution of this problem.

Now consider the dual problem

$$\left.\begin{array}{rl} \text{maximize} & (-b^T, -x^{*T} C^T) y \\ \text{subject to} & (-A^T, -C^T) y \leq C^T 1 \\ & y \geq 0 \end{array}\right\} \tag{3.37}$$

of this linear programming problem.

From the (strong) duality theorem of linear programming, it holds that

$$(-b^T, -x^{*T} C^T) y^* = 1^T Cx^*$$

and thus

$$(1^T + y_2^{*T}) Cx^* = -b^T y_1^*, \quad y^{*T} = (y_1^{*T}, y_2^{*T}).$$

Substituting

$$w = 1 + y_2^{*T},$$

we have

$$\sum_{i=1}^{k} w_i c_i x = w^T Cx = (1 + y_2^*)^T Cx, \qquad \forall x \in X.$$

Next observe that the dual problem of the linear programming problem

$$\left.\begin{array}{rl} \text{minimize} & (1 + y_2^*)^T Cx \\ \text{subject to} & Ax \leq b, \; x \geq 0 \end{array}\right\} \tag{3.38}$$

becomes

$$\left.\begin{array}{ll} \text{maximize} & -b^T u \\ \text{subject to} & -A^T u \leq C^T(1 + y_2^*) \\ & u \geq 0. \end{array}\right\} \tag{3.39}$$

$x = x^*$ and $u = y_1^*$ are feasible solutions for the corresponding linear programming problems and each value of the objective functions is equal. Hence, for any $x \in X$, it follows that

$$\sum_{i=1}^{k} w_i c_i x^* = (1 + y_2^*)^T C x^* \leq \min_{x \in X} (1 + y_2^*)^T C x = (1 + y_2^*)^T C x = \sum_{i=1}^{k} w_i c_i x.$$

This implies that x^* is an optimal solution of the weighting problem for $w = (w_1, \ldots, w_k) \geq 0$. This completes the proof of the theorem. □

The weighting coefficients of the weighting problem give the trade-off rate information between the objective functions. They show how many units of value of one objective function have to be given up in order to obtain one additional unit of value of the other objective function. This fact may be intuitively understood as follows.

Geometrically, in the k-dimensional $z = (z_1, \ldots, z_k)$ space,

$$W = w_1 z_1(x) + w_2 z_2(x) + \cdots + w_k z_k(x) = c \quad \text{(constant)} \tag{3.40}$$

represents the hyperplane (note that in the case of two objectives it is a line, in the case of three objectives, a plane) with the normal vector $w = (w_1, \ldots, w_k)$. Solving the weighting problem for the given weighting coefficient $w^* > 0$ yields the minimum c such that this hyperplane has at least one common point with the feasible region Z in the $z = (z_1, \ldots, z_k)$ space, and the corresponding Pareto optimal solution x^* is obtained as in Figure 3.5.

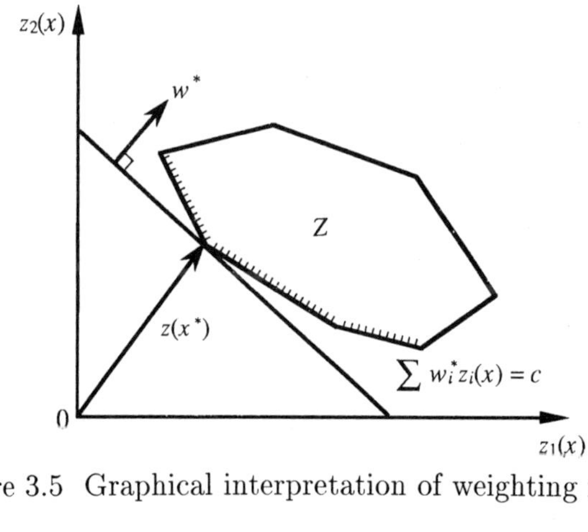

Figure 3.5 Graphical interpretation of weighting method

The hyperplane for this minimum c is the supporting hyperplane of the feasible region Z at the point $z(x^*)$ of the Pareto optimal surface. The condition for the small displacement from the point $z(x^*)$ belonging to this supporting hyperplane is $\Delta W = 0$, i.e.,

$$w_1^* \Delta z_1 + w_2^* \Delta z_2 + \cdots + w_k^* \Delta z_k = 0. \tag{3.41}$$

For fixed values of $\Delta z_j = 0$, $j = 2, 3, \ldots, k$, $j \neq i$ except z_1 and z_i, we have

$$w_1^* \Delta z_1 + w_i^* \Delta z_i = 0. \tag{3.42}$$

Hence, it holds that

$$-\frac{\partial z_1}{\partial z_i} = \frac{w_i^*}{w_1^*}. \tag{3.43}$$

Therefore, the ratios of the weighting coefficients w_i^*/w_1^* give the trade-off rates between the objective functions at $z(x^*)$.

To illustrate the weighting method, consider the problem of Example 3.3. The corresponding weighting problem becomes as follows:

$$\left.\begin{array}{lll} \text{minimize} & wz(x) \triangleq & w_1(-x_1 - 2x_2) + w_2(3x_1 + 2x_2) \\ \text{subject to} & & 2x_1 + 6x_2 \leq 27 \\ & & 8x_1 + 6x_2 \leq 45 \\ & & 3x_1 + \ x_2 \leq 15 \\ & & x_1 \geq 0,\ x_2 \geq 0. \end{array}\right\} \tag{3.44}$$

For this problem, for example, if we choose $w_1 = 0.6$ and $w_2 = 0.4$, we obtain

$$wz(x) = 0.6x_1 - 0.4x_2.$$

As depicted in Figure 3.6, it can be easily understood that solving the corresponding weighting problem yields the extreme point $E(0, 4.5)$ as a Pareto optimal solution. Also, as two extreme cases, if we set $w_1 = 1$, $w_2 = 0$ and $w_1 = 0$, $w_2 = 1$, from Figure 3.6, the optimal solutions of the corresponding weighting problems become the extreme point $D(3, 3.5)$ and the extreme point $A(0, 0)$, respectively. In these cases, although the condition $w > 0$ of Theorem 3.1 is not satisfied, from Figure 3.6, it can be seen that these two extreme points are Pareto optimal solutions.

(2) The constraint method

The constraint method for characterizing Pareto optimal solutions is to solve the constraint problem formulated by taking one objective function as the objective function and letting all the other objective functions be inequality constraints

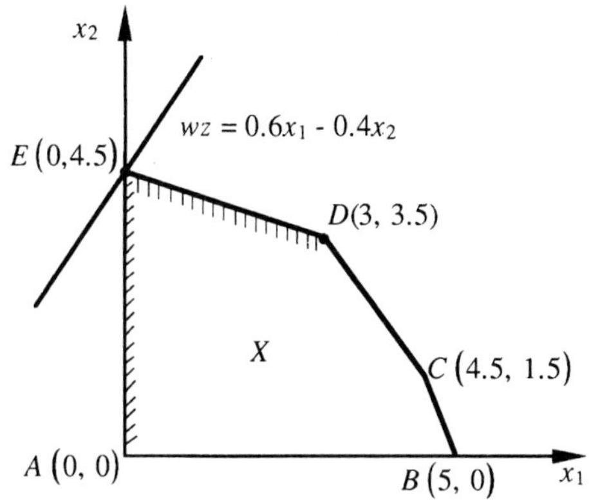

Figure 3.6 Weighting method for Example 3.3

(Haimes et al. 1971, Haimes and Hall 1974). The constraint problem is defined by

$$\left.\begin{array}{ll} \text{minimize} & z_j(x) \\ \text{subject to} & z_i(x) \leq \varepsilon_i, \ i = 1, \ldots, k; \ i \neq j \\ & x \in X. \end{array}\right\} \tag{3.45}$$

The relationships between the optimal solution x^* to the constraint problem and the Pareto optimal concept of the MOLP can be characterized by the following theorems.

Theorem 3.3

If $x^* \in X$ is a unique optimal solution to the constraint problem for some ε_i, $i = 1, \ldots, k$; $i \neq j$, then x^* is a Pareto optimal solution to the MOLP.

Proof

If a unique optimal solution x^* to the constraint problem is not a Pareto optimal solution of the MOLP, then there exists $x \in X$ such that $z_l(x) < z_l(x^*)$ for some l and $z_i(x) \leq z_i(x^*)$, $i = 1, \ldots, k; i \neq l$. This means either

$$z_i(x) \leq z_i(x^*) \leq \varepsilon_i, \quad i = 1, \ldots, k; \ i \neq j, \quad z_j(x) < z_j(x^*),$$

or

$$z_i(x) \leq z_i(x^*) \leq \varepsilon_i, \quad i = 1, \ldots, k; \ i \neq j, \quad z_j(x) = z_j(x^*),$$

which contradicts the assumption that x^* is a unique optimal solution of the constraint problem for some ε_i, $i = 1, \ldots, k$; $i \neq j$. □

As can be easily understood from the proof of this theorem, if the uniqueness of a solution is not guaranteed, only weak Pareto optimality is guaranteed.

Theorem 3.4

If $x^* \in X$ is a Pareto optimal solution of the MOLP, then x^* is an optimal solution of the constraint problem for some ε_i, $i = 1, \ldots, k$; $i \neq j$.

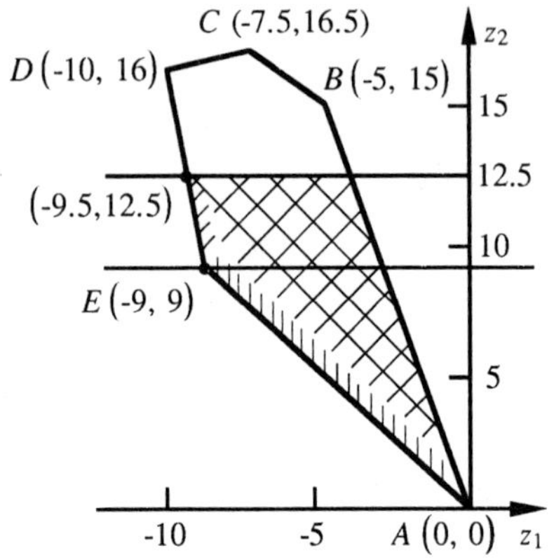

Figure 3.7 Constraint method for Example 3.3

Proof

If a Pareto optimal solution $x^* \in X$ of the MOLP is not an optimal solution to the constraint problem for ε_i, $i = 1, \ldots, k;\ i \neq j$, then there exists $x \in X$ such that

$$z_j(x) < z_j(x^*) \quad z_i(x) \leq \varepsilon_i = z_i(x^*), \quad i = 1, \ldots, k;\ i \neq j,$$

which contradicts the fact that x^* is a Pareto optimal solution to the MOLP. □

To illustrate the constraint method, consider Example 3.3. The constraint problem for j =1 becomes

$$\left.\begin{array}{lrl} \text{minimize} & z_1 = & -x_1 - 2x_2 \\ \text{subject to} & & 2x_1 + 6x_2 \leq 27 \\ & & 8x_1 + 6x_2 \leq 45 \\ & & 3x_1 + \ x_2 \leq 15 \\ & z_2 = & 3x_1 + 2x_2 \leq \varepsilon_2 \\ & & x_1 \geq 0,\ x_2 \geq 0. \end{array}\right\} \qquad (3.46)$$

Here, for example, if we choose $\varepsilon_2 = 9$, as illustrated in Figure 3.7, it can be understood that the optimal solution for this constraint problem occurs at the extreme point $E(-9, 9)$, and hence, yields a Pareto optimal solution. Also, if we choose $\varepsilon_2 = 12.5$, we can obtain a Pareto optimal solution such that $(z_1, z_2) = (-9.5, 12.5)$ as in Figure 3.7.

(3) The weighted minimax method

The weighted minimax method for characterizing Pareto optimal solutions is to solve the following weighted minimax problem (Bowman 1976):

$$\left.\begin{array}{ll} \text{minimize} & \max_{i=1,\ldots,k} w_i z_i(x) \\ \text{subject to} & x \in X \end{array}\right\} \tag{3.47}$$

or equivalently

$$\left.\begin{array}{ll} \text{minimize} & v \\ \text{subject to} & w_i z_i(x) \leq v, \; i = 1, \ldots, k \\ & x \in X \end{array}\right\} \tag{3.48}$$

by introducing the auxiliary variable v. Here, without loss of generality, it can be assumed that $z_i(x) > 0$, $i = 1, \cdots, k$, for all $x \in X$. Because, for the objective functions not satisfying $z_i(x) > 0, x \in X$, using the individual minimum $z_i^{\min} = \min_{x \in X} z_i(x)$ of the corresponding functions and setting

$$\hat{z}_i(x) = z_i(x) - z_i^{\min}, \tag{3.49}$$

it holds that $\hat{z}_i(x) > 0$, $i = 1, \ldots, k$, for all $x \in X$.

Geometrically, in the weighted minimax problem, the contours of $\max\{w_i z_i\} = c$ (constant) in the objective function space become rectangles corresponding to the given weighting coefficients. Hence, solving the weighted minimax problem yields the Pareto optimal solutions such that these rectangles support the feasible region

$$Z = \{z(x) \mid x \in X\} \tag{3.50}$$

as depicted in Figure 3.8.

The relationships between the optimal solution x^* of the weighted minimax problem and the Pareto optimal concept of the MOLP can be characterized by the following theorems.

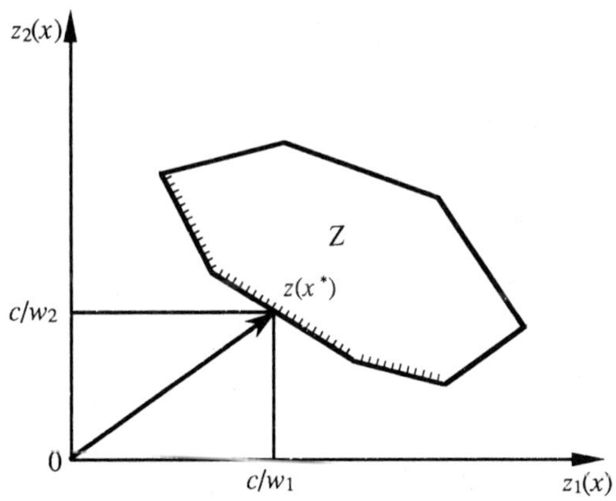

Figure 3.8 Graphical interpretation of weighted minimax method

Theorem 3.5

If $x^* \in X$ is a unique optimal solution of the weighted minimax problem for some $w = (w_1, \ldots, w_k) \geq 0$, then x^* is a Pareto optimal solution of the MOLP.

Proof

If a unique optimal solution x^* of the weighted minimax problem for some $w = (w_1, \ldots, w_k) \geq 0$ is not a Pareto optimal solution, then there exists $x \in X$ such that $z_j(x) < z_j(x^*)$ for some j and $z_i(x) \leq z_i(x^*)$, $i = 1, \ldots, k$; $i \neq j$. In view of $w = (w_1, \ldots, w_k) \geq 0$, it follows

$$w_i z_i(x) \leq w_i z_i(x^*), \quad i = 1, \ldots, k.$$

Hence,

$$\max_{x \in X} w_i z_i(x) \leq \max_{x \in X} w_i z_i(x^*).$$

This contradicts the fact that x^* is a unique optimal solution to the weighted minimax problem for $w = (w_1, \ldots, w_k) \geq 0$. □

From the proof of this theorem, if the uniqueness of a solution is not guaranteed, only weak Pareto optimality is guaranteed.

Theorem 3.6

If $x^* \in X$ is a Pareto optimal solution of the MOLP, then x^* is an optimal solution of the minimax problem for some $w = (w_1, \ldots, w_k) > 0$.

Proof

For a Pareto optimal solution $x^* \in X$ of the MOLP, choose $w^* = (w_1^*, \ldots, w_k^*) > 0$ such that $w_i^* z_i(x^*) = v$, $i = 1, \ldots, k$. Then in view of $z_i(x) > 0$, $i = 1, \ldots, k$, for all $x \in X$, we have $w^* = (w_1^*, \ldots, w_k^*) > 0$. Now assume that x^* is not an optimal solution of the minimax problem, then there exists $x \in X$ such that

$$w_i^* z_i(x) < w_i^* z_i(x^*) = v^*, \ i = 1, \ldots, k.$$

Noting $w^* = (w_1^*, \ldots, w_k^*) > 0$, this implies the existence of $x \in X$ such that

$$z_i(x) < z_i(x^*), \ i = 1, \ldots, k,$$

which contradicts the assumption that x^* is a Pareto optimal solution. □

From the proof, Pareto optimality in Theorem 3.6 can be replaced by weak Pareto optimality.

To illustrate the weighted minimax method, consider the problem of Example 3.3. Observe that the individual minimums of $z_1(x)$ and $z_2(x)$ are $\min_{x \in X} z_1(x) = -10$ and $\min_{x \in X} z_2(x) = 0$, respectively. If we substitute $\hat{z}_1(x) = z_1(x) - (-10)$ and choose $w_1 = 0.5$,$w_2 = 0.5$, then the weighted minimax problem becomes

$$\left.\begin{array}{ll} \text{minimize} & \max\,(-0.5x_1 - x_2 + 5, 1.5x_1 + x_2) \\ \text{subject to} & 2x_1 + 6x_2 \leq 27 \\ & 8x_1 + 6x_2 \leq 45 \\ & 3x_1 + \ \ x_2 \leq 15 \\ & x_1 \geq 0, \, x_2 \geq 0 \end{array}\right\} \tag{3.51}$$

or equivalently

$$\left.\begin{array}{ll} \text{minimize} & v \\ \text{subject to} & 2x_1 + 6x_2 \leq 27 \\ & 8x_1 + 6x_2 \leq 45 \\ & 3x_1 + \ \ x_2 \leq 15 \\ & -0.5x_1 - \ \ x_2 + 5 \leq v \\ & 1.5x_1 + \ \ x_2 \leq v \\ & x_1 \geq 0,\ x_2 \geq 0. \end{array}\right\} \tag{3.52}$$

Noting that $\hat{z}_1 = z_1 + 10$, as illustrated in Figure 3.9, it can be understood that the optimal solution of this problem is the point which translates the extreme point $E(-9, 9)$ by 10 along the z_1 axis. Hence, the optimal solution of the original problem is $(1, 9)$ which is a Pareto optimal solution.

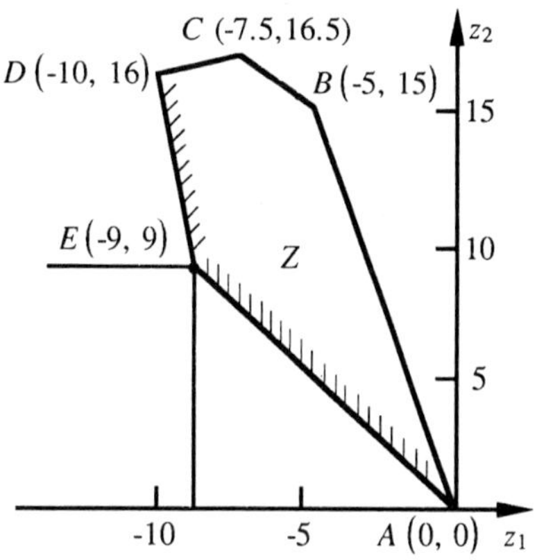

Figure 3.9 Weighted minimax method for Example 3.3

From Theorems 3.3 and 3.5, if the uniqueness of the optimal solution x^* for the scalarizing problem is not guaranteed, it is necessary to perform the Pareto optimality test of x^*. The Pareto optimality test for x^* can be performed by solving the following linear programming problem with the decision variables $x = (x_1, \ldots, x_n)^T$ and $\varepsilon = (\varepsilon_1, \ldots, \varepsilon_k)^T$:

$$\left.\begin{array}{ll} \text{maximize} & \sum_{i=1}^{k} \varepsilon_i \\ \text{subject to} & z_i(x) + \varepsilon_i = z_i(x^*),\ i = 1, \ldots, k \\ & x \in X,\ \varepsilon \geq 0. \end{array}\right\} \tag{3.53}$$

For the optimal solutions $\bar{x}$ and $\bar{\varepsilon}$ of this linear programming problem, the following theorem holds.

Theorem 3.7

For the optimal solution $\bar{x}$ and $\bar{\varepsilon}$ of the Pareto optimality test problem,

(1) *If all $\bar{\varepsilon}_i = 0$, then x^* is a Pareto optimal solution of the MOLP.*

(2) *If at least one $\bar{\varepsilon}_i \geq 0$, then x^* is not a Pareto optimal solution of the MOLP. Instead of x^*, $\bar{x}$ is the Pareto optimal solution corresponding to the scalarization problem.*

Proof

(1) If x^* is not a Pareto optimal solution to the MOLP, then there exists $x \in X$ such that $z_j(x) < z_j(x^*)$ for some j and $z_i(x) \leq z_i(x^*)$, $i = 1, \ldots, k$; $i \neq j$, which contradicts the assumption that all $\bar{\varepsilon}_i = 0$.

(2) If at least one $\bar{\varepsilon}_i \geq 0$ and $\bar{x}$ is not a Pareto optimal solution of the MOLP, then there exists $x \in X$ such that $z_j(x) < z_j(\bar{x})$ for some j and $z_i(x) \leq z_i(\bar{x})$, $i = 1, \ldots, k$; $i \neq j$. Hence, there exists $x \in X$ such that $z(x) + \varepsilon' = z(\bar{x})$ for some $\varepsilon' \geq 0$. This contradicts the optimality of $\bar{\varepsilon}$. □

Recently, as a generalization of such existing scalarization methods for characterizing Pareto optimal solutions of multiobjective optimization problems, a fully integrated scalarization method, called the hyperplane method was introduced by the authors (1989, 1990). The details of the hyperplane method can be found in the Appendix.

3.2.3 Linear goal programming

The term goal programming first appeared in a 1961 text by Charnes and Cooper to deal with multiobjective linear programming problems that assumed the decision maker (DM) could specify goals or aspiration levels for the objective functions. Subsequent works on goal programming have been numerous, including texts on goal programming by Ijiri (1965), Lee (1972), and Ignizio (1976, 1982), and survey papers by Charnes and Cooper (1977) and Ignizio (1983).

The key idea behind goal programming is to minimize the deviations from goals or aspiration levels set by the DM. Goal programming therefore, in most cases, seems to yield a satisficing solution in the same spirit as March and Simon (1958) rather than an optimizing one.

As discussed in the previous subsection, in general, the multiobjective linear programming (MOLP) problem can be formulated as follows:

$$\underset{x \in X}{\text{minimize}} \quad z(x) \triangleq (z_1(x), \ldots, z_k(x))^T \tag{3.54}$$

where $z_1(x) = c_1 x, \ldots, z_k(x) = c_k x$ are k distinct objective functions of the decision vector x and

$$X \triangleq \{x \in R^n \mid Ax \leq b,\ x \geq 0\} \tag{3.55}$$

is the linearly constrained feasible set.

For linear goal programming, however, a set of k goals is specified by the DM for the k objective functions $z_i(x)$ and the MOLP is converted into the problem of coming "as close as possible" to the set of specified goals which may not be simultaneously attainable.

The general formulation of goal programming thus becomes

$$\underset{x \in X}{\text{minimize}} \quad d\,(z(x), \hat{z}) \tag{3.56}$$

where $\hat{z} = (\hat{z}_1, \ldots, \hat{z}_k)$ is the goal vector specified by the DM and $d\,(\cdot, \cdot)$ represents the distance between $z(x)$ and $\hat{z}$ according to some selected norm.

The simplest version of (3.56), where the absolute value or ℓ_1 norm is used, becomes

$$\underset{x \in X}{\text{minimize}} \quad d_1(z(x), \hat{z}) \triangleq \sum_{i=1}^{k} |c_i x - \hat{z}_i|. \tag{3.57}$$

More generally, using the ℓ_1 norm with weights (the weighted ℓ_1 norm), it becomes

$$\underset{x \in X}{\text{minimize}} \quad d_1^w(z(x), \hat{z}) \triangleq \sum_{i=1}^{k} w_i |c_i x - \hat{z}_i| \tag{3.58}$$

where w_i is the nonnegative weight to the ith objective function.

This linear goal programming problem can easily be converted to an equivalent linear programming problem by introducing the auxiliary variables

$$d_i^+ = \frac{1}{2}\left\{|z_i(x) - \hat{z}_i| + (z_i(x) - \hat{z}_i)\right\} \tag{3.59}$$

and

$$d_i^- = \frac{1}{2}\left\{|z_i(x) - \hat{z}_i| - (z_i(x) - \hat{z}_i)\right\} \tag{3.60}$$

for each $i = 1, \ldots, k$. Thus, the equivalent linear goal programming formulation to the problem (3.58) becomes

$$\text{minimize} \quad \sum_{i=1}^{k} w_i(d_i^+ + d_i^-) \tag{3.61}$$

$$\text{subject to} \quad z_i(x) - d_i^+ + d_i^- = \hat{z}_i \quad i = 1, \ldots, k \tag{3.62}$$

$$Ax \leq b, \; x \geq 0 \tag{3.63}$$

$$d_i^+ \cdot d_i^- = 0, \; i = 1, \ldots, k \tag{3.64}$$

$$d_i^+ \geq 0, \; d_i^- \geq 0, \; i = 1, \ldots, k. \tag{3.65}$$

It is appropriate to consider here the practical significance of d_i^+ and d_i^-. From the definition of d_i^+ and d_i^-, it can be easily understood that

$$d_i^+ = \begin{cases} z_i(x) - \hat{z}_i & \text{if } z_i(x) \geq \hat{z}_i \\ 0 & \text{if } z_i(x) \leq \hat{z}_i \end{cases} \tag{3.66}$$

and

$$d_i^- = \begin{cases} \hat{z}_i - z_i(x) & \text{if } \hat{z}_i \geq z_i(x) \\ 0 & \text{if } \hat{z}_i \leq z_i(x). \end{cases} \tag{3.67}$$

Thus, d_i^+ and d_i^- represent, respectively, the over-achievement and under-achievement of the ith goal and, hence, are called deviational variables. Obviously, over-achievement and under-achievement can never occur simultaneously. When $d_i^+ > 0$, then d_i^- must be zero, and vice versa. This fact is reflected by (3.64) which is automatically satisfied at every iteration of the simplex method of linear programming because d_i^+ and d_i^- never become basic variables simultaneously. Nonlinear equality constraints (3.64) are always satisfied in the simplex method, and consequently, it is clear that the simplex method can be applied to solve this type of linear goal programming problem.

Depending on the decision situations, the DM is sometimes concerned only with either the over-achievement or under-achievement of a specified goal. Such a situation can be incorporated into the goal programming formulation by assigning over- and under-achievement weights w_i^+ and w_i^- to d_i^+ and d_i^-, respectively. For example, if each $z_i(x)$ is a cost-type objective function with its goal $\hat{z}_i$, over-achievement is not desirable. For this case, we set $w_i^+ = 1$ and $w_i^- = 0$, and the problem (3.61) $\sim$ (3.65) is modified as follows:

$$\text{minimize} \quad \sum_{i=1}^{k} w_i^+ d_i^+ \tag{3.68}$$

$$\text{subject to} \quad z_i(x) - d_i^+ + d_i^- = \hat{z}_i \quad i = 1, \ldots, k \tag{3.69}$$

$$Ax \leq b, \ x \geq 0 \tag{3.70}$$

$$d_i^+ \cdot d_i^- = 0, \ i = 1, \ldots, k \tag{3.71}$$

$$d_i^+ \geq 0, \ d_i^- \geq 0, \ i = 1, \ldots, k. \tag{3.72}$$

Conversely, for benefit-type objective functions, under-achievement is not desirable. For this case, we set $w_i^- = 1$ and $w_i^+ = 0$ to replace $\sum_{i=1}^{k} w_i^+ d_i^+$ by $\sum_{i=1}^{k} w_i^- d_i^-$ as the objective function in (3.61). This particular goal programming is called one-sided goal programming.

The linear goal programming formulation can also be modified into a more general form by introducing the preemptive priorities, P_i, in place of, or together with, the numerical weights $w_i^+, w_i^- \geq 0$. When the objective functions $z_1(x), \ldots, z_k(x)$ are divided into L ordinal ranking classes, $1 \leq L \leq k$, having the preemptive priorities $P_1, \ldots, P_L$ in decreasing order, it may be convenient to write

$$P_l \gg P_{l+1}, \quad l = 1, \ldots, L-1 \tag{3.73}$$

to mean that no real number t, however large, can produce

$$tP_{l+1} \gg P_l, \quad l = 1, \ldots, L-1. \tag{3.74}$$

By incorporating such preemptive priorities P_l together with over- and under-achievement weights w_i^+ and w_i^-, the general linear goal programming formulation takes on the following form:

$$\text{minimize} \quad \sum_{l=1}^{L} P_l \Big(\sum_{i \in I_l} (w_i^+ d_i^+ + w_i^- d_i^-) \Big) \tag{3.75}$$

$$\text{subject to} \quad z_i(x) - d_i^+ + d_i^- = \hat{z}_i \quad i = 1, \ldots, k \tag{3.76}$$

$$Ax \leq b, \ x \geq 0 \tag{3.77}$$

$$d_i^+ \cdot d_i^- = 0, \ i = 1, \ldots, k \tag{3.78}$$

$$d_i^+ \geq 0, \ d_i^- \geq 0, \ i = 1, \cdots, k \tag{3.79}$$

where $I_l \neq \emptyset$ is the index set of objective functions in the lth priority class. Observe that when there are k distinct ordinal ranking classes (i.e., $L = k$) with $c_i x$ belonging to the ith priority class, the objective function in (3.75) then becomes simply

$$\sum_{i=1}^{k} P_i (w_i^+ d_i^+ + w_i^- d_i^-). \tag{3.80}$$

To solve this type of linear goal programming problem, we begin by trying to achieve the goals of all objective functions in the first priority class. Having done that, we try to satisfy the goals in the second priority class, keeping the goals in the first class satisfied. The process is repeated until either a unique solution is obtained at some stage or all priority classes are considered. This is equivalent to solving, at most, L linear programming problems sequentially, for which the simplex method can easily be applied with some modifications.

Further details concerning the algorithm, extensions, and applications can be found in the text of Lee (1972) and Ignizio (1976, 1982).

Example 3.4 (Production planning problem with goals)

To illustrate the linear goal programming method, assume that the manager of company C established the following goals for the production planning problem incorporating environmental quality in Example 3.3. Here, Q and R are used instead of P_1 and P_2 to avoid confusion in the notation.

P_1 : To achieve at least 8 million yen of total profit.

P_2 : To keep the pollution level below 10 units.

P_3 : To produce at least 2 tons of Q and 3 tons of R. However, assign the weight ratio of 1:2 for Q and R by considering the profit contribution ratio of these two products.

Then the corresponding linear goal programming problem can be formulated as follows:

$$\left.\begin{array}{ll} \text{minimize} & P_1 d_1^- + P_2 d_2^+ + P_3(d_3^- + 2d_4^-) \\ \text{subject to} & x_1 + 2x_2 - d_1^+ + d_1^- = 8 \\ & 3x_1 + 2x_2 - d_2^+ + d_2^- = 10 \\ & x_1 \qquad - d_3^+ + d_3^- = 2 \\ & \qquad x_2 - d_4^+ + d_4^- = 3 \\ & 2x_1 + 6x_2 \leq 27 \\ & 8x_1 + 6x_2 \leq 45 \\ & 3x_1 + x_2 \leq 15 \\ & x_1 \geq 0,\ x_2 \geq 0 \\ & d_i^+ \geq 0,\ d_i^- \geq 0,\ i = 1, 2, 3, 4. \end{array}\right\} \tag{3.81}$$

To obtain a graphical solution for this simple example in the $x_1 - x_2$ plane, the three priority goals are plotted as straight lines together with the constrained set in Figure 3.10. Although only the decision variables x_1 and x_2 are used in this plot, the effect of increasing either d_i^+ or d_i^- is reflected by the arrow signs. The region which satisfies both the constrained set and the first priority goal (i.e., $d_i^+ = 0$ and $d_i^- \geq 0$) is shown as the cross-hatched region. To achieve the second priority goal without degrading the achievement of the first priority goal, the area of feasible solution should be limited to the crisscross-hatched area in Figure 3.10. However, as can been seen from Figure 3.10, concerning the third priority goals, both d_3^- and d_4^- cannot be set zero. In this way, the final solution

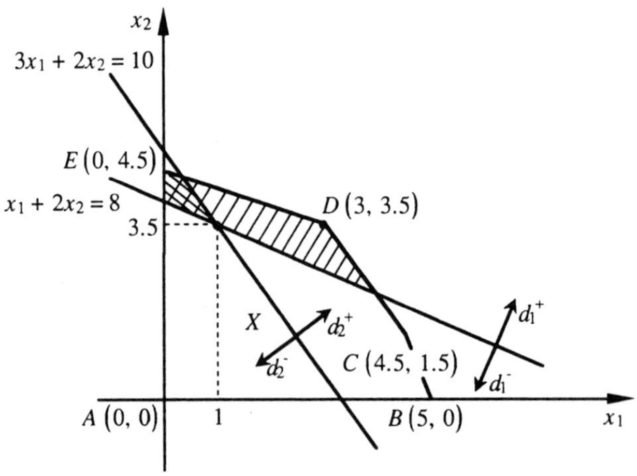

Figure 3.10 Graphical solution for Example 3.4

of this problem occurs at the point $x = (x_1, x_2) = (1, 3.5)$ in which only the first and second priority goals are satisfied.

A well-known extension of the goal programming approach is obtained if the goal vector $\hat{z} = (\hat{z}_1, \ldots, \hat{z}_k)$ is replaced by the so-called ideal or utopia vector $z^{\min} = (z_1^{\min}, \ldots, z_k^{\min})$, where $z_i^{\min} = \max_{x \in X} z_i(x),\ 1, \ldots, k$. The resulting problem can be interpreted as an attempt to minimize deviation from the ideal or utopia vector (point). Realizing that the ideal vector is generally infeasible in most of the MOLP with conflicting objective functions, Yu (1973) and Zeleny (1973) introduced the concept of compromise solution. This will be discussed further in the next chapter.

3.3 Interactive multiobjective linear programming

The STEP method (STEM) proposed by Benayoun et al. (1971) seems to be known as one of the first interactive multiobjective linear programming techniques, but there have been some modifications and extensions (see, for example, Fichefet 1976; Choo and Atkins 1980). Essentially, the STEM algorithm consists of two major steps. Step 1 seeks a Pareto optimal solution that is near to the ideal point in the minimax sense. Step 2 requires the decision maker (DM) to compare the objective vector with the ideal vector and to indicate which objectives can be sacrificed, and by how much, in order to improve the current levels of unsatisfactory objectives. The STEM algorithm is quite simple to understand and implement, in the sense that the DM is required to give only the amounts to be sacrificed of some satisfactory objectives until all objectives become satisfactory. However, the DM will never arrive at the final solution if the DM is not willing to sacrifice any of the objectives. Moreover, in many practical situations, the DM will probably want to indicate directly the aspiration level for each objective rather than just specify the amount by which satisfactory objectives can be sacrificed.

Wierzbicki (1979a) developed a relatively practical interactive method called the reference point method (RPM) by introducing the concept of a reference point suggested by the DM which reflects in some sense the desired values of the objective functions. The basic idea behind the RPM is that the DM can specify reference values for the objective functions and change the reference objective levels interactively due to learning or improved understanding during the solution process. In this procedure, when the DM specifies a reference point, the corresponding scalarization problem is solved for generating the Pareto optimal solution which is, in a sense, close to the reference point or better than that if the reference point is attainable. Then the DM either chooses the current Pareto optimal solution or modifies the reference point to find a satisficing solution.

Since then, some similar interactive multiobjective programming methods have been developed along this line (see, for example, Steuer and Choo 1983; Nakayama and Furukawa 1985). However, it is important to point out here that for dealing with the fuzzy goals of the DM for each of the objective functions of the multiobjective linear programming problem, Sakawa, Yano, and Yumine (1987) developed the extended fuzzy version of the RPM that supplies the DM with the trade-off information even if the fuzzy goals of the DM are not considered. Although the details of the method will be discussed in the next chapter, it would certainly be appropriate to discuss here the RPM with trade-off information rather than the RPM proposed by Wierzbicki.

Consider the following multiobjective linear programming (MOLP) problem:

$$\left.\begin{array}{rl} \text{minimize} & z_1(x) \triangleq c_1 x \\ \text{minimize} & z_2(x) \triangleq c_2 x \\ & \cdots\cdots \\ \text{minimize} & z_k(x) \triangleq c_k x \\ \text{subject to} & x \in X \triangleq \{x \in R^n \mid Ax \leq b,\ x \geq 0\} \end{array}\right\} \tag{3.82}$$

where $z_1(x) = c_1 x, \ldots, z_k(x) = c_k x$ are k distinct objective functions of the decision vector x and X is the linearly constrained feasible set.

For each of the multiple conflicting objective functions $z(x) = (z_1(x), \ldots, z_k(x))^T$, assume that the DM can specify the so-called reference point $\bar{z} = (\bar{z}_1, \ldots, \bar{z}_k)^T$ which reflects in some sense the desired values of the objective functions of the DM. Also assume that the DM can change the reference point interactively due to learning or improved understanding during the solution process. When the DM specifies the reference point $\bar{z} = (\bar{z}_1, \ldots, \bar{z}_k)^T$, the corresponding Pareto optimal solution, which is, in the minimax sense, nearest to the reference point or better than that if the reference point is attainable, is obtained by solving the following minimax problem :

$$\left.\begin{array}{rl} \text{minimize} & \max\limits_{i=1,\ldots,k} \{z_i(x) - \bar{z}_i\} \\ \text{subject to} & x \in X, \end{array}\right\} \tag{3.83}$$

or equivalently

$$\left.\begin{array}{rl} \text{minimize} & v \\ \text{subject to} & z_i(x) - \bar{z}_i \leq v,\ i = 1, \ldots, k \\ & x \in X. \end{array}\right\} \tag{3.84}$$

The case of the two-objective functions in the $z_1 - z_2$ plane is shown geometrically in Figure 3.11. For the two reference points $\bar{z}^1 = (\bar{z}_1^1, \bar{z}_2^1)^T$ and $\bar{z}^2 = (\bar{z}_1^2, \bar{z}_2^2)^T$ specified by the DM, solving the corresponding minimax problems yields the corresponding Pareto optimal solutions $z^1(x^1)$ and $z^2(x^2)$.

The relationships between the optimal solutions of the minimax problem and the Pareto optimal concept of the MOLP can be characterized by the following two theorems.

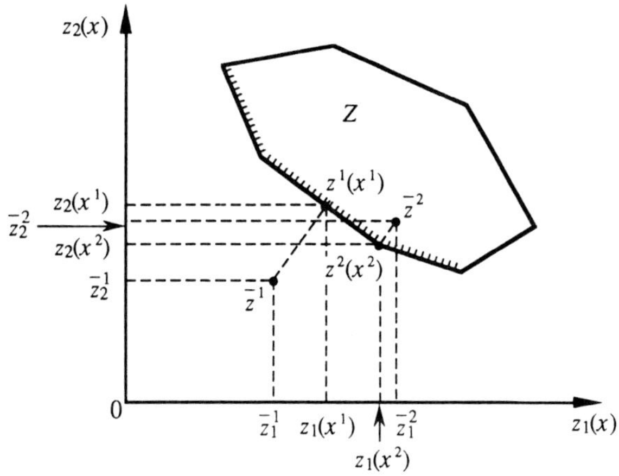

Figure 3.11 Graphical interpretation of minimax method

Theorem 3.8

If $x^* \in X$ is a unique optimal solution of the minimax problem for any reference point $\bar{z}$, then x^* is a Pareto optimal solution of the MOLP.

Proof

If a unique optimal solution x^* of the minimax problem is not a Pareto optimal solution of the MOLP, then there exists $x \in X$ such that $z_i(x) \leq z_i(x^*)$, $i = 1, \ldots, k$; $i \neq j$, and $z_j(x) < z_j(x^*)$ for some j. Hence, it follows that

$$\max_{x \in X}\{z_i(x) - \bar{z}_i\} \leq \max_{x \in X}\{z_i(x^*) - \bar{z}_i\}.$$

This contradicts the assumption that x^* is a unique optimal solution of the minimax problem. □

From the proof of this theorem, if the uniqueness of a solution is not guaranteed, only weak Pareto optimality is guaranteed.

Theorem 3.9

If x^* is a Pareto optimal solution of the MOLP, then x^* is an optimal solution of the minimax problem for some reference point $\bar{z}$.

Proof

For a Pareto optimal solution $x^* \in X$ of the MOLP, choose a reference point $\bar{z} = (\bar{z}_1, \ldots, \bar{z}_k)^T$ such that $z_i(x^*) - \bar{z}_i = v^*$, $i = 1, \ldots, k$. For this reference point, if x^* is not an optimal solution of the minimax problem, then there exists $x \in X$ such that

$$z_i(x) - \bar{z}_i < z_i(x^*) - \bar{z}_i = v^*, \ i = 1, \ldots, k.$$

This implies the existence of $x \in X$ such that

$$z_i(x) < z_i(x^*), \ i = 1, \ldots, k,$$

which contradicts the fact that x^* is a Pareto optimal solution. □

If x^*, an optimal solution to the minimax problem, is not unique, then, as discussed in section 3.2, the Pareto optimality test for x^* can be performed by solving the following problem:

$$\left.\begin{array}{ll} \text{maximize} & \sum_{i=1}^{k} \varepsilon_i \\ \text{subject to} & z_i(x) + \varepsilon_i = z_i(x^*),\ i = 1, \ldots, k \\ & x \in X,\ \varepsilon \geq 0. \end{array}\right\} \tag{3.85}$$

For the optimal solutions $\bar{x}$ and $\bar{\varepsilon}$ of this linear programming problem with decision variables x and ε, as was shown in Theorem 3.7, (1) If all $\bar{\varepsilon}_i = 0$, then x^* is a Pareto optimal solution of the MOLP, and (2) If at least one $\bar{\varepsilon}_i \geq 0$, then not x^* but $\bar{x}$ is a Pareto optimal solution of the MOLP.

Now, given the Pareto optimal solution for the reference point specified by the DM by solving the corresponding minimax problem, the DM must either be satisfied with the current Pareto optimal solution or modify the reference point. To help the DM express a degree of preference, trade-off information between a standing objective function $z_1(x)$ and each of the other objective functions is very useful. Such a trade-off between $z_1(x)$ and $z_i(x)$ for each $i = 2, \ldots, k$ is easily obtainable since it is closely related to the strict positive simplex multipliers of the minimax problem. Let the simplex multipliers associated with the constraints of the minimax problem be denoted by π_i, $i = 1, \ldots, k$. If all $\pi_i > 0$ for each i, it can be proved that the following expression holds (for details, see the Appendix):

$$-\frac{\partial z_1(x)}{\partial z_i(x)} = \frac{\pi_1}{\pi_i},\ i = 2, \ldots, k. \tag{3.86}$$

Geometrically, however, we can understand it as follows:

In the $(z_1, \ldots, z_k, w)$ space, the tangent hyperplane at some point on the Pareto surface can be described by

$$H(z_1, \ldots, z_k, w) = a_1 z_1 + \cdots + a_k z_k + bw = c.$$

The necessary and sufficient condition for the small displacement from this point belonging to this tangent hyperplane is $\Delta H = 0$, i.e.,

$$a_1 \Delta z_1 + \cdots + a_k \Delta z_k + b \Delta w = 0.$$

For fixed values of $\Delta z_j = 0$ $(j = 2, ..., k,\ j \neq i)$ and $\Delta w = 0$ except Δz_1 and Δz_i, we have

$$a_1 \Delta z_1 + a_i \Delta z_i = 0.$$

Similarly, for fixed values of $\Delta z_i = 0$ $(i = 1, \ldots, k, i \neq j)$ except Δz_j and Δw, we have

$$a_j \Delta z_j + b \Delta w = 0.$$

It follows from the last two relations that

$$-\frac{\Delta z_i}{\Delta z_1} = \frac{a_1}{a_i} = \frac{(-a_1/b)}{(-a_i/b)} = \frac{(\Delta w/\Delta z_1)}{(\Delta w/\Delta z_i)}.$$

Consequently, it holds that

$$-\frac{\partial z_i}{\partial z_1} = \frac{(\partial w/\partial z_1)}{(\partial w/\partial z_i)}.$$

Using the simplex multipliers π_i, $i = 1, ..., k$, associated with all the active constraints of the minimax problem, since $\partial w/\partial z_i = \pi_i$, we obtain (3.86).

It should be stressed here that in order to obtain the trade-off rate information from (3.86), all constraints of the minimax problem must be active. Therefore, if there are inactive constraints, it is necessary to replace $\bar{z}_i$ for inactive constraints by $z_i(x^*)$ and to solve the corresponding minimax problem to obtain the simplex multipliers.

We can now construct the interactive algorithm to derive the satisficing solution for the DM from the Pareto optimal solution set. The steps marked with an asterisk involve interaction with the DM. Observe that this interactive multiobjective linear programming method can be interpreted as the reference point method (RPM) with trade-off information.

Interactive multiobjective linear programming

Step 0: Calculate the individual minimum $z_i^{\min} = \min_{x \in X} z_i(x)$ and maximum $z_i^{\max} = \max_{x \in X} z_i(x)$ of each objective function under the given constraints.

Step 1*: Ask the DM to select the initial reference point by considering the individual minimum and maximum. If the DM finds it difficult or impossible to identify such a point, ideal point $z_i^{\min} = \min_{x \in X} z_i(x)$ can be used for that purpose.

Step 2: For the reference point specified by the DM, solve the corresponding minimax problem to obtain the Pareto optimal solution together with the trade-off rate information between the objective functions.

Step 3*: If the DM is satisfied with the current levels of the Pareto optimal solution, stop. Then the current Pareto optimal solution is the satisficing solution for the DM. Otherwise, ask the DM to update the current reference point by considering the current values of the objective functions together with the trade-off rates between the objective functions and return to step 2.

It should be stressed to the DM that any improvement of one objective function can be achieved only at the expense of at least one of the other objective functions.

Example 3.5 (Production planning with environmental quality)
To demonstrate interactive multiobjective linear programming, consider the following production planning problem with environmental quality as previously discussed in Example 3.3:

$$\left.\begin{array}{lrl} \text{minimize} & z_1(x) = & -x_1 - 2x_2 \\ \text{minimize} & z_2(x) = & 3x_1 + 2x_2 \\ \text{subject to} & & 2x_1 + 6x_2 \leq 27 \\ & & 8x_1 + 6x_2 \leq 45 \\ & & 3x_1 + x_2 \leq 15 \\ & & x_1 \geq 0,\ x_2 \geq 0. \end{array}\right\}$$

Recall that these two objective functions are the minimization of both the minus profit (z_1) and the amount of pollution (z_2).

First, observe that individual minimums and maximums for each of the objective functions are

$$z_1^{\min} = -10,\ z_1^{\max} = 0, \quad z_2^{\min} = 0,\ z_2^{\max} = 16.5.$$

Considering these values, assume that the DM specified the reference point as

$$\bar{z}_1 = -10, \quad \bar{z}_2 = 0.$$

For this reference point, as can be easily seen from Figure 3.12, solving the corresponding minimax problem yields the Pareto optimal solution

$$z_1 = -5, \quad z_2 = 5 \qquad (x_1 = 0,\ x_2 = 2.5)$$

and the trade-off rate between the objective functions

$$-\frac{\partial z_2}{\partial z_1} = 1.$$

On the basis of such information, assume that the DM updated the reference point to

$$\bar{z}_1 = -10, \quad \bar{z}_2 = 10$$

improving the satisfaction levels of the profit at the expense of the amount of pollution. For the updated reference point, solving the corresponding minimax problem yields the Pareto optimal solution

$$z_1 = -9.25, \quad z_2 = 10.75 \qquad (x_1 = 0.75,\ x_2 = 4.25)$$

and the trade-off rate

$$-\frac{\partial z_2}{\partial z_1} = 7.$$

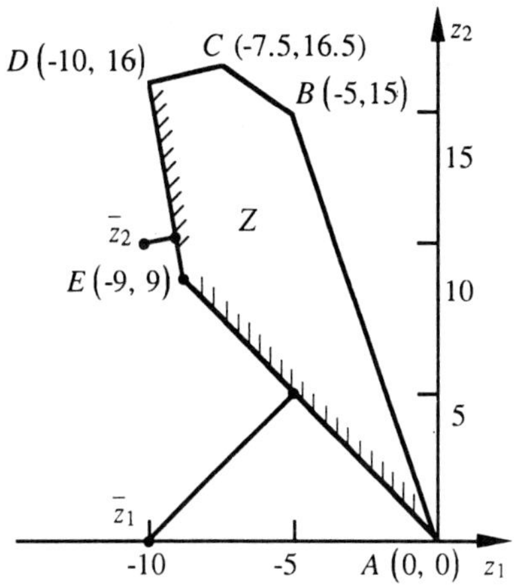

Figure 3.12 Interactive multiobjective linear programming for Example 3.5

If the DM is satisfied with the current values of the objective functions, the procedure stops. Otherwise, a similar procedure continues in this fashion until the satisficing solution of the DM is derived.

3.4 Fuzzy multiobjective linear programming

In 1978, H.-J. Zimmermann extended his fuzzy linear programming approach to the following multiobjective linear programming problem with k linear objective functions $z_i(x) = c_i x$, $i = 1, \ldots, k$:

$$\left.\begin{array}{rl} \text{minimize} & z(x) \triangleq (z_1(x), z_2(x), \ldots, z_k(x))^T \\ \text{subject to} & Ax \leq b, \; x \geq 0 \end{array}\right\} \tag{3.87}$$

where $c_i = (c_{i1}, \ldots, c_{in})$, $i = 1, \ldots, k$, $x = (x_1, \ldots, x_n)^T$, $b = (b_1, \ldots, b_m)^T$ and $A = [a_{ij}]$ is an $m \times n$ matrix.

For each of the objective functions $z_i(x) = c_i x$, $i = 1, \ldots, k$, of this problem, assume that the decision maker (DM) has a fuzzy goal such as "the objective function $z_i(x)$ should be substantially less than or equal to some value p_i." Then the corresponding linear membership function $\mu_i^L(z_i(x))$ is defined as

$$\mu_i^L(z_i(x)) = \begin{cases} 0 & ; \; z_i(x) \geq z_i^0 \\ \dfrac{z_i(x) - z_i^0}{z_i^1 - z_i^0} & ; \; z_i^0 \geq z_i(x) \geq z_i^1 \\ 1 & ; \; z_i(x) \leq z_i^1 \end{cases} \tag{3.88}$$

where z_i^0 or z_i^1 denotes the value of the objective function $z_i(x)$ such that the degree of membership function is 0 or 1 respectively.

Figure 3.13 illustrates the graph of the possible shape of the linear membership function.

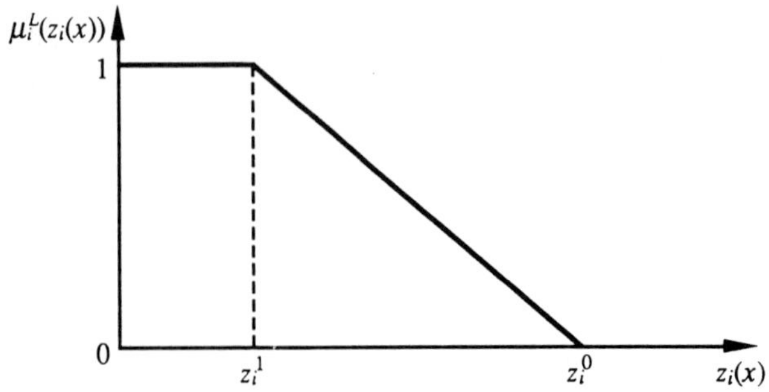

Figure 3.13 Linear membership function

Using such linear membership functions $\mu_i^L(z_i(x))$, $i = 1, \ldots, k$, and following the fuzzy decision of Bellman and Zadeh (1970), the original multiobjective linear programming problem can be interpreted as

$$\left.\begin{array}{ll} \text{maximize} & \min\limits_{i=1,\ldots,k} \{\mu_i^L(z_i(x))\} \\ \text{subject to} & Ax \leq b, \; x \geq 0. \end{array}\right\} \tag{3.89}$$

By introducing the auxiliary variable λ, it can be reduced to the following conventional linear programming problem:

$$\left.\begin{array}{ll} \text{maximize} & \lambda \\ \text{subject to} & \lambda \leq \mu_i^L(z_i(x)), \; i = 1, 2, \ldots, k \\ & Ax \leq b, \; x \geq 0. \end{array}\right\} \tag{3.90}$$

By assuming the existence of the optimal solution x^{io} of the individual objective function minimization problem under the constraints defined by

$$\min_{x \in X} z_i(x), \; i = 1, \ldots, k, \tag{3.91}$$

Zimmermann (1978) suggested a way to determine the linear membership function $\mu_i^L(z_i(x))$. To be more specific, using the individual minimum

$$z_i^{\min} = z_i(x^{io}) = \min_{x \in X} z_i(x), \; i = 1, \ldots, k, \tag{3.92}$$

together with

$$z_i^{\mathrm{m}} = \max(z_i(x^{1o}), \ldots, z_i(x^{i-1,o}), z_i(x^{i+1,o}), \ldots, z_i(x^{ko})), \; i = 1, \ldots, k, \tag{3.93}$$

he determined the linear membership function as in (3.88) by choosing $z_i^1 = z_i^{\min}$ and $z_i^0 = z_i^{\mathrm{m}}$. For this membership function, it can be easily shown that if the optimal solution of (3.89) or (3.90) is unique, it is also a Pareto optimal solution of the MOLP.

In the case where not only fuzzy goals but also fuzzy constraints exist, using linear membership functions for fuzzy constraints, similar discussion can be made. Zimmermann called the fuzzy decision the minimum operator, and for other aggregation patterns than the minimum operator, he considered the product fuzzy decision. He called the product fuzzy decision the product operator, and proposed using the product operator. In this case, the problem to be solved becomes

$$\left.\begin{array}{ll} \text{maximize} & \prod_{i=1}^{k} \mu_i^L(z_i(x)) \\ \text{subject to} & Ax \leq b, \ x \geq 0. \end{array}\right\} \tag{3.94}$$

Unfortunately, with the product operator, even if we use the linear membership functions, the objective function of this problem becomes a nonlinear function, and hence, the linear programming method cannot be applied.

Example 3.6 (Production planning with environmental quality)
To illustrate the fuzzy multiobjective linear programming method proposed by Zimmermann, consider the following production planning problem with pollution as previously discussed in Example 3.3:

$$\left.\begin{array}{lrr} \text{minimize} & z_1(x) = & -x_1 - 2x_2 \\ \text{minimize} & z_2(x) = & 3x_1 + 2x_2 \\ \text{subject to} & & 2x_1 + 6x_2 \leq 27 \\ & & 8x_1 + 6x_2 \leq 45 \\ & & 3x_1 + x_2 \leq 15 \\ & & x_1 \geq 0, \, x_2 \geq 0. \end{array}\right\}$$

Note that these two objective functions are the minimization of both the minus profit (z_1) and the amount of pollution (z_2).

The individual minimum and maximum of these objective functions become

$$z_1^{\min} = -10, \quad z_1^{\max} = 0, \quad z_2^{\min} = 0, \quad z_2^{\max} = 16.5.$$

Assume that the DM subjectively determined the corresponding linear membership functions $\mu_i^L(z_i), i = 1, 2$ as follows:

$$\begin{cases} \text{fuzzy goal for } z_1\text{:} & \mu_1^L(-8) = 0, \quad \mu_1^L(-10) = 1 \\ \text{fuzzy goal for } z_2\text{:} & \mu_2^L(14) = 0, \quad \mu_2^L(9) = 1 \end{cases}$$

where linear functions are assumed from $\mu_i = 0$ to $\mu_i = 1$ for $i = 1, 2$.

To be more specific, the fuzzy goals of the DM are assumed to be expressed by the following membership functions:

$$\mu_1^L(z_1(x)) = \begin{cases} 0 & ; \quad z_1(x) \geq -8 \\ \dfrac{-x_1 - 2x_2 + 8}{-2} & ; \quad -8 \geq z_1(x) \geq -10 \\ 1 & ; \quad z_1(x) \leq -10, \end{cases}$$

$$\mu_2^L(z_2(x)) = \begin{cases} 0 & ; \quad z_2(x) \geq 14 \\ \dfrac{3x_1 + 2x_2 - 14}{-5} & ; \quad 14 \geq z_2(x) \geq 9 \\ 1 & ; \quad z_2(x) \leq 9. \end{cases}$$

Then the equivalent linear programming problem formulation becomes

$$\left.\begin{array}{ll} \text{maximize} & \lambda \\ \text{subject to} & 0.51x_1 + x_2 - \lambda \geq 4 \\ & 0.6x_1 + 0.4x_2 + \lambda \leq 2.8 \\ & 2x_1 + 6x_2 \leq 27 \\ & 8x_1 + 6x_2 \leq 45 \\ & 3x_1 + x_2 \leq 15 \\ & x_1 \geq 0,\ x_2 \geq 0. \end{array}\right\} \tag{3.95}$$

Solving this problem by the simplex method of linear programming yields the optimal solution

$$x_1 = 0.78947, \quad x_2 = 4.2368, \quad \lambda = 0.63158.$$

This means that the overall satisfaction of the fuzzy goals of the DM is 0.63158, the corresponding total profit $(-z_1)$ is 9.26307, and the amount of pollution (z_2) is 10.84201. This point $(x_1, x_2) = (0.78947, 4.2368)$ or $(z_1, z_2) = (-9.26307, 10.84201)$ lies on the edge ED in Figure 3.3 or Figure 3.4. From this, it can be understood that it is also a Pareto optimal solution of the original MOLP.

In 1981, by considering the rate of increased membership satisfaction must not always be constant as in the case of the linear membership function proposed by Zimmermann, Leberling introduced special nonlinear functions and showed that the resulting nonlinear programming problem can be equivalently converted to a conventional linear programming problem. For each of the objective functions $z_i(x)$, $i = 1, \ldots, k$, of the MOLP, Leberling assumed the following nonlinear tangent hyperbolic membership function for representing the fuzzy goals of the DM:

$$\mu_i^H(z_i(x)) = \frac{1}{2}\tanh((z_i(x) - b_i)\alpha_i) + \frac{1}{2} \tag{3.96}$$

where $\alpha_i < 0$ is a shape parameter and b_i represents the value of $z_i(x)$ such that $\mu_i^H(z_i(x)) = 0.5$. Using the individual minimum $z_i^{\min}$ and z_i^{m} as was defined in (3.92) and (3.93), he set $b_i = (z_i^{\mathrm{m}} + z_i^0)/2$.

The hyperbolic membership function $\mu_i^H(z_i(x))$ introduced by Leberling has the following properties:

(1) $\mu_i^H(z_i(x))$ is a strictly monotone increasing function with respect to x.

(2) $\mu_i^H(z_i(x))$ is convex for $z_i(x) \geq \frac{1}{2}(z_i^{\mathrm{m}} + z_i^0)$ and concave for $z_i(x) \leq \frac{1}{2}(z_i^{\mathrm{m}} + z_i^0)$ with respect to $z_i(x)$.

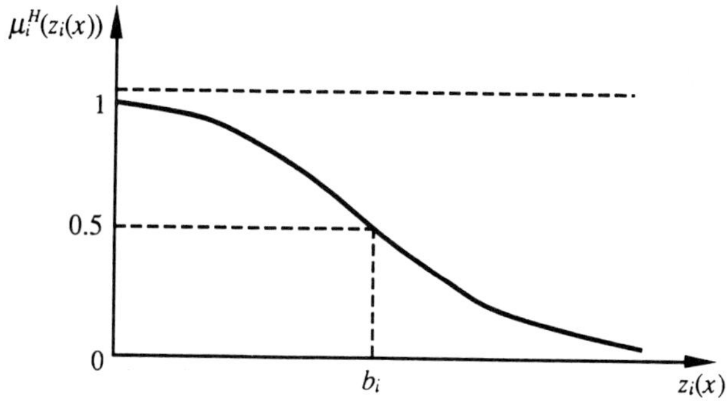

Figure 3.14 Hyperbolic membership function

(3) $\mu_i^H(z_i(x))$ satisfies $0 < \mu_i^H(z_i(x)) < 1$ for all x and approaches asymptotically $\mu_i^H(x) = 0$ and $\mu_i^H(x) = 1$.

Figure 3.14 depicts a possible shape of $\mu_i^H(z_i(x))$ with respect to $z_i(x)$.

By adopting the hyperbolic membership function for representing the fuzzy goals of the DM together with the fuzzy decision of Bellman and Zadeh (1970), the problem to be solved can be expressed as

$$\left.\begin{array}{ll} \text{maximize} & \lambda \\ \text{subject to} & \lambda \leq \mu_i^H(z_i(x)) \quad i = 1, \ldots, k \\ & Ax \leq b \\ & x \geq 0. \end{array}\right\} \tag{3.97}$$

Observe that $0 < \lambda^0 < 1$ holds for the optimal solution (λ^0, x^0) of this problem due to $0 < \mu_i^H(z_i(x)) < 1$.

In this formulation, however, $\mu_i^H(z_i(x))$ is a nonlinear function, and hence, we cannot directly apply the linear programming method. To circumvent such difficulty, Leberling transformed the problem in the following way:

In view of $\lambda^0 > 0$, this problem can be rewritten as

$$\left.\begin{array}{ll} \text{maximize} & \lambda \\ \text{subject to} & \lambda \leq \frac{1}{2}\tanh((z_i(x) - b_i)\alpha_i) + \frac{1}{2} \\ & \quad i = 1, \ldots, k \\ & Ax \leq b \\ & x \geq 0, \quad \lambda \geq 0, \end{array}\right\} \tag{3.98}$$

or equivalently

$$\left.\begin{array}{ll} \text{maximize} & \lambda \\ \text{subject to} & \tanh((z_i(x) - b_i)\alpha_i) \geq 2\lambda - 1 \\ & \quad i = 1, \ldots, k \\ & Ax \leq b \\ & x \geq 0, \quad \lambda \geq 0. \end{array}\right\} \tag{3.99}$$

Observing that the hyperbolic function $\tanh(x)$ and hyperbolic inverse function $\tanh^{-1}(x)$ are strictly monotone increasing functions with respect to x, we have

$$\left.\begin{array}{ll} \text{maximize} & \lambda \\ \text{subject to} & (z_i(x) - b_i)\alpha_i \geq \tanh^{-1}(2\lambda - 1) \\ & \quad i = 1, \ldots, k \\ & Ax \leq b \\ & x \geq 0, \quad \lambda \geq 0. \end{array}\right\} \tag{3.100}$$

Now, if we define

$$x_{n+1} = \tanh^{-1}(2\lambda - 1), \tag{3.101}$$

then

$$\lambda = \frac{1}{2}\tanh(x_{n+1}) + \frac{1}{2}. \tag{3.102}$$

Since $\tanh(x)$ is a strictly monotone increasing function with respect to x, the maximization of λ is equivalent to the maximization of x_{n+1}. Hence, (3.100) can be transformed to the following ordinary linear programming problem:

$$\left.\begin{array}{ll} \text{maximize} & x_{n+1} \\ \text{subject to} & \alpha_i z_i(x) - x_{n+1} \geq \alpha_i b_i, \quad i = 1, \ldots, k \\ & Ax \leq b \\ & x \geq 0. \end{array}\right\} \tag{3.103}$$

Let (x^*_{n+1}, x^*) be an optimal solution of this problem. Then from (3.102), the optimal solution (λ^*, x^*) of the problem (3.103) can be obtained by

$$(\lambda^*, x^*) = \Big(\frac{1}{2}\tanh(x^*_{n+1}) + \frac{1}{2}, x^*\Big). \tag{3.104}$$

Concerning the Pareto optimality of an optimal solution of (3.97), Leberling proved the following theorem.

Theorem 3.10

(1) *If (λ^*, x^*) is an optimal solution of (3.97), then x^* is a weak Pareto optimal solution of the MOLP.*

(2) *If* (λ^*, x^*) *is a unique optimal solution of (3.97), then* x^* *is a Pareto optimal solution of the MOLP.*

Proof

(1) If x^* is not a weak Pareto optimal solution of the MOLP, then there exists a feasible solution $\overline{x}$ $(\neq x^*)$ such that $z_i(x^*) > z_i(\overline{x})$, $i = 1, \ldots, k$. Observing that $\mu_i^H(z_i(x))$ is a strictly monotone decreasing function with respect to $z_i(x)$, this implies $\mu_i^H(z_i(x^*)) < \mu_i^H(z_i(\overline{x}))$, $i = 1, \ldots, k$. Hence, we have

$$\lambda^* = \min_{i=1,\ldots,k} \mu_i^H(z_i(x^*)) < \min_{i=1,\ldots,k} \mu_i^H(z_i(\overline{x})) = \overline{\lambda}$$

which contradicts the optimality of (λ^*, x^*).

(2) If x^* is not a Pareto optimal solution of the MOLP, then there exists a feasible solution $\hat{x}$ $(\neq x^*)$ such that $z_j(x^*) > z_j(\hat{x})$ for some j and $z_i(x^*) \geq z_i(\hat{x})$, $i = 1, ..., k; i \neq j$. Noting that $\mu_i^H(z_i(x))$ is a strictly monotone decreasing function with respect to $z_i(x)$, this implies $\mu_j^H(z(x^*)) < \mu_j^H(z(\hat{x}))$ for some j and $\mu_i^H(z(x^*)) \leq \mu_i^H(z(\hat{x}))$, $i = 1, ..., k; i \neq j$. Hence, we have

$$\lambda^* = \min_{i=1,\ldots,k} \mu_i^H(z_i(x^*)) \leq \min_{i=1,\ldots,k} \mu_i^H(z_i(\hat{x})) = \hat{\lambda}$$

which contradicts the unique optimality of (λ^*, x^*) for either $\lambda^* < \hat{\lambda}$ or $\lambda^* = \hat{\lambda}$.

This completes the proof of the theorem. □

In 1981, as another extension of the linear membership function of Zimmermann, E. L. Hannan proposed a different approach from Leberling. For each of the objective functions of the MOLP, assuming that the DM could specify the degree of membership for several values of $z_i(x)$, he introduced the following piecewise linear membership function:

$$\mu_i^{PL}(z_i(x)) = \sum_{i=1}^{N_i} \alpha_{ij}|z_i(x) - g_{ij}| + \beta_i z_i(x) + \gamma_i \tag{3.105}$$

where

$$\alpha_{ij} = (t_{i,j+1} - t_{ij})/2, \quad \beta_i = (t_{i,N_i+1} + t_{i1})/2, \quad \gamma_i = (s_{i,N_i+1} + s_{i1})/2$$
$$i = 1, \ldots, k$$
$$j = 1, \ldots, N_i \ (N_i \text{ is a number of divided points}).$$

In this piecewise linear membership function, it is assumed that $\mu_i^{PL}(z_i(x)) = t_{ir} z_i(x) + s_{ir}$ for each segment $g_{i,r-1} \leq z_i(x) \leq g_{ir}$ where t_{ir} is the slope and s_{ir} is the y-intercept of the line segment initiated at $g_{i,r-1}$ and terminated at g_{ir}. Observe that, since the values of $\mu_i^{PL}(z_i(x))$ represent the grade of membership, $0 \leq \mu_i^{PL}(z_i(x)) \leq 1$ for all $z_i(x)$.

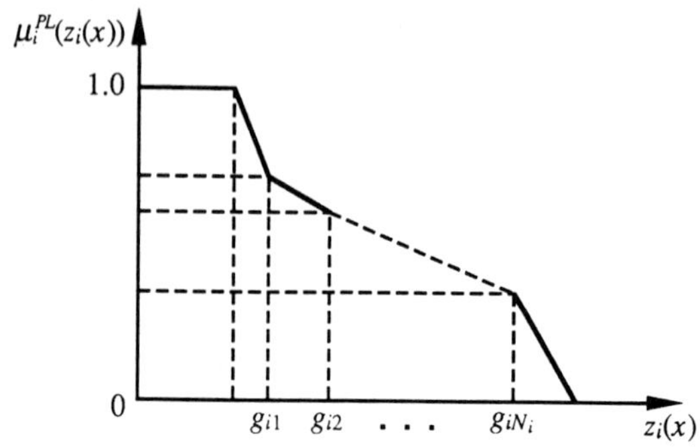

Figure 3.15 Piecewise linear membership function

Figure 3.15 illustrates a possible shape of a piecewise linear membership function.

By adopting the piecewise linear membership function to represent the fuzzy goal of the DM for the MOLP together with the fuzzy decision of Bellman and Zadeh (1970), the problem to be solved becomes

$$\left.\begin{array}{rl} \text{maximize} & \lambda \\ \text{subject to} & \lambda \leq \mu_i^{PL}(z_i(x)), \quad i = 1, \ldots, k \\ & Ax \leq b \\ & x \geq 0,\ \lambda \geq 0. \end{array}\right\} \tag{3.106}$$

To formulate this problem as a goal programming problem discussed in subsection 3.2.3, he introduced the following nonnegative deviational variables d_{ij}^+ and d_{ij}^-

$$\left.\begin{array}{l} z_i(x) - d_{i1}^+ + d_{i1}^- = g_{i1} \\ \cdots\cdots\cdots\cdots\cdots\cdots \\ z_i(x) - d_{iN_i}^+ + d_{iN_i}^- = g_{iN_i} \end{array} \quad i = 1, \ldots, k \right\} \tag{3.107}$$

where d_{ij}^+ and d_{ij}^- denote the deviational variables for the ith point and g_{ij} represent the values of $z_i(x)$ for the ith objective function at the jth point.

Hence, the piecewise linear membership function $\mu_i^{PL}(z_i(x))$ can be expressed as

$$\mu_i^{PL}(z_i(x)) = \sum_{j=1}^{N_i} \alpha_{ij}(d_{ij}^+ + d_{ij}^-) + \beta_i z_i(x) + \gamma_i, \ i = 1, \ldots, k. \tag{3.108}$$

Therefore, if we adopt the piecewise linear membership function together with the fuzzy decision of Bellman and Zadeh (1970), the problem to be solved can be converted to the following ordinary linear programming problem:

$$\left.\begin{array}{ll} \text{maximize} & \lambda \\ \text{subject to} & \lambda \leq \mu_i^{PL}(z_i(x)),\ i = 1, \ldots, k \\ & z_i(x) - d_{ij}^+ + d_{ij}^- = g_{ij} \\ & \qquad i = 1, \ldots, k;\ j = 1, \ldots, N_i \\ & Ax \leq b,\ x \geq 0,\ \lambda \geq 0 \\ & d_{ij}^+ \geq 0,\ d_{ij}^- \geq 0 \\ & \qquad i = 1, \ldots, k;\quad j = 1, \ldots, N_i. \end{array}\right\} \tag{3.109}$$

Instead of the fuzzy decision, Hannan (1981) also suggested specifying the goals $\hat{\mu}_i\ (0 < \hat{\mu}_i \leq 1)$ for each of the linear membership functions and the priorities among them. Then the MOLP can be converted to the following general fuzzy linear goal programming problem:

$$\left.\begin{array}{ll} \text{maximize} & \sum_{l=1}^{L} P_l \left(\sum_{i \in I_l} e_i^- \right) \\ \text{subject to} & \mu_i^{PL}(z_i(x)) - e_i^+ + e_i^- = \hat{\mu}_i,\ i = 1, \ldots, k \\ & z_i(x) - d_{ij}^+ + d_{ij}^- = g_{ij} \\ & \qquad i = 1, \ldots, k;\quad j = 1, \ldots, N_i \\ & Ax \leq b,\ x \geq 0,\ \lambda \geq 0 \\ & d_{ij}^+ \geq 0,\ d_{ij}^- \geq 0,\ i = 1, \ldots, k;\ j = 1, \ldots, N_i \\ & e_i^+ \geq 0,\ e_i^- \geq 0,\ i = 1, \ldots, k \end{array}\right\} \tag{3.110}$$

where $I_l \neq \emptyset$ is the set of lth priority class and e_i^+ and e_i^- are deviational variables.

So far, we have discussed fuzzy multiobjective linear programming in the same line as Zimmermann. However, suppose that the interaction with the DM establishes that the first membership function should be linear, the second hyperbolic, the third piecewise linear, and so forth. In such a situation, following the fuzzy decision of Bellman and Zadeh (1970), the resulting problem becomes a nonlinear programming problem and cannot be solved by a linear programming method.

In 1983, to quantify the fuzzy goals of the DM by eliciting the corresponding membership functions, Sakawa (1983a, 1983b) proposed using five types of membership functions: linear, exponential, hyperbolic, hyperbolic inverse, and piecewise linear functions. Through the use of these membership functions including nonlinear ones, the fuzzy goals of the DM are quantified. Then following the fuzzy decision of Bellmann and Zadeh (1970), the problem becomes a nonlinear programming problem. However, it can be reduced to a set of linear inequalities if some variable is fixed. Based on this idea, Sakawa (1983a, 1983b) proposed a new method combining the use of the bisection method and the linear programming method.

To elicit a membership function $\mu_i(z_i(x))$ from the DM for each of the objective functions $z_i(x), i = 1, 2, ..., k$, he suggested calculating the individual minimum $z_i^{\min}$ and maximum $z_i^{\max}$ of each objective function $z_i(x)$ under the given constraints in advance. Taking into account the calculated individual minimum and maximum of each objective function together with the rate of increase of membership of satisfaction, the DM is asked to select a membership function in a subjective manner from among five types of functions: linear, exponential, hyperbolic, hyperbolic inverse, and piecewise linear functions. The parameter values are determined through interaction with the DM in the following ways, where z_i^a represents the value of $z_i(x)$ such that the degree of membership function $\mu_i(z_i(x))$ is $a \in [0, 1]$ within $z_i^{\min}$ and $z_i^{\max}$.

(1) Linear membership function

For each objective function, the corresponding linear membership function is defined by

$$\mu_i(z_i(x)) = [z_i(x) - z_i^0]/[z_i^1 - z_i^0]. \tag{3.111}$$

The linear membership function can be determined by asking the DM to specify the two points z_i^0 and z_i^1 within $z_i^{\max}$ and $z_i^{\min}$. Figure 3.16 illustrates the graph of the linear membership function.

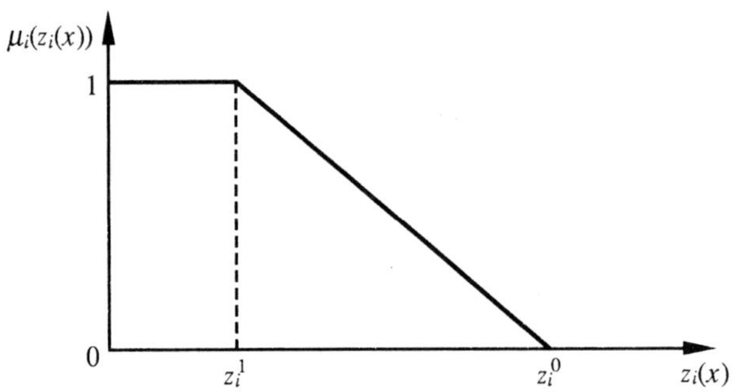

Figure 3.16 Linear membership function

(2) Exponential membership function

For each objective function, the corresponding exponential membership function is defined by

$$\mu_i(z_i(x)) = a_i[1 - \exp\{-\alpha_i(z_i(x) - z_i^0)/(z_i^1 - z_i^0)\}], \tag{3.112}$$

where $a_i > 1, \alpha_i > 0$ or $a_i < 0, \alpha_i < 0$. The exponential membership function can be determined by asking the DM to specify the three points z_i^0, $z_i^{0.5}$, and z_i^1 within $z_i^{\max}$ and $z_i^{\min}$, where α_i is a shape parameter. Figure 3.17 illustrates the graph of the exponential membership function.

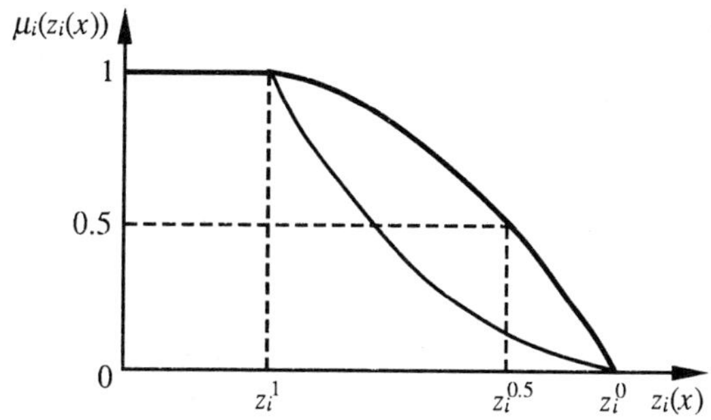

Figure 3.17 Exponential membership function

(3) Hyperbolic membership function

For each objective function, the corresponding hyperbolic membership function is defined by

$$\mu_i(z_i(x)) = \frac{1}{2}\tanh((z_i(x) - b_i)\alpha_i) + \frac{1}{2}, \tag{3.113}$$

where $\alpha_i < 0$. The hyperbolic membership function can be determined by asking the DM to specify the two points $z_i^{0.25}$ and $z_i^{0.5}$ within $z_i^{\max}$ and $z_i^{\min}$, where α_i is a shape parameter and b_i is associated with the point of inflection. Figure 3.18 illustrates the graph of the hyperbolic membership function.

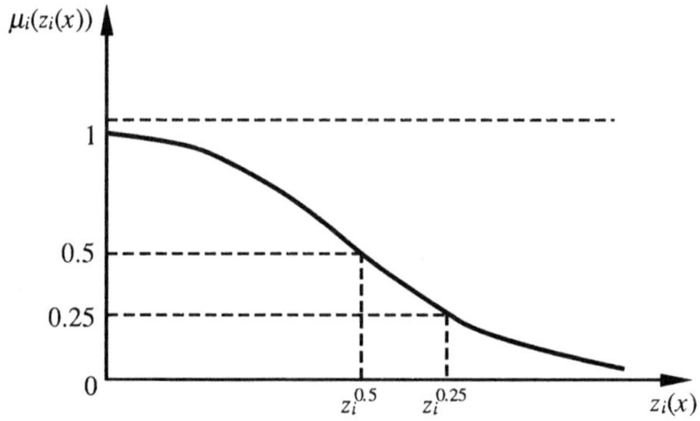

Figure 3.18 Hyperbolic membership function

(4) Hyperbolic inverse membership function

For each objective function, the corresponding hyperbolic inverse membership function is defined by

$$\mu_i(z_i(x)) = a_i \tanh^{-1}((z_i(x) - b_i)\alpha_i) + \frac{1}{2}, \tag{3.114}$$

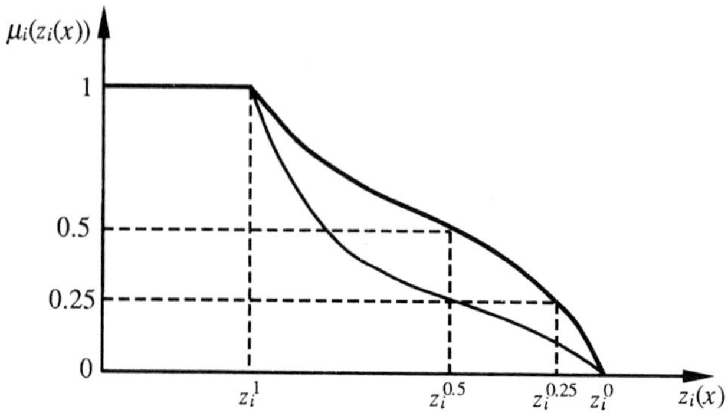

Figure 3.19 Hyperbolic inverse membership function

where $a_i > 0$ and $\alpha_i < 0$. The hyperbolic inverse membership function can be determined by asking the DM to specify the three points z_i^0, $z_i^{0.25}$, and $z_i^{0.5}$ within $z_i^{\max}$ and $z_i^{\min}$, where α_i is a shape parameter and b_i is associated with the point of inflection. Figure 3.19 illustrates the graph of the hyperbolic inverse membership function.

(5) Piecewise linear membership function

For each objective function, the corresponding piecewise linear membership function is defined by

$$\mu_i(z_i(x)) = t_{ir} z_i(x) + s_{ir} \quad \text{for} \quad g_{ir-1} \leq z(x) \leq g_{ir}. \tag{3.115}$$

It is assumed that $\mu_i(z_i(x)) = t_{ir} z_i(x) + s_{ir}$ for each segment $g_{ir-1} \leq z(x) \leq g_{ir}$, where t_{ir} is the slope and s_{ir} is the y-intercept for the section of the line initiated at g_{ir-1} and terminated at g_{ir}. The piecewise linear membership function can be determined by asking the DM to specify the degree of membership in each of several values of objective functions within $z_i^{\max}$ and $z_i^{\min}$. Figure 3.20 illustrates the graph of the piecewise linear membership function.

After determining the membership functions for each of the objective functions and adopting the fuzzy decision of Bellman and Zadeh (1970), the resulting problem to be solved is

$$\underset{x \in X}{\text{maximize}} \left\{ \min_{i=1,\ldots,k} (\mu_i(z_i(x))) \right\} \tag{3.116}$$

or equivalently

$$\left.\begin{aligned} &\text{maximize} && \lambda \\ &\text{subject to} && \lambda \leq \mu_i(z_i(x)), \quad i = 1, \ldots, k \\ & && Ax \leq b \\ & && \quad x \geq 0,\ \lambda \geq 0. \end{aligned}\right\} \tag{3.117}$$

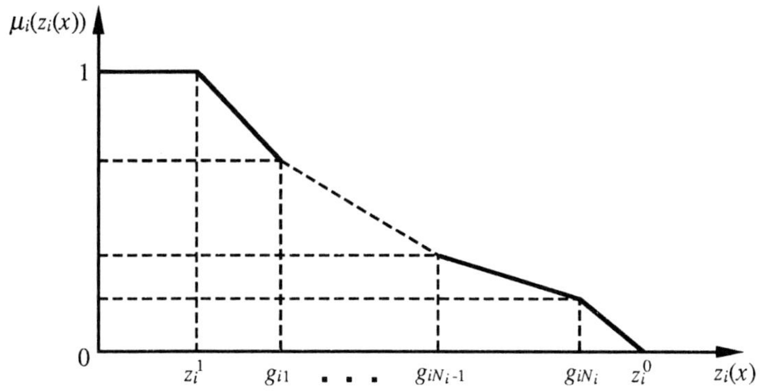

Figure 3.20 Piecewise linear membership function

Unfortunately, however, with the five types of membership functions, the resulting problem is a nonlinear programming problem. To solve the formulated problem on the basis of linear programming, Sakawa (1983a, 1983b) made the following transformation of each constraint for the five types of membership functions.

(1) $\lambda \leq [z_i(x) - z_i^0]/[z_i^1 - z_i^0]$
Since $z_i^0 > z_i^1$, it follows that

$$z_i(x) \leq \lambda(z_i^1 - z_i^0) + z_i^0. \tag{3.118}$$

(2) $\lambda \leq a_i[1 - \exp\{-\alpha_i(z_i(x) - z_i^0)/(z_i^1 - z_i^0)\}]$

In the case of $a_i > 1$ and $-\alpha_i > 0$, it follows that

$$\exp\{-\alpha_i(z_i(x) - z_i^0)/(z_i^1 - z_i^0)\} \leq (a_i - \lambda)/a_i.$$

By taking the logarithm and rearranging, it holds that

$$z_i(x) \leq z_i^0 - \{(z_i^1 - z_i^0)/\alpha_i\}\log\{(a_i - \lambda)/a_i\}. \tag{3.119}$$

Similarly, in the case of $a_i < 0$ and $\alpha_i < 0$, the same result follows.

(3) $\lambda \leq \frac{1}{2}\tanh((z_i(x) - b_i)\alpha_i) + \frac{1}{2}$
Since tanh and $\tanh^{-1}$ are strictly increasing functions, it follows that

$$(z_i(x) - b_i)\alpha_i \geq \tanh^{-1}(2\lambda - 1).$$

For $\alpha_i < 0$, it holds that

$$z_i(x) \leq b_i + \frac{1}{\alpha_i}\tanh^{-1}(2\lambda - 1). \tag{3.120}$$

(4) $\lambda \leq a_i\tanh^{-1}((z_i(x) - b_i)\alpha_i) + \frac{1}{2}$

Since tanh and $\tanh^{-1}$ are strictly increasing and $a_i > 0$, $\alpha_i < 0$, it follows that

$$z_i(x) \leq b_i + \frac{1}{\alpha_i} \tanh\left((\lambda - \frac{1}{2})/a_i \right). \tag{3.121}$$

(5) $\lambda \leq t_{ir} z_i(x) + s_{ir}$ for $f(g_{ir-1}) \leq \lambda \leq f(g_{ir})$.

Since $t_{ir} < 0$, it follows that

$$z_i(x) \leq (\lambda - s_{ir})/t_{ir} \quad \text{for} \quad f(g_{ir-1}) \leq \lambda \leq f(g_{ir}). \tag{3.122}$$

For the five types of membership functions, if we convert each constraint $\lambda \leq \mu_i(z_i(x))$ into the form described above, we arrive at the following problem:

$$\left.\begin{array}{ll} \text{maximize} & \lambda \\ \text{subject to} & z_i(x) \leq \lambda(z_i^1 - z_i^0) + z_i^0 \\ & z_i(x) \leq z_i^0 - ((z_i^1 - z_i^0)/\alpha_i)\log((a_i - \lambda)/a_i) \\ & z_i(x) \leq b_i + (1/\alpha_i)\tanh^{-1}(2\lambda - 1) \\ & z_i(x) \leq b_i + (1/\alpha_i)\tanh((\lambda - \frac{1}{2})/a_i) \\ & z_i(x) \leq (\lambda - s_{ir})/t_{ir} \text{ for } f(g_{ir-1}) \leq \lambda \leq f(g_{ir}) \\ & Ax \leq b, \quad 0 \leq \lambda \leq 1. \end{array}\right\} \tag{3.123}$$

It is important to note here that, in this formulation, if the value of λ is fixed, it can be reduced to a set of linear inequalities. Obtaining the optimal solution λ^* to the above problem is equivalent to determining the maximum value of λ so that there exists an admissible set satisfying the constraints of equations (3.123). Since λ satisfies $0 \leq \lambda \leq 1$, we have the following method for solving this problem by combined use of the bisection method and phase one of linear programming technique.

Step 1: Set $\lambda = 0$ and test whether an admissible set satisfying the constraints of (3.123) exists or not using phase one of the simplex method. If an admissible set exists, proceed. Otherwise, the DM must reassess the membership function.

Step 2: Set $\lambda = 1$ and test whether an admissible set satisfying the constraints of (3.123) exists or not using phase one of the simplex method. If an admissible set exists, the maximum degree of membership $\lambda^* = 1$ is achieved. Otherwise, go to the next step since the maximum λ that satisfies the constraints of (3.123) exists between 0 and 1.

Step 3: For the initial value of $\lambda_1 = 0.5$, update the value of λ using the bisection method as follows:

$$\begin{cases} \lambda_{n+1} = \lambda_n + 1/2^{n+1} & \text{if an admissible set exists for } \lambda_n, \\ \lambda_{n+1} = \lambda_n - 1/2^{n+1} & \text{if no admissible set exists for } \lambda_n. \end{cases}$$

For each λ_n, $n = 1, 2, ...,$ test whether an admissible set of (3.123) exists or not using the simplex method and determine the maximum value of λ satisfying the constraints of (3.123).

In this way, we can determine the optimal solution λ^*. Then the DM selects an appropriate standing objective from among the objectives $z_i(x)$, $i = 1, ..., k$. For notational convenience in the following without loss of generality, let it be $z_1(x)$. Then the following linear programming problem is solved for $\lambda = \lambda^*$:

$$\left.\begin{array}{l} \text{minimize } \; z_1(x) \\ \text{subject to the constraints of (3.123).} \end{array}\right\} \quad (3.124)$$

It can be easily shown that an optimal solution to this problem, say x^*, satisfies weak Pareto optimality.

In order to test the Pareto optimality, we solve the following linear programming problem:

$$\left.\begin{array}{ll} \text{maximize} & \sum_{i=1}^{k} \varepsilon_i \\ \text{subject to} & z_i(x) + \varepsilon_i = z_i(x^*) \\ & x \in X, \;\; \varepsilon \geq 0. \end{array}\right\} \quad (3.125)$$

Let $\bar{x}$ be an optimal solution to this problem. Then as discussed in Theorem 3.7, (1) If all $\varepsilon_i = 0$, then x^* is a Pareto optimal solution, and (2) If at least one $\varepsilon_i > 0$, then $\bar{x}$ is a Pareto optimal solution.

Further details including an illustrative numerical example and an application can be found in Sakawa (1983a, 1983b).

3.5 Interactive fuzzy multiobjective linear programming

In the fuzzy approaches to multiobjective linear programming problems proposed by Zimmermann and his successors, however, it has been implicitly assumed that the fuzzy decision of Bellman and Zadeh (1970) is the proper representation of the fuzzy preferences of the decision maker (DM). Therefore, these approaches are preferable only when the DM feels that the fuzzy decision is appropriate when combining the fuzzy goals and/or constraints. However, such situations seem to occur rarely in practice and consequently it becomes evident that an interaction with the DM is necessary.

In this section, assuming that the DM has a fuzzy goal for each of the objective functions in multiobjective linear programming problems, we present an interactive fuzzy multiobjective linear programming method incorporating the desirable features of the interactive approaches into the fuzzy approaches.

In general, the multiobjective linear programming (MOLP) problem is represented as the following vector-minimization problem:

$$\left.\begin{array}{ll} \text{minimize} & z(x) \triangleq (z_1(x), z_2(x), \ldots, z_k(x))^T \\ \text{subject to} & x \in X \triangleq \{x \in R^n \mid Ax \leq b, \; x \geq 0\} \end{array}\right\} \quad (3.126)$$

where x is an n-dimensional vector of decision variables, $z_1(x) = c_1x, \ldots, z_k(x) = c_kx$ are k conflicting linear objective functions, and X is the feasible set of linearly constrained decisions.

Fundamental to the MOLP is the concept of Pareto optimal solutions, also known as a noninferior solution.

However, considering the imprecise nature inherent in human judgments in multiobjective linear programming problems, the DM may have a fuzzy goal expressed as " $z_i(x)$ should be substantially less than or equal to some value p_i."

In a minimization problem, a fuzzy goal stated by the DM may be to achieve "substantially less than or equal to p_i." This type of statement can be quantified by eliciting a corresponding membership function. Figure 3.21 illustrates the possible shape of the membership function representing the fuzzy goal to achieve substantially less than or equal to p_i.

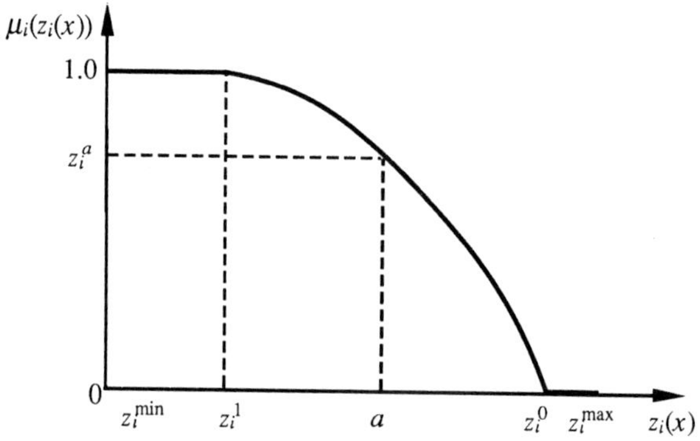

Figure 3.21 Membership function for minimization problem

To elicit a membership function $\mu_i(z_i(x))$ from the DM for each of the objective functions $z_i(x)$, $i = 1, ..., k$, we first calculate the individual minimum $z_i^{\min} = \min_{x \in X} z_i(x)$ and maximum $z_i^{\max} = \max_{x \in X} z_i(x)$ of each objective function $z_i(x)$ under the given constraints.

Taking into account the calculated individual minimum and maximum of each objective function together with the rate of increase of membership of satisfaction, the DM must determine the subjective membership function $\mu_i(z_i(x))$, which is a strictly monotone decreasing function with respect to $z_i(x)$. Here, it is assumed that $\mu_i(z_i(x)) = 0$ or $\rightarrow 0$ if $z_i(x) \geq z_i^0$ and $\mu_i(z_i(x)) = 1$ or $\rightarrow 1$ if $z_i(x) \leq z_i^1$, where z_i^a represents the value of $z_i(x)$ such that the value of membership function $\mu_i(z_i(x))$ is $a \in [0, 1]$ within $z_i^{\min}$ and $z_i^{\max}$.

So far, we have restricted ourselves to a minimization problem and consequently assumed that the DM has a fuzzy goal such as " $z_i(x)$ should be substantially less than or equal to p_i." In the fuzzy approaches, however, we can further treat a more general multiobjective linear programming problem in which the DM has two types of fuzzy goals expressed in words such as " $z_i(x)$ should be in the vicinity of r_i " (called fuzzy equal) or " $z_i(x)$ should be substantially less than or equal to p_i or greater than or equal to q_i " (called fuzzy min or fuzzy max).

Such a generalized multiobjective linear programming problem (GMOLP) may now be expressed as

$$\left.\begin{array}{lll} \text{fuzzy min} & z_i(x) & i \in I_1 \\ \text{fuzzy max} & z_i(x) & i \in I_2 \\ \text{fuzzy equal} & z_i(x) & i \in I_3 \\ \text{subject to} & x \in X & \end{array}\right\} \tag{3.127}$$

where $I_1 \cup I_2 \cup I_3 = \{1, 2, ..., k\}$, $I_i \cap I_j = \emptyset, i, j = 1, 2, 3, i \neq j$.

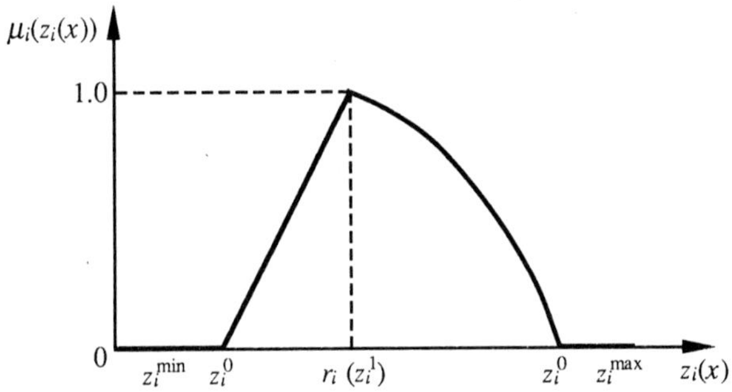

Figure 3.22 Fuzzy equal membership function

Here "fuzzy min $z_i(x)$" or "fuzzy max $z_i(x)$" represents the fuzzy goal of the DM such as " $z_i(x)$ should be substantially less than or equal to p_i or greater than or equal to q_i," and "fuzzy equal $z_i(x)$" represents the fuzzy goal such as " $z_i(x)$ should be in the vicinity of r_i."

Concerning the membership function for the fuzzy goal of the DM such as " $z_i(x)$ should be in the vicinity of r_i," it is obvious that a strictly monotone increasing function $d_{iL}(z_i)$ $(i \in I_{3L})$ and a strictly monotone decreasing function $d_{iR}(z_i)$ $(i \in I_{3R})$, $(I_3 = I_{3L} \cup I_{3R})$ corresponding to the left and right sides of r_i must be determined through interaction with the DM.

As an example, Figure 3.22 illustrates the possible shape of the fuzzy equal membership functions where the left function is linear and the right function is exponential.

When the fuzzy equal is included in the fuzzy goals of the DM, it is desirable that $z_i(x)$ should be as close to r_i as possible. Consequently, the notion of Pareto optimal solutions defined in terms of objective functions cannot be applied. For this reason, we introduce the concept of M-Pareto optimal solutions which is defined in terms of membership functions instead of objective functions. M refers to membership.

Definition 3.4 (M-Pareto optimal solution)

$x^* \in X$ is said to be an M-Pareto optimal solution to the GMOLP if and only if there does not exist another $x \in X$ such that $\mu_i(z_i(x)) \geq \mu_i(z_i(x^*))$ for all i and $\mu_j(z_j(x)) \neq \mu_j(z_j(x^*))$ for at least one j.

Having elicited the membership functions $\mu_i(z_i(x))$, $i = 1, ..., k$, from the DM for each of the objective functions $z_i(x)$, $i = 1, ..., k$, the MOLP and/or the GMOLP can be converted into the fuzzy multiobjective optimization problem (FMOP) defined by

$$\underset{x \in X}{\text{maximize}} \quad (\mu_1(z_1(x)), \mu_2(z_2(x)), ..., \mu_k(z_k(x))). \tag{3.128}$$

By introducing a general aggregation function

$$\mu_D(\mu(z(x))) = \mu_D(\mu_1(z_1(x)), \mu_2(z_2(x)), ..., \mu_k(z_k(x))), \tag{3.129}$$

a general fuzzy multiobjective decision making problem (FMDMP) can be defined by

$$\underset{x \in X}{\text{maximize}} \quad \mu_D(\mu(z(x))). \tag{3.130}$$

Observe that the value of $\mu_D(\mu(z(x)))$ can be interpreted as representing an overall degree of satisfaction with the DM's multiple fuzzy goals.

The fuzzy decision or the minimum operator of Bellman and Zadeh (1970)

$$\min_{i=1,...,k} (\mu_1(z_1(x)), \mu_2(z_2(x)), ..., \mu_k(z_k(x))) \tag{3.131}$$

can be viewed only as one special example of $\mu_D(\mu(z(x)))$.

In the conventional fuzzy approaches discussed thus far, it has been implicitly assumed that the minimum operator is the proper representation of the DM's fuzzy preferences, and hence, the FMDMP has been interpreted as

$$\underset{x \in X}{\text{maximize}} \min_{i=1,...,k} \quad (\mu_1(z_1(x)), \mu_2(z_2(x)), ..., \mu_k(z_k(x))) \tag{3.132}$$

or equivalently

$$\left.\begin{array}{ll} \text{maximize} & v \\ \text{subject to} & v \leq \mu_i(z_i(x)), \ i = 1, 2, \ldots, k \\ & x \in X. \end{array}\right\} \tag{3.133}$$

However, it should be emphasized here that this approach is preferable only when the DM feels that the minimum operator is appropriate. In other words, in general decision situations, the DM does not always use the minimum operator when combining the fuzzy goals and/or constraints. Probably the most crucial problem in the FMDMP is the identification of an appropriate aggregation function which well represents the DM's fuzzy preferences. If $\mu_D(\cdot)$ can be explicitly identified, then the FMDMP reduces to a standard mathematical programming problem. However, this rarely happens, and as an alternative, an interaction with the DM is necessary for finding the satisficing solution of the FMDMP.

In the interactive multiobjective linear programming method proposed by Sakawa, Yano, and Yumine (1987), the membership functions $\mu(z(x)) = (\mu_1(z_1(x)), \ldots, \mu_k(z_k(x)))^T$ for each of the objective functions $z(x) = (z_1(x), \ldots, z_k(x))^T$ are determined first. To generate a candidate for the satisficing solution which is also (M-) Pareto optimal, the DM is then asked to specify the aspiration levels of achievement for the membership values of all membership functions, called the reference membership levels. The reference membership values can be viewed as natural extensions of the reference point of Wierzbicki (1979a) in objective function spaces.

For the DM's reference membership levels $\bar{\mu} = (\bar{\mu}_1, \ldots, \bar{\mu}_k)^T$, the corresponding (M-) Pareto optimal solution, which is nearest to the requirements in the minimax sense or better than that if the reference membership levels are attainable, is obtained by solving the following minimax problem

$$\underset{x \in X}{\text{minimize}} \quad \max_{i=1,\ldots,k} \{\bar{\mu}_i - \mu_i(z_i(x))\} \tag{3.134}$$

or equivalently

$$\left.\begin{array}{ll} \text{minimize} & v \\ \text{subject to} & \bar{\mu}_i - \mu_i(z_i(x)) \leq v, \; i = 1, 2, \ldots, k \\ & x \in X. \end{array}\right\} \tag{3.135}$$

If all of the membership functions $\mu_i(z_i(x))$, $i = 1, \ldots, k$, are linear, the minimax problem becomes a linear programming problem, and hence, we can obtain an optimal solution by directly applying the simplex method of linear programming.

However, with the strictly monotone decreasing or increasing membership functions, which may be nonlinear, the resulting minimax problem becomes a nonlinear programming problem. For notational convenience, denote the strictly monotone decreasing function for the fuzzy min and the right function of the fuzzy equal by $d_{iR}(z_i)$ $(i \in I_1 \cup I_{3R})$ and the strictly monotone increasing function for the fuzzy max and the left function of the fuzzy equal by $d_{iL}(z_i)$ $(i \in I_2 \cup I_{3L})$. Then in order to solve the formulated problem on the basis of the linear programming method, convert each constraint $\bar{\mu}_i - \mu_i(z_i(x)) \leq v$, $i = 1, \ldots, k$, of the minimax problem (3.135) into the following form using the strictly monotone property of $d_{iL}(\cdot)$ and $d_{iR}(\cdot)$:

$$\left.\begin{array}{ll} \text{minimize} & v \\ \text{subject to} & z_i(x) \leq d_{iR}^{-1}(\bar{\mu}_i - v), \ i \in I_1 \cup I_{3R} \\ & z_i(x) \geq d_{iL}^{-1}(\bar{\mu}_i - v), \ i \in I_2 \cup I_{3L} \\ & x \in X. \end{array}\right\} \tag{3.136}$$

It is important to note here that, if the value of v is fixed, it can be reduced to a set of linear inequalities. Obtaining the optimal solution v^* to the above problem is equivalent to determining the minimum value of v so that there exists an admissible set satisfying the constraints of (3.136). Since v satisfies $\bar{\mu}_{\max} - 1 \leq v \leq \bar{\mu}_{\max}$, where $\bar{\mu}_{\max}$ denotes the maximum value of $\bar{\mu}_i$, $i = 1, ..., k$, we have the following method for solving this problem by combined use of the bisection method and the simplex method of linear programming. Here, when $\bar{\mu}_i - v \leq 0$, set $\bar{\mu}_i - v = 0$ in view of the constraints $\bar{\mu}_i - v \leq \mu_i(z_i(x))$ for $0 \leq \mu_i(z_i(x)) \leq 1$, $i = 1, \ldots, k$.

Step 1: Set $v = \bar{\mu}_{\max}$ and test whether an admissible set satisfying the constraints of (3.136) exists or not using phase one of the simplex method. If an admissible set exists, proceed. Otherwise, the DM must reassess the membership function.

Step 2: Set $v = \bar{\mu}_{\max} - 1$ and test whether an admissible set satisfying the constraints of (3.136) exists or not using phase one of the simplex method. If an admissible set exists, set $v^* = \bar{\mu}_{\max} - 1$. Otherwise, go to the next step since the minimum v which satisfies the constraints of (3.136) exists between $\bar{\mu}_{\max} - 1$ and $\bar{\mu}_{\max}$.

Step 3: For the initial value of $v = \bar{\mu}_{\max} - 0.5$, update the value of v using the bisection method as follows:

$$\begin{cases} v_{n+1} = v_n - 1/2^{n+1} & \text{if an admissible set exists for } v_n, \\ v_{n+1} = v_n + 1/2^{n+1} & \text{if no admissible set exists for } v_n. \end{cases}$$

For each v_n, $n = 1, 2, ...$, test whether an admissible set of (3.136) exists or not using the sensitivity analysis technique for changes in the right-hand side of the simplex method and determine the minimum value of v satisfying the constraints of (3.136).

In this way, we can determine the optimal solution v^*. Then the DM selects an appropriate standing objective from among the objectives $z_i(x)$, $i = 1, ..., k$. For notational convenience in the following without loss of generality, let it be $z_1(x)$ and $1 \in I_1$. Then the following linear programming problem is solved for

$v = v^*$:

$$\left.\begin{array}{ll} \text{minimize} & z_1(x) \\ \text{subject to} & z_i(x) \le d_{iR}^{-1}(\bar{\mu}_i - v^*),\ i\,(\neq 1) \in I_1 \cup I_3 \\ & z_i(x) \ge d_{iL}^{-1}(\bar{\mu}_i - v^*),\ i\,(\neq 1) \in I_2 \cup I_3 \\ & x \in X. \end{array}\right\} \quad (3.137)$$

The relationships between the optimal solutions of the minimax problem and the (M-) Pareto optimal concept of the MOLP can be characterized by the following theorem.

Theorem 3.11

(1) *If $x^* \in X$ is a unique optimal solution to the minimax problem for some $\bar{\mu}_i$, $i = 1, ..., k$, then x^* is a (M-) Pareto optimal solution to the (G) MOLP.*

(2) *If x^* is a (M-) Pareto optimal solution to the (G) MOLP with $0 < \mu_i(z_i(x^*)) < 1$ holding for all i, then there exists $\bar{\mu}_i$, $i = 1, ..., k$, such that x^* is an optimal solution to the minimax problem.*

The proof of this theorem follows directly from the definitions of optimality and (M-) Pareto optimality by making use of contradiction arguments.

It must be noted here that, for generating (M-) Pareto optimal solutions using this theorem, uniqueness of solution must be verified. In the ad hoc numerical approach, however, to test the (M-) Pareto optimality of current optimal solution x^*, we solve the following Pareto optimality test problem as discussed in Theorem 3.7:

$$\left.\begin{array}{ll} \text{maximize} & \sum_{i=1}^{k} \varepsilon_i \\ \text{subject to} & \mu_i(z_i(x)) - \varepsilon_i = \mu_i(z_i(x^*)),\ i = 1, \ldots, k \\ & x \in X,\ \varepsilon = (\varepsilon_1, ..., \varepsilon_k)^T \ge 0. \end{array}\right\} \quad (3.138)$$

For the optimal solution $\bar{x}$ and $\bar{\varepsilon}$ to this linear programming problem, as discussed in Theorem 3.7, (1) If all $\bar{\varepsilon}_i = 0$, then x^* is a (M-) Pareto optimal solution to the (G) MOLP, and (2) If at least one $\bar{\varepsilon}_i > 0$, not x^* but $\bar{x}$ is a (M-) Pareto optimal solution of the (G) MOLP.

The DM must either be satisfied with the current (M-) Pareto optimal solution or act on this solution by updating the reference membership levels. In order to help the DM express a degree of preference, as was discussed in the previous subsection, trade-off information between a standing membership function $\mu_1(z_1(x))$ and each of the other membership functions is very useful. Such trade-off information is easily obtainable since it is closely related to the simplex multipliers of problem (3.137).

Let the simplex multipliers corresponding to the constraints $z_i(x), i = 2, \ldots, k$ of the linear problem (3.137) be denoted by $\pi_i^* = \pi_i(x^*)$, $i = 2, \ldots, k$, where x^*

is an optimal solution of (3.137). If x^* is a nondegenerate solution of (3.137) and all the constraints of (3.137) are active, then by using the results in Haimes and Chankong (1979), the trade-off information between the objective functions can be represented by

$$-\frac{\partial z_1(x)}{\partial z_i(x)} = -\pi_i^*, \; i = 2, \ldots, k. \tag{3.139}$$

Hence, by the chain rule, the trade-off information between the membership functions is given by

$$-\frac{\partial \mu_1(z_1(x))}{\partial \mu_i(z_i(x))} = \frac{\partial \mu_1(z_1(x))}{\partial z_1(x)} \frac{\partial z_1(x)}{\partial z_i(x)} \left\{ \frac{\partial \mu_i(z_i(x))}{\partial z_i(x)} \right\}^{-1}, \; i = 2, \ldots, k. \tag{3.140}$$

Therefore, for each $i = 2, \ldots, k$, we have the following expression:

$$-\frac{\partial \mu_1(z_1(x))}{\partial \mu_i(z_i(x))} = \pi_i^* \frac{\partial \mu_1(z_1(x))/\partial z_1(x)}{\partial \mu_i(z_i(x))/\partial z_i(x)}, \; i = 2, \ldots, k. \tag{3.141}$$

It should be stressed here that in order to obtain the trade-off rate information from (3.141), all the constraints of problem (3.137), must be active. Therefore, if there are inactive constraints, it is necessary to replace $\bar{\mu}_i$ for inactive constraints by $\bar{\mu}_i(z_i(x^*))$ and solve the corresponding problem to obtain the simplex multipliers.

We can now construct the interactive algorithm in order to derive the satisficing solution for the DM from the (M-) Pareto optimal solution set where the steps marked with an asterisk involve interaction with the DM. This interactive fuzzy multiobjective programming method can also be interpreted as the fuzzy version of the reference point method (RPM) with trade-off information.

Interactive fuzzy multiobjective linear programming

Step 0: Calculate the individual minimum and maximum of each objective function under the given constraints.

Step 1*: Elicit a membership function from the DM for each of the objective functions.

Step 2: Set the initial reference membership levels to 1.

Step 3: For the reference membership values, solve the corresponding minimax problem to obtain the (M-) Pareto optimal solution and the membership function value together with the trade-off rate information between the membership functions.

Step 4*: If the DM is satisfied with the current levels of the (M-) Pareto optimal solution, stop. Then the current (M-) Pareto optimal solution is the satisficing solution of the DM. Otherwise, ask the DM to update the current reference membership levels by considering the current values of the membership functions together with the trade-off rates between the membership functions and return to step 3.

It should be stressed to the DM that any improvement of one membership function can be achieved only at the expense of at least one of the other membership functions.

Example 3.7 (Production problem with environmental quality)
To demonstrate interactive fuzzy multiobjective programming, consider the production problem with environmental quality in Example 3.3. In addition to the fuzzy goals of the DM such as (1) maximization of the profit and (2) minimization of pollution considering the profit contribution ratio of the two products, assume that the DM also has a fuzzy goal such as (3) production of product P_2 should be twice that of product P_1.

Then the corresponding GMOLP can be formulated as follows:

$$\left.\begin{array}{rl} \text{fuzzy max} & x_1 + 2x_2 \\ \text{fuzzy min} & 3x_1 + 2x_2 \\ \text{fuzzy equal} & x_1 - 2x_2 \\ \text{subject to} & x \in X \end{array}\right\} \tag{3.142}$$

where X is the feasible region as in Example 3.3.

First, observe that the individual minimums and maximums of each objective function are

$$z_1^{\min} = 0, \quad z_1^{\max} = 10, \quad z_2^{\min} = 0, \quad z_2^{\max} = 16.5, \quad z_3^{\min} = -9, \quad z_3^{\max} = 5.$$

Considering these values, assume that the DM determines the following linear membership functions for the fuzzy goals

$$\left\{\begin{array}{lll} \text{fuzzy max} & \mu_1(5) = 0, & \mu_1(8) = 1 \\ \text{fuzzy min} & \mu_2(14) = 0, & \mu_2(9) = 1 \\ \text{fuzzy equal} & \mu_{3L}(-7) = 0, & \mu_{3L}(0) = \mu_{3R}(0) = 1, \quad \mu_{3R}(4) = 0 \end{array}\right.$$

where linear functions are assumed from $\mu_i = 0$ to $\mu_i = 1$ for $i = 1, 2, 3$.

Then, for the initial reference membership value 1, solving the corresponding minimax problem yields the M-Pareto optimal solution

$$z_1 = 6.833, \quad z_2 = 10.94, \quad z_3 = -2.722 \qquad (x_1 = 2.056,\ x_2 = 2.389),$$

the corresponding membership values

$$\mu_1 = 0.6111, \quad \mu_2 = 0.6111, \quad \mu_3 = 0.6111,$$

and the trade-off rates between the membership functions

$$-\frac{\partial \mu_2}{\partial \mu_1} = 1.2, \quad -\frac{\partial \mu_3}{\partial \mu_1} = 0.8571.$$

On the basis of such information, the DM updates the reference membership values to

$$\bar{\mu}_1 = 0.7, \quad \bar{\mu}_2 = 0.8, \quad \bar{\mu}_3 = 0.5,$$

improving the satisfaction levels for the profit and the amount of pollution at the expense of the ratio of two products.

For the updated reference membership values, the corresponding minimax problem yields the M-Pareto optimal solution

$$z_1 = 6.983, \quad z_2 = 10.19, \quad z_3 = -3.772 \qquad (x_1 = 1.606, \ x_2 = 2.689),$$

the membership values

$$\mu_1 = 0.6611, \quad \mu_2 = 0.7611, \quad \mu_3 = 0.4611,$$

and the trade-off rates

$$-\frac{\partial \mu_2}{\partial \mu_1} = 1.2, \quad -\frac{\partial \mu_3}{\partial \mu_1} = 0.8571.$$

If the DM is satisfied with the current values of the membership functions, the procedure stops. Otherwise, a similar procedure continues until the satisficing solution for the DM is obtained.

CHAPTER 4

FUZZY NONLINEAR PROGRAMMING

In 1974, Tanaka et al. first introduced the concept of fuzzy mathematical programming following the fuzzy decision proposed by Bellman and Zadeh (1970) on a general level. Under some assumptions about a fuzzy goal and a fuzzy constraint for the decision maker, they proved that the fuzzy mathematical programming problem of finding the maximizing decision of the fuzzy decision can be reduced to the conventional nonlinear programming problem. Since then, fuzzy mathematical programming has been developed not only on a general level but on a more practical level with many successful applications.

In this chapter, after overviewing the basic concepts of nonlinear programming via a simple numerical example, fuzzy nonlinear programming first proposed by Tanaka et al. on a general level is presented because of its historical importance. Then fundamental notions and some well-recognized methods of multiobjective nonlinear programming are presented. Finally, fuzzy multiobjective nonlinear programming and interactive fuzzy multiobjective nonlinear programming, both proposed by the authors, are explained in detail by putting special emphasis on Pareto optimality.

4.1 Nonlinear programming and fuzzy nonlinear programming

4.1.1 Nonlinear programming

To introduce the basic notions of nonlinear programming, first recall the production planning problem discussed in Example 3.1.

Example 3.1 (Production planning problem)

Company C would like to maximize the total profit producing two products P_1 and P_2 utilizing three different materials M_1, M_2, and M_3. The company knows that to produce 1 ton of product P_1 requires 2 tons of material M_1, 8 tons of material M_2, and 3 tons of material M_3, while to produce 1 ton of product P_2 requires 6 tons of material M_1, 6 tons of material M_2, and 1 ton of material M_3. The total amounts of available materials are limited to 27 tons, 45 tons, and 15 tons for M_1, M_2, and M_3, respectively. It also knows that product P_1 yields a profit of 1 million yen per ton, while P_2 yields 2 million yen. Given these limited materials, the company is trying to figure out how many units of products P_1 and P_2 should be produced to maximize the total profit.

Denote the number of tons produced of products P_1 and P_2 by x_1 and x_2 respectively. This production planning problem can be formulated as the following linear programming problem:

$$\left.\begin{array}{lrl} \text{minimize} & z = & -x_1 - 2x_2 \\ \text{subject to} & & 2x_1 + 6x_2 \leq 27 \\ & & 8x_1 + 6x_2 \leq 45 \\ & & 3x_1 + \ x_2 \leq 15 \\ & & x_1 \geq 0,\ x_2 \geq 0. \end{array}\right\} \tag{4.1}$$

Example 4.1 (Nonlinear production planning problem)

In practice, however, it was found that the unit profit contributions of products P_1 and P_2 are not constants as in the linear profit function. They depend on the amounts of products P_1 and P_2 and are $4 - x_1$, $11 - x_2$ for P_1 and P_2, respectively.

The resulting profit function is defined by

$$(4 - x_1)x_1 + (11 - x_2)x_2. \tag{4.2}$$

In such a situation, it is more appropriate to formulate the production planning problem as the following nonlinear programming problem than as a linear one:

$$\left.\begin{array}{lrl} \text{minimize} & z = & x_1^2 + \ x_2^2 - 4x_1 - 11x_2 \\ \text{subject to} & & 2x_1 + 6x_2 \leq 27 \\ & & 8x_1 + 6x_2 \leq 45 \\ & & 3x_1 + \ x_2 \leq 15 \\ & & x_1 \geq 0,\ x_2 \geq 0. \end{array}\right\} \tag{4.3}$$

Although the constrained set, having five extreme points, is the same as Example 3.1, the nonlinear objective function has contours of constant value which are not parallel lines, as in the linear case, but concentric circles. This simple nonlinear programming problem involving only two variables can be solved graphically in the $x_1 - x_2$ plane as shown in Figure 4.1.

The minimum value of z corresponds to the contour of lowest value having at least one point in common with the constrained set, as can be seen from Figure 4.1. Therefore, the optimal solution occurs at the point $(1.5, 4)$ where the contour $z = -31.75$ is a tangent to the boundary of the constrained set. Hence, the optimal solution is $x_1 = 1.5, x_2 = 4$ with $z = -31.75$. Observe that this is not an extreme point of the constrained set although it is a boundary point.

If the objective function of the problem is changed to

$$z = (x_1 - 1)^2 + (x_2 - 1)^2, \tag{4.4}$$

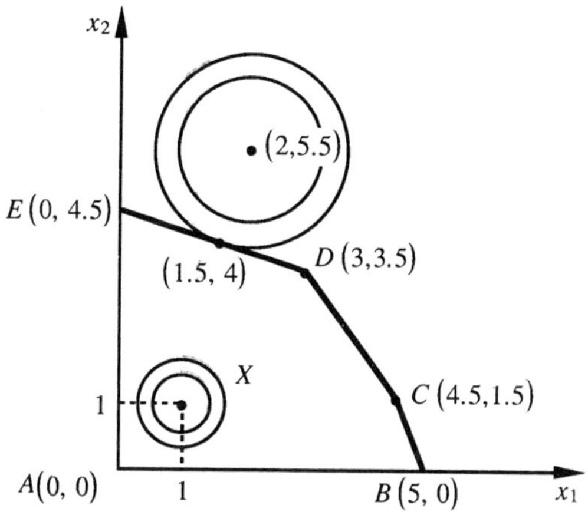

Figure 4.1 Graphical solution for Example 4.1

the optimal solution is obviously $x_1 = 1, x_2 = 1$, which is not even a boundary point but an interior point of the constrained set.

It is significant to realize here that the optimal solution of a nonlinear programming problem will in general not occur at an extreme point of the constrained set and may not even be on the boundary. Therefore, nonlinear programming problems could not be solved by just examining only the finite set of extreme points as in the simplex method of linear programming. As a result, they are far more difficult to solve than linear programming problems.

Also, the possible existence of local minima which might not be optimal overall is another characteristic of nonlinear programming problems which can cause serious difficulty. For example, if the objective function of the previous problem has two minima and at least one of them is an interior point of the constrained set, then the constrained problem would have two local minima.

These properties, discussed thus far via simple numerical examples, are direct consequences of the consideration of nonlinearities and takes us into the field of nonlinear programming.

The preceding simple nonlinear production planning problem can be immediately generalized to the following nonlinear programming (NLP) problem[§]:

minimize the single-valued nonlinear objective function

$$f(x_1, x_2, \ldots, x_n) \tag{4.5}$$

[§] For convenience, we start from the problem with inequality constraints, but similar discussions can be made for the problems with both equality and inequality constraints.

subject to the m nonlinear inequality constraints

$$\left.\begin{array}{c} g_1(x_1, x_2, \ldots, x_n) \leq 0 \\ g_2(x_1, x_2, \ldots, x_n) \leq 0 \\ \cdots\cdots\cdots\cdots\cdots\cdots \\ g_m(x_1, x_2, \ldots, x_n) \leq 0, \end{array}\right\} \tag{4.6}$$

or in more compact vector form,

$$\left.\begin{array}{ll} \text{minimize} & f(x) \\ \text{subject to} & g_i(x) \leq 0, \ i = 1, \ldots, m \end{array}\right\} \tag{4.7}$$

or

$$\left.\begin{array}{ll} \text{minimize} & f(x) \\ \text{subject to} & x \in X \triangleq \{x \in R^n \mid g_i(x) \leq 0, \ i = 1, ..., m\} \end{array}\right\} \tag{4.7$'$}$$

where x is an n-dimensional vector of decision variables, $f(x) = f(x_1, x_2, \ldots, x_n)$ and $g_i(x) = g_i(x_1, x_2, \ldots, x_n)$, $i = 1, \ldots, m$, are given real-valued nonlinear functions of n real variables $x_1, x_2, \ldots, x_n$, and X is the feasible set of the nonlinearly constrained set.

Several basic notions and definitions involved in nonlinear programming follow.

(1) Local and global minimum: A point x^* is said to be a local minimum point of the NLP if there exists a real number $\delta > 0$ such that $f(x) \geq f(x^*)$ for all $x \in X$ satisfying $\|x - x^*\| < \delta$. If $f(x) > f(x^*)$ for all $x \in X$, $x \neq x^*$, within a distance δ of x^*, then x^* is said to be a strict local minimum point of the NLP. A point x^* is said to be a global minimum point of the NLP if $f(x) \geq f(x^*)$ for all $x \in X$. If $f(x) > f(x^*)$ for all $x \in X$, $x \neq x^*$, then x^* is said to be a strict global minimum point of the NLP.

(2) Convex set: A nonempty set S is said to be convex if the line segment joining any two points of the set also belongs to the set, i.e.,

$$\lambda x_1 + (1 - \lambda) x_2 \in S \ \ \forall x_1, x_2 \in S \ \text{ and } \forall \lambda \in [0, 1]. \tag{4.8}$$

(3) Convex function: A function $f(x)$ defined on a nonempty convex set S in R^n is said to be convex if

$$f(\lambda x_1 + (1 - \lambda) x_2) \leq \lambda f(x_1) + (1 - \lambda) f(x_2) \ \forall x_1, x_2 \in S \ \text{ and } \forall \lambda \in [0, 1]. \tag{4.9}$$

A function $f(x)$ is said to be strictly convex if the above inequality holds as a strict inequality for $x_1 \neq x_2$ and $\lambda \in (0, 1)$. A function $f(x)$ is said to be concave (strictly concave) if $-f(x)$ is convex (strictly convex).

(4) Convex programming: A nonlinear programming problem to minimize a convex function (or maximize a concave function) is said to be convex programming.

Geometrically, a function is convex if the line joining two points on its graph lies nowhere below the graph. It is easy to see that a linear function is both convex and concave. Conversely, if a function is both convex and concave, it is linear.

The following properties follow directly from the definitions of convex sets and convex functions.

(1) Intersection of convex sets: The intersection of any number of convex sets is also a convex set.

(2) Level set of convex function: If $f(x)$ is a convex function defined on a convex set S, then the level set S_α defined by

$$S_\alpha = \{x \in S \mid f(x) \leq \alpha\} \tag{4.10}$$

is also convex for all real numbers α.

As an immediate consequence of these properties, if all the functions $f(x)$ and $g_i(x)$, $i = 1, \ldots, m$, are convex, the NLP becomes a convex programming problem. Observe that, since a linear function is convex, a linear programming problem is also a convex programming problem.

The desirable feature of convex programming follows.

Theorem 4.1 (Convex programming)

Any local minimum of a convex programming problem is a global minimum.

To introduce the most important theoretical results in the field of nonlinear programming, first introduced by Kuhn and Tucker (1951), define the following basic notions.

(1) Active constraint: An inequality constraint $g_i(x) \leq 0$ is said to be active at a point x^* if $g_i(x^*) = 0$.

(2) Regular point: A point x^* satisfying $g_i(x^*) \leq 0$ is said to be a regular point if the gradient vectors $\nabla g_i(x)$ of all active constraints (i.e., for all $g_i(x) \leq 0$ such that $g_i(x^*) = 0$) are linearly independent.

(3) Lagrangian function: The Lagrangian function associated with the NLP is defined as

$$L(x, \lambda) = f(x) + \sum_{i=1}^{m} \lambda_i g_i(x), \quad \lambda_i \geq 0 \tag{4.11}$$

where the λ_i are Lagrange multipliers for the inequality constraint $g_i(x) \leq 0$.

(4) **Nondegenerate**: An active constraint is said to be nondegenerate if the corresponding Lagrange multiplier for the active constraint is strictly positive.

By imposing the constraint qualification of a regular point assumption, the first-order Kuhn-Tucker necessary conditions for optimality can be expressed as follows.

Theorem 4.2 (Kuhn-Tucker necessary conditions)

Let x^* be a regular point of the constraints of the NLP and assume all the functions $f(x)$ and $g_i(x)$ of the NLP are differentiable. If x^* is a local minimum of the NLP, then there exist Lagrange multipliers λ_i, $i = 1, ..., m$, such that

$$\nabla_x L(x, \lambda) = \nabla f(x) + \sum_{i=1}^{m} \lambda_i \nabla g_i(x) = 0, \tag{4.12}$$

$$\lambda_i g_i(x) = 0, \quad i = 1, \ldots, m, \tag{4.13}$$

$$\lambda_i \geq 0, \quad i = 1, \ldots, m. \tag{4.14}$$

Although the Kuhn-Tucker conditions have been shown to be necessary for both convex and nonconvex problems, they are also sufficient for convex programming as shown in the following theorem.

Theorem 4.3 (Kuhn-Tucker sufficient conditions)

Let all the functions $f(x)$ and $g_i(x)$ of the NLP be convex and differentiable. Suppose x^* satisfies the Kuhn-Tucker conditions. Then x^* is a global minimum of the NLP.

By assuming that all the functions of the NLP are twice continuously differentiable and using the Hessian matrix of $L(x, \lambda)$, the second-order necessary or sufficient conditions for optimality can be expressed as follows.

Theorem 4.4 (Second-order necessary conditions)

Let x^* be a regular point of the constraints of the NLP. If x^* is a local minimum of the NLP, then there exist Lagrange multipliers λ_i^*, $i = 1, ..., m$, such that the Kuhn-Tucker conditions (4.12) to (4.14) hold and the Hessian matrix

$$\nabla_x^2 L(x^*, \lambda^*) = \nabla^2 f(x^*) + \sum_{i=1}^{m} \lambda_i^* \nabla^2 g_i(x^*) \tag{4.15}$$

is positive semidefinite on the tangent subspace T of the active constraint at x^*, where

$$T = \{y \mid \nabla g_i(x^*) y = 0, \text{ for all } i \text{ with } \lambda_i > 0\}. \tag{4.16}$$

Theorem 4.5 (Second-order sufficient conditions)

Let x^* be a regular point of the constraints of the NLP. Assume that there exist Lagrange multipliers λ_i^*, $i = 1, ..., m$, such that the Kuhn-Tucker conditions (4.12) to (4.14) hold and the Hessian matrix

$$\nabla_x^2 L(x^*, \lambda^*) = \nabla^2 f(x^*) + \sum_{i=1}^{m} \lambda_i^* \nabla^2 g_i(x^*)$$

is positive definite on T. Then x^* is a local minimum of the NLP.

The following sensitivity theorem is very useful for developing the trade-off information in multiobjective programming problems.

Theorem 4.6 (Sensitivity theorem)

Let $f(x)$ and $g_i(x)$ be twice continuously differentiable and consider the following nonlinear programming problem:

$$\left.\begin{array}{ll} \text{minimize} & f(x) \\ \text{subject to} & g_i(x) \leq \varepsilon_i, \ i = 1, \ldots, m. \end{array}\right\} \tag{4.17}$$

Let x^* be a local optimal solution of this problem for $\varepsilon_i = 0$, $i = 1, ..., m$, satisfying the following conditions:

(1) *x^* is a regular point of the constraints of the problem.*

(2) *The second-order sufficiency conditions are satisfied at x^*.*

(3) *There are no degenerate active constraints at x^*.*

Then there exists a continuously differentiable vector-valued function $x(\cdot)$ defined in a neighborhood of 0 in R^m such that $x(0) = x^*$ and such that for every ε near 0, $x(\varepsilon)$ is a strict local minimum of the problem. Furthermore,

$$-\frac{\partial f[x(0)]}{\partial \varepsilon_i} = \lambda_i, \quad i = 1, \ldots, m. \tag{4.18}$$

Optimality conditions for convex programming in the absence of differentiability can often be derived by using the following separation theorem of convex sets.

Theorem 4.7 (Separation theorem)

Let A and B be two nonempty disjoint convex subsets in R^n. Then there exists a hyperplane which separates them, i.e., there exists a nonzero vector c in R^n and a scalar α such that

$$\left.\begin{array}{l} cx \leq \alpha, \quad \forall x \in A \\ cx \geq \alpha, \quad \forall x \in B. \end{array}\right\}$$

Further details of the theory and algorithms of nonlinear programming can be found in standard texts including Mangasarian (1969), Zangwill (1969), Avriel (1976), Bazaraa and Shetty (1979), Gill, Murray, and Wright (1981) and Luenberger (1973, 1984).

4.1.2 Fuzzy nonlinear programming

Similar to fuzzy linear programming as proposed by Zimmermann (1976), in contrast to the conventional nonlinear programming problem discussed above, it is possible to soften the rigid requirements of the decision maker (DM) to strictly minimize the objective function and strictly satisfy the constraints. In such a situation, the nonlinear programming problem may be softened into the following fuzzy version

$$\left.\begin{aligned} \widetilde{\text{minimize}} \quad & f(x) \\ \text{subject to} \quad & g_i(x) \preceq 0, \; i = 1, \ldots, m \end{aligned}\right\} \tag{4.19}$$

where the symbols " $\widetilde{\text{minimize}}$ " and " $\preceq$ " denote a relaxed or fuzzy version of those of " minimize " and " $\leq$ " meaning that " the objective function should be minimized as well as possible " and " the constraints should be possibly well satisfied, " respectively.

Such fuzzy requirements can be quantified by eliciting membership functions $\mu_i(g_i(x))$, $i = 0, \ldots, m$, from the DM for all the functions $f(x)$ and $g_i(x)$, $i = 1, \ldots, m$, where, for notational convenience, $\mu(f(x))$ is denoted by $\mu_0(g_0(x))$. By taking account of the rate of increased membership satisfaction, the DM must determine the subjective membership function $\mu_i(g_i(x))$ which is a strictly monotone decreasing function with respect to g_i in the following way:

$$\mu_i(g_i(x)) = \begin{cases} 1 & ; \quad g_i(x) \leq g_i^1 \\ d_i(x) & ; \quad g_i^1 \leq g_i(x) \leq g_i^0 \\ 0 & ; \quad g_i(x) \geq g_i^0. \end{cases} \tag{4.20}$$

As shown in Figure 4.2, g_i^1 or g_i^0 represents the value of g_i such that the grade of the membership function $\mu_i(g_i(x))$ is 1 or 0, and the grades of membership for the intermediate function values are expressed by a strictly monotone decreasing function $d_i(x)$ with respect to g_i.

Using these membership functions, if we follow the fuzzy decision of Bellman and Zadeh (1970), the problem of finding the maximizing decision is to choose x^* such that

$$\mu_D(x^*) = \max \min_{i=0,\ldots,m} \{\mu_i(g_i(x))\}. \tag{4.21}$$

By introducing the auxiliary variable λ, this problem can be transformed into the following equivalent conventional nonlinear programming problem:

$$\left.\begin{aligned} \text{maximize} \quad & \lambda \\ \text{subject to} \quad & \lambda \leq \mu_i(g_i(x)), \; i = 0, ..., m. \end{aligned}\right\} \tag{4.22}$$

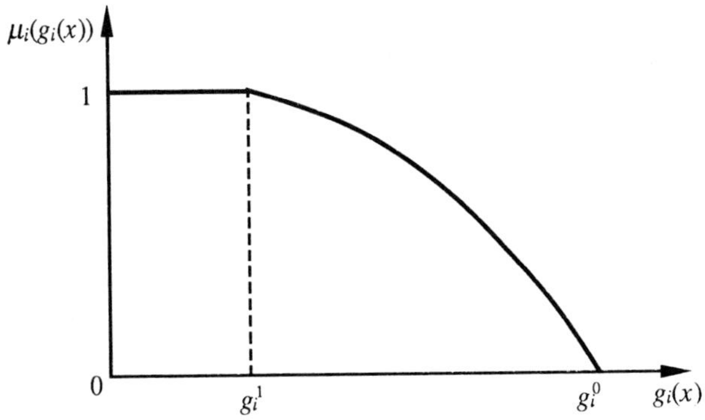

Figure 4.2 Strictly monotone decreasing membership function

Therefore, it is evident that any existing algorithms of nonlinear programming can be applied to solve this problem and obtain the maximizing decision.

A generalization of the above results to multiobjective nonlinear programming along this line will be further discussed in Section 4.4.

In the remainder of this subsection, however, it is quite appropriate to present the concept of fuzzy mathematical programming first introduced by Tanaka et al. (1974) in the framework of the fuzzy decision of Bellman and Zadeh (1970). It is acceptable because of its historic importance.

In the original fuzzy decision of Bellman and Zadeh, they use the set of alternatives $X = R^n$ to introduce the concept of fuzzy mathematical programming on a general level.

Now assume a fuzzy goal G on X characterized by its membership function

$$\mu_G : X \to [0, 1] \tag{4.23}$$

and a fuzzy constraint C on X characterized by its membership function

$$\mu_C : X \to [0, 1]. \tag{4.24}$$

Then as discussed in Section 2.4, the fuzzy decision D resulting from the fuzzy goal G and fuzzy constraint C is defined as

$$\mu_D(x) = \min(\mu_G(x), \mu_C(x)). \tag{4.25}$$

The maximizing decision, if it exists, is given by

$$\max_{x \in X} \mu_D(x) = \max_{x \in X} \min(\mu_G(x), \mu_C(x)). \tag{4.26}$$

In the same spirit as Bellman and Zadeh (1970), Tanaka et al. (1974) formulated the fuzzy mathematical programming problem as finding the maximum fuzzy decision (maximizing decision). Realizing that there does not always exist $x \in X$ which yield the maximizing decision, Tanaka et al. expressed the fuzzy mathematical programming problem as follows:

$$\sup_{x\in X} \mu_D(x) = \sup_{x\in X}[\min(\mu_G(x), \mu_C(x))]. \tag{4.27}$$

Although it seems to be very difficult to derive the maximizing decision for the general forms of the membership functions $\mu_G(x)$ and $\mu_C(x)$, with some assumptions about the membership functions, Tanaka et al. proved that this problem can be reduced to the following conventional mathematical programming problem:

$$\left.\begin{array}{ll} \sup & \mu_G(x) \\ \text{subject to} & \mu_G(x) \leq \mu_C(x). \end{array}\right\} \tag{4.28}$$

More generally, if we consider m fuzzy constraints $C_1, \ldots, C_m$ on X and simply take $C = C_1 \cap \cdots \cap C_m$, this relation also holds true.

Although the proof of the equivalence between (4.27) and (4.28) is somewhat mathematical, we are giving the proof because of its historic significance.

Proof of equivalence between (4.27) and (4.28)

Assume that the membership functions $\mu_G(x)$ and $\mu_C(x)$ are continuous and C is normal, i.e., there is x such that $\mu_C(x) = 1$.

They first proved the following relation

$$\sup_{x\in X} \mu_D(x) = \sup_{\alpha\in[0,1]} \min\left[\alpha, \sup_{x\in C_\alpha} \mu_G(x)\right] \tag{4.29}$$

where C_α is an α-level set of C defined by

$$C_\alpha = \{x \in X \mid \mu_C(x) \geq \alpha\}.$$

From the decomposition theorem of the α-level set, C can be represented by

$$C = \bigcup_{\alpha\in[0,1]} \alpha C_\alpha$$

or

$$\mu_C(x) = \sup_{\alpha\in[0,1]} \alpha\, \mu_{C_\alpha}(x).$$

Hence, the fuzzy decision becomes

$$\begin{aligned} \mu_D(x) &= \min(\mu_G(x), \mu_C(x)) \\ &= \min\left[\mu_G(x), \sup_{\alpha\in[0,1]} \alpha\, \mu_{C_\alpha}(x)\right] \\ &= \min\left[\mu_G(x), \sup_{\alpha\in[0,1]} \left(\min(\alpha, \mu_{C_\alpha}(x))\right)\right] \\ &= \sup_{\alpha\in[0,1]} \min[\alpha, \mu_G(x), \mu_{C_\alpha}(x)]. \end{aligned}$$

As a result, the maximizing decision can be expressed as

$$\begin{aligned}\sup_{x\in X}\mu_D(x) &= \sup_{x\in X}\sup_{\alpha\in[0,1]}\min\left[\alpha,\mu_G(x),\mu_{C_\alpha}(x)\right]\\ &= \sup_{\alpha\in[0,1]}\sup_{x\in X}\min\left[\alpha,\mu_G(x),\mu_{C_\alpha}(x)\right]\\ &= \sup_{\alpha\in[0,1]}\min\left[\alpha,\sup_{x\in X}\min(\mu_G(x),\mu_{C_\alpha}(x))\right].\end{aligned}$$

Noting that

$$\mu_{C_\alpha}(x)=\begin{cases}1 & x\in C_\alpha\\ 0 & x\notin C_\alpha,\end{cases}$$

it follows that

$$\sup_{x\in X}\min\left[\mu_G(x),\mu_{C_\alpha}(x)\right]=\sup_{x\in C_\alpha}\mu_G(x).$$

Thus, we have the relation (4.29), i.e.,

$$\sup_{x\in X}\mu_D(x)=\sup_{\alpha\in[0,1]}\min\left[\alpha,\sup_{x\in C_\alpha}\mu_G(x)\right].$$

Here, for notational convenience, if we set

$$g(\alpha)=\sup_{x\in C_\alpha}\mu_G(x) \tag{4.30}$$

and

$$h(\alpha)=\min\left(\alpha,g(\alpha)\right), \tag{4.31}$$

this relation can be rewritten as

$$\sup_{x\in X}\mu_D(x)=\sup_{\alpha\in[0,1]}h(\alpha). \tag{4.32}$$

Now consider what kind of properties are required for $g(\alpha)$ to reduce the problem to the conventional mathematical programming problem.

It is obvious that $g(\alpha)$ has the following two properties:

(1) $g(0)=\sup_{x\in X}\mu_G(x)>0$,

(2) $g(\alpha)\geq g(\beta)$ for $\alpha\leq\beta$.

Actually, (1) follows directly from $C_0=\{x\in X \mid \mu_C(x)\geq 0\}=X$, and (2) follows from $g(\alpha)=\sup_{x\in C_\alpha}\mu_G(x)\geq\sup_{x\in C_\beta}\mu_G(x)=g(\beta)$.

If $g(\alpha)$ is continuous on the closed interval $[0,1]$, then there always exists a fixed point $\bar{\alpha}$ such that

$$g(\bar{\alpha})=\bar{\alpha}$$

and

$$\sup_{x\in X}\mu_D(x)=\bar{\alpha}=g(\bar{\alpha}).$$

From $g(0) > 0$, it follows that $\bar{\alpha} > 0$ and

$$h(\bar{\alpha}) = \min(\bar{\alpha}, g(\bar{\alpha})) = \bar{\alpha}.$$

Also, noting that $g(\alpha)$ is monotonic nonincreasing for $\alpha < \bar{\alpha}$, it follows that

$$g(\alpha) \geq g(\bar{\alpha}) = \bar{\alpha} > \alpha.$$

Hence,

$$h(\alpha) = \min(\alpha, g(\alpha)) < \bar{\alpha} = h(\bar{\alpha}),$$

which implies

$$\alpha < \bar{\alpha} \Rightarrow h(\alpha) < h(\bar{\alpha}).$$

On the other hand, since $g(\alpha) \leq g(\bar{\alpha}) = \bar{\alpha} < \alpha$ for $\alpha > \bar{\alpha}$, it follows that

$$h(\alpha) = \min(\alpha, g(\alpha)) < g(\alpha) \leq \bar{\alpha} = h(\bar{\alpha}),$$

which implies

$$\alpha > \bar{\alpha} \Rightarrow h(\alpha) \leq h(\bar{\alpha}).$$

From these two relations, it holds that

$$\sup_{\alpha \in [0,1]} h(\alpha) = h(\bar{\alpha}) = \bar{\alpha}.$$

Hence, (4.32) can be rewritten as

$$\sup_{x \in X} \mu_D(x) = \bar{\alpha}. \tag{4.33}$$

From the definition of $g(\alpha)$ together with $g(\bar{\alpha}) = \bar{\alpha}$, it holds that

$$\sup_{x \in X} \mu_D(x) = \sup_{x \in C_{\bar{\alpha}}} \mu_G(x) = \bar{\alpha}. \tag{4.34}$$

Now, for compact notation, define

$$A = \{x \in X \mid \mu_C(x) \geq \mu_G(x)\} \tag{4.35}$$

and consider a point x_0 of the level set $C_{\bar{\alpha}}$. Then it follows that

$$\mu_C(x_0) \geq \bar{\alpha} = \sup_{x \in C_{\bar{\alpha}}} \mu_G(x) \geq \mu_G(x_0),$$

which means x_0 belongs to the set A, i.e.,

$$C_{\bar{\alpha}} \subset A.$$

Hence,

$$\sup_{x \in A} \mu_D(x) \geq \sup_{x \in C_{\bar{\alpha}}} \mu_G(x).$$

On the other hand, from the definition of the set A, it follows that

$$\begin{aligned}\sup_{x\in X}\mu_D(x) &= \sup_{x\in X}\min\left(\mu_G(x),\mu_C(x)\right)\\ &\geq \sup_{x\in A}\min\left(\mu_G(x),\mu_C(x)\right)\\ &= \sup_{x\in A}\mu_G(x).\end{aligned}$$

From these two results, we have

$$\sup_{x\in X}\mu_D(x) \geq \sup_{x\in A}\mu_G(x) \geq \sup_{x\in C_{\bar{\alpha}}}\mu_G(x). \tag{4.36}$$

In view of (4.34), we finally arrive at the following relation:

$$\sup_{x\in X}\mu_D(x) = \sup_{x\in A}\mu_G(x). \tag{4.37}$$

This completes the proof of the equivalence. □

Some extensions along this line can also be found in Orlovski (1977), Negoita and Ralescu (1977), and Flachs and Pollatscheck (1978).

4.2 Multiobjective nonlinear programming

4.2.1 Problem formulation and solution concepts

Recall the production planning problem with environmental quality together with the nonlinear production planning problem discussed respectively in Examples 3.2 and 4.1.

Example 4.2 (Two-objective nonlinear production planning)
In the production planning problem with environmental quality in Example 3.2, as was discussed in Example 4.1, the profit function is a nonlinear function of two variables x_1 and x_2 such as

$$(4-x_1)x_1+(11-x_2)x_2 \tag{4.38}$$

where x_1 and x_2 denote the number of tons produced of products P_1 and P_2, respectively.

Also, the amount of pollution is not a linear function of two variables x_1 and x_2, as in Example 3.2, but a nonlinear function of x_1 and x_2 such as

$$(x_1+3)x_1+(x_2+2)x_2. \tag{4.39}$$

In such a situation, the production planning problem considering environmental quality should be formulated as the following two-objective nonlinear programming problem:

$$\left.\begin{array}{lrl} \text{minimize} & z_1 = & x_1^2 + x_2^2 - 4x_1 - 11x_2 \\ \text{minimize} & z_2 = & x_1^2 + x_2^2 + 3x_1 + 2x_2 \\ \text{subject to} & & 2x_1 + 6x_2 \leq 27 \\ & & 8x_1 + 6x_2 \leq 45 \\ & & 3x_1 + x_2 \leq 15 \\ & & x_1 \geq 0,\ x_2 \geq 0. \end{array}\right\} \quad (4.40)$$

This simple two-objective nonlinear programming problem is generalized to the following multiobjective nonlinear programming (MONLP) problem:

minimize k nonlinear objective functions

$$\left.\begin{array}{c} f_1(x_1, x_2, \ldots, x_n) \\ f_2(x_1, x_2, \ldots, x_n) \\ \cdots\cdots\cdots \\ f_k(x_1, x_2, \ldots, x_n) \end{array}\right\} \quad (4.41)$$

subject to the nonlinear inequality constraints

$$\left.\begin{array}{c} g_1(x_1, x_2, \ldots, x_n) \leq 0 \\ g_2(x_1, x_2, \ldots, x_n) \leq 0 \\ \cdots\cdots\cdots \\ g_m(x_1, x_2, \ldots, x_n) \leq 0 \end{array}\right\} \quad (4.42)$$

or in a vector-minimization form

$$\left.\begin{array}{ll} \text{minimize} & f(x) \triangleq (f_1(x), f_2(x), \ldots, f_k(x))^T \\ \text{subject to} & x \in X \triangleq \{x \in R^n \mid g_j(x) \leq 0,\ j = 1, ..., m\}, \end{array}\right\} \quad (4.43)$$

where $x = (x_1, x_2, \ldots, x_n)^T$ is an n-dimensional vector of decision variables, $f_i(x) = f_i(x_1, x_2, \ldots, x_n)$, $i = 1, \ldots, k$, and $g_j(x) = g_j(x_1, x_2, \ldots, x_n)$, $j = 1, \ldots, m$, are given real-valued nonlinear functions of n real variables $x_1, x_2, \ldots, x_n$ and X is the feasible set of the nonlinearly constrained set.

As discussed in multiobjective linear programming (MOLP), a direct application of the optimality for single-objective nonlinear programming to the MONLP leads us to the following complete optimality concept.

Definition 4.1 (Complete optimal solution)

x^* is said to be a complete optimal solution to the MONLP if and only if there exists $x^* \in X$ such that $f_i(x^*) \leq f_i(x)$, $i = 1, \cdots, k$, for all $x \in X$.

However, when the objective functions of the MONLP conflict with each other, a complete optimal solution does not always exist, and hence, the Pareto or weak Pareto optimality concept is defined as follows.

Definition 4.2 (Pareto optimal solution)

x^* is said to be a Pareto optimal solution to the MONLP if and only if there does not exist another $x \in X$ such that $f_i(x) \leq f_i(x^*)$ for all i and $f_j(x) \neq f_j(x^*)$ for at least one j.

Definition 4.3 (Weak Pareto optimal solution)

x^* is said to be a weak Pareto optimal solution to the MONLP if and only if there does not exist another $x \in X$ such that $f_i(x) < f_i(x^*)$, $i = 1, \ldots, k$.

In practice, however, since only local optimal solutions are guaranteed in solving a single-objective nonlinear programming problem by any available nonlinear programming technique, unless the problem is convex, the local Pareto optimality concept is also defined for the MONLP. The concept of local Pareto optimal solutions was first introduced by Geoffrion (1968).

Definition 4.4 (Local Pareto optimal solution)

$x^* \in X$ is said to be a local Pareto optimal solution to the MONLP if and only if there exists a real number $\delta > 0$ such that x^* is Pareto optimal in $X \cap N(x^*, \delta)$, i.e., there does not exist another $x \in X \cap N(x^*, \delta)$ such that $f_i(x) \leq f_i(x^*)$ for all i and $f_j(x) \neq f_j(x^*)$ for at least one j, where $N(x^*, \delta)$ denotes the δ neighborhood of x^* defined by $\{x \in R^n \mid \|x - x^*\| < \delta\}$.

4.2.2 Scalarization methods

Similar to the linear case, there are several possible ways of scalarizing the MONLP for characterizing Pareto optimal solutions. In this subsection, we discuss the weighting method, the constraint method, and the weighted minimax method because of their importance. However, for the MONLP, it should be noted that only local Pareto optimality is guaranteed in the following theorems if only local optimality is guaranteed due to nonconvexity.

(1) The weighting method

The weighting method for obtaining Pareto optimal solutions is to solve the following weighting problem formulated by taking the weighted sum of all the objective functions of the original MONLP (Kuhn and Tucker 1951):

$$\left.\begin{array}{ll} \text{minimize} & wf(x) \triangleq \sum_{i=1}^{k} w_i f_i(x) \\ \text{subject to} & x \in X, \end{array}\right\} \tag{4.44}$$

where $w = (w_1, \ldots, w_k)$ is the vector of weighting coefficients assigned to the objective functions, and assumed to be

$$w = (w_1, \ldots, w_k) \geq 0. \tag{4.45}$$

The relationships between the optimal solution x^* of the weighting coefficient problem and the Pareto optimality concept of the MONLP can be characterized by the following theorems.

Theorem 4.8

If $x^* \in X$ is an optimal solution of the weighting problem for some $w > 0$, then x^* is a Pareto optimal solution of the MONLP.

The proof of this theorem is the same as for the linear case and the condition of this theorem can be replaced by a unique optimal solution of the weighting problem for $w \geq 0$.

Theorem 4.9

Assume that all $f_i(x)$ and $g_j(x)$ are convex in the MONLP. If $x^* \in X$ is a Pareto optimal solution of the MONLP, then x^* is an optimal solution of the weighting problem for some $w = (w_1, \ldots, w_k) \geq 0$.

Proof

Let $x^* \in X$ be a Pareto optimal solution and define the following two subsets in R^k:

$$\begin{aligned} U &= \{u \in R^k \mid u > 0\} \\ V &= \{v \in R^k \mid v \leq f(x^*) - f(x), \forall x \in X\}. \end{aligned}$$

Then, obviously, $U \cap V = \emptyset$ and U is a convex set. Also, V is a convex set due to the convexity assumption.

Since $u > 0$, by the separation theorem of convex sets (Theorem 4.7), there exists a row vector $w \geq 0$ ($w \neq 0$) in R^k such that

$$wv \leq 0 < wu, \quad \forall u \in U, \quad \forall v \in V.$$

In view of $wv = w(f(x^*) - f(x)) \leq 0$ for all v such that $v = f(x^*) - f(x)$, we have

$$wf(x) \geq wf(x^*) \quad \forall x \in X,$$

which implies that x^* is an optimal solution to the weighting problem for $w \geq 0$. This completes the proof of the theorem. □

This theorem, which is proven through the separation theorem of convex sets, requires the convexity assumption and hence cannot be applied for nonconvex problems.

As for the MOLP, the weighting coefficients of the weighting problem give the trade-off rate information between the objective functions as follows:

$$-\frac{\partial f_1}{\partial f_i} = \frac{w_i^*}{w_1^*}, \quad i = 2, \ldots, k. \tag{4.46}$$

(2) The constraint method

The constraint method for characterizing Pareto optimal solutions is to solve the following constraint problem formulated by taking one objective function as the objective function and letting all other objective functions be inequality constraints (Haimes et al. 1971, Haimes and Hall 1974):

$$\left.\begin{array}{ll} \text{minimize} & f_j(x) \\ \text{subject to} & f_i(x) \leq \varepsilon_i,\ i = 1, \ldots, k;\ i \neq j \\ & x \in X. \end{array}\right\} \tag{4.47}$$

Similar to the linear case, the relationships between the optimal solution x^* to the constraint problem and the Pareto optimal concept of the MONLP can be characterized by the following theorems.

Theorem 4.10

If $x^* \in X$ is a unique optimal solution to the constraint problem for some $\varepsilon_i,\ i = 1, \ldots, k;\ i \neq j$, then x^* is a Pareto optimal solution to the MONLP.

Theorem 4.11

If $x^* \in X$ is a Pareto optimal solution of the MONLP, then x^* is an optimal solution of the constraint problem for some $\varepsilon_i,\ i = 1, \ldots, k;\ i \neq j$.

The proofs of these theorems are the same as the MOLP, and hence, no convexity assumption is required for the constraint method.

For the constraint problem, consider the Lagrange multipliers associated with the active constraints. The Lagrangian function $L(x, \lambda)$ for the constraint problem with respect to the ε-constraints is formulated as

$$L(x, \lambda) = f_j(x) + \sum_{\substack{i=1 \\ i \neq j}}^{k} \lambda_{ji}(f_i(x) - \varepsilon_i), \tag{4.48}$$

where $\lambda_{ji},\ i = 1, 2, \ldots, k;\ i \neq j$, denote the corresponding Lagrange multipliers.

Assume that the Kuhn-Tucker conditions hold. Then it follows that

$$\lambda_{ji}(f_i(x) - \varepsilon_i) = 0,\ \ \lambda_{ji} \geq 0,\ i = 1, \ldots, k;\ \ i \neq j. \tag{4.49}$$

Also, from (4.48), under appropriate conditions, we have

$$\lambda_{ji} = -\frac{\partial L}{\partial \varepsilon_i},\quad i = 1, \ldots, k;\ i \neq j. \tag{4.50}$$

Observing (4.49), it follows that

$$\lambda_{ji} = -\frac{\partial f_j(x)}{\partial \varepsilon_i},\quad i = 1, \ldots, k;\ i \neq j. \tag{4.51}$$

Furthermore, from (4.49), $\lambda_{ji} > 0$ if the corresponding ε constraint is active, i.e., $f_i(x) - \varepsilon_i = 0$. Therefore,

$$\lambda_{ji} = -\frac{\partial f_j(x)}{\partial f_i(x)},\quad i = 1, \ldots, k;\ i \neq j. \tag{4.52}$$

This means that the Lagrange multipliers λ_{ji}, $i = 1, \ldots, k$; $i \neq j$, represent the trade-off rates between $f_j(x)$ and $f_i(x)$ on the Pareto surface.

(3) The weighted minimax method

The weighted minimax method for characterizing Pareto optimal solutions is to solve the following weighted minimax problem (Bowman 1976):

$$\left.\begin{array}{ll} \text{minimize} & \max_{i=1,\ldots,k} w_i f_i(x) \\ \text{subject to} & x \in X \end{array}\right\} \tag{4.53}$$

or equivalently

$$\left.\begin{array}{ll} \text{minimize} & v \\ \text{subject to} & w_i f_i(x) \leq v, \; i = 1, \ldots, k \\ & x \in X \end{array}\right\} \tag{4.54}$$

where v is a scalar variable. Here, without loss of generality, it can be assumed that $f_i(x) > 0$, $i = 1, \ldots, k$, for all $x \in X$. Because, for the objective functions not satisfying $f_i(x) > 0$, $x \in X$, using the individual minimum $f_i^{\min} = \min_{x \in X} f_i(x)$ of the corresponding functions and setting $\hat{f}_i(x) = f_i(x) - f_i^{\min}$, it holds that $\hat{f}_i(x) > 0$, $i = 1, \ldots, k$, for all $x \in X$.

Similar to the linear case, the relationships between the optimal solution x^* of the weighted minimax problem and the Pareto optimal concept of the MONLP can be characterized by the following theorems.

Theorem 4.12

If $x^* \in X$ is a unique optimal solution of the weighted minimax problem for some $w = (w_1, \ldots, w_k) \geq 0$, then x^* is a Pareto optimal solution of the MONLP.

Theorem 4.13

If $x^* \in X$ is a Pareto optimal solution of the MONLP, then x^* is an optimal solution of the minimax problem for some $w = (w_1, \ldots, w_k) > 0$.

The proofs of these theorems are the same as that of the MOLP, and if the uniqueness of a solution is not guaranteed, only weak Pareto optimality is guaranteed in Theorem 4.5.

Assume that $x^*(w^*)$ and $v^*(w^*)$, a local optimal solution to the weighted minimax problem, are unique for the given $w^* > 0$ and all constraints for v^* are active. Then the relation between w_i^* and $f_i(x^*)$ is $w_1^* f_1(x^*) = w_2^* f_2(x^*) = \cdots = w_k^* f_k(x^*) = v^*$. Hence, the ratio of the weighting vector is given by

$$w_1^* : w_2^* : \cdots : w_k^* = 1/f_1(x^*) : 1/f_2(x^*) : \cdots : 1/f_k(x^*) \tag{4.55}$$

From Theorems 4.10 and 4.12, if the uniqueness of the optimal solution x^* for the scalarizing problem is not guaranteed, it is necessary to perform the Pareto optimality test of x^* by solving the following nonlinear programming problem:

$$\left.\begin{array}{ll} \text{maximize} & \sum_{i=1}^{k} \varepsilon_i \\ \text{subject to} & f_i(x) + \varepsilon_i = f_i(x^*),\ i = 1, \ldots, k \\ & x \in X,\ \varepsilon = (\varepsilon_1, \ldots, \varepsilon_k)^T \geq 0. \end{array}\right\} \quad (4.56)$$

For the optimal solutions $\bar{x}$ and $\bar{\varepsilon}$ of this nonlinear programming problem, the following theorem holds.

Theorem 4.14

For the optimal solution $\bar{x}$ and $\bar{\varepsilon}$ of the Pareto optimality test problem,

(1) *If all $\bar{\varepsilon}_i = 0$, then x^* is a Pareto optimal solution of the MONLP.*

(2) *If at least one $\bar{\varepsilon}_i > 0$, then not x^* but $\bar{x}$ is the Pareto optimal solution.*

The proof of this theorem is the same as for the linear case.

It is now appropriate to compare the characteristics of the scalarization methods discussed thus far. From the computational point of view, the weighting problem is the easiest one to solve under suitable convex assumptions for the MONLP. But if the convexity assumptions are absent, not all of the Pareto optimal solutions can be obtained by solving the weighting problem for any $w > 0$. On the other hand, by solving the weighted minimax problem for any $w > 0$, or the constraint problem for an appropriately selected value of ε, all of the Pareto optimal solutions to the MONLP involving nonconvexities can be obtained. Since the range for ε is not known numerically in advance, there is a possibility that the constraint problem will be infeasible for some value of ε. Moreover, in the constraint problem, the objective function $f_j(\cdot)$ is treated specially and not all objective functions are treated equally. With this observation, in some sense, it seems that the weighted minimax problem may be much more favorable when compared to the others.

Recently, as a generalization of such existing scalarization methods for characterizing Pareto optimal solutions of multiobjective optimization problems, a fully integrated scalarization method, called the hyperplane method, was introduced by the authors (Yano and Sakawa 1989a; Sakawa and Yano 1990c). The details of the hyperplane method can be found in the Appendix.

4.2.3 Nonlinear goal programming

The extension of the idea of linear goal programming to nonlinear multiobjective programming problems where objective functions $f_1(x), \ldots, f_k(x)$ and constraints $g_1(x), \ldots, g_m(x)$ may be nonlinear is straightforward. For nonlinear goal programming, similar to the linear case, when the goal vector $\hat{f} = (\hat{f}_1, \ldots, \hat{f}_k)^T$ is specified by the DM for the vector-valued objective function $f(x) = (f_1(x), \ldots, f_k(x))^T$, the MONLP is interpreted as the problem of coming "as close as possible" to the set of specified goals which may not be simultaneously attainable. Therefore, if we use the general weighted ℓ_p norm to measure the distance between $f(x)$ and $\hat{f}$, a nonlinear goal programming problem can be formulated as follows:

$$\underset{x \in X}{\text{minimize}} \quad \|f(x) - \hat{f}\|_p^w, \quad w \geq 0, \ 1 \leq p \leq \infty \tag{4.57}$$

where the weighted ℓ_p norm $\|\cdot\|_p^w$ for $1 \leq p < \infty$ and $p = \infty$ are defined respectively as follows:

$$\|f(x) - \hat{f}\|_p^w = \left(\sum_{i=1}^{k} w_i |f_i(x) - \hat{f}_i|^p \right)^{1/p}, \ 1 \leq p < \infty \tag{4.58}$$

$$\|f(x) - \hat{f}\|_\infty^w = \max_{i=1,\ldots,k} \{w_i |f_i(x) - \hat{f}_i|\}. \tag{4.59}$$

The idea of linear goal programming for the MOLP with $p = 1$ originally proposed by Charnes and Cooper (1961) can be immediately extended to the MONLP by considering the following nonlinear programming problem with $p = 1$:

$$\underset{x \in X}{\text{minimize}} \quad \|f(x) - \hat{f}\|_1^w \ ; w \geq 0. \tag{4.60}$$

Introducing the auxiliary variables

$$d_i^+ = \frac{1}{2}\left\{|f_i(x) - \hat{f}_i| + (f_i(x) - \hat{f}_i)\right\} \tag{4.61}$$

and

$$d_i^- = \frac{1}{2}\left\{|f_i(x) - \hat{f}_i| - (f_i(x) - \hat{f}_i)\right\} \tag{4.62}$$

for each $i = 1, \ldots, k$, this problem can be converted to the following equivalent nonlinear goal programming problem:

$$\underset{x \in X}{\text{minimize}} \quad \sum_{i=1}^{k} w_i (d_i^+ + d_i^-) \tag{4.63}$$

$$\text{subject to} \quad f_i(x) - d_i^+ + d_i^- = \hat{f}_i, \quad i = 1, \ldots, k \tag{4.64}$$

$$d_i^+ \cdot d_i^- = 0, \ i = 1, \ldots, k \tag{4.65}$$

$$d_i^+ \geq 0, \ d_i^- \geq 0, \ i = 1, \ldots, k. \tag{4.66}$$

Here, from (4.61) and (4.62), observe that

$$d_i^+ = \begin{cases} f_i(x) - \hat{f}_i & \text{if } f_i(x) \geq \hat{f}_i \\ 0 & \text{if } f_i(x) \leq \hat{f}_i \end{cases} \tag{4.67}$$

and

$$d_i^- = \begin{cases} \hat{f}_i - f_i(x) & \text{if } \hat{f}_i \geq f_i(x) \\ 0 & \text{if } \hat{f}_i \leq f_i(x). \end{cases} \tag{4.68}$$

Thus d_i^+ and d_i^- represent the over-achievement and the under-achievement of the ith goal respectively and, hence, are called deviational variables.

With respect to alternative situations regarding the decision-making progress, either over- or under-achievement of a specific goal is the only concern of the DM. Assimilation into the goal programming formulation is achieved by assigning w_i^+ and w_i^- to d_i^+ and d_i^-, representing over- and under-achievement weights, respectively. If each $f_i(x)$ is a cost-type objective function with a goal $\hat{f}_i$, then over-achievement is undesirable. In this case, $w_i^+ = 1$ and $w_i^- = 0$ are set, and the nonlinear goal programming formulation is modified to the following so-called one-sided goal programming formulation:

$$\begin{aligned} \underset{x \in X}{\text{minimize}} \quad & \sum_{i=1}^{k} w_i^+ d_i^+ && (4.69) \\ \text{subject to} \quad & f_i(x) - d_i^+ + d_i^- = \hat{f}_i \quad i = 1, \ldots, k && (4.70) \\ & d_i^+ \cdot d_i^- = 0, \; i = 1, \ldots, k && (4.71) \\ & d_i^+ \geq 0, \; d_i^- \geq 0, \; i = 1, \ldots, k. && (4.72) \end{aligned}$$

Conversely, for benefit-type objective functions, under-achievement is undesirable. In this case, $w_i^- = 1$ and $w_i^+ = 0$ are set to replace $\sum_{i=1}^{k} w_i^+ d_i^+$ by $\sum_{i=1}^{k} w_i^- d_i^-$ as the objective function.

A more general form of the nonlinear goal programming formulation can be brought into existence by the introduction of preemptive priorities replacing or in addition to the numerical weights P_i, in place of or together with the numerical weights $w_i^+, w_i^- \geq 0$. Given the situation where objective functions $f_1(x), \ldots, f_k(x)$ are divided into L ordinal ranking classes, $1 \leq L \leq k$, having the preemptive priorities $P_1, \ldots, P_L$ in decreasing order and with the use of preemptive priorities P_l together with over- and under-achievement weights w_i^+ and w_i^-, the general nonlinear goal programming formulation becomes

$$\begin{aligned} \underset{x \in X}{\text{minimize}} \quad & \sum_{l=1}^{L} P_l \left(\sum_{i \in I_l} (w_i^+ d_i^+ + w_i^- d_i^-) \right) && (4.73) \\ \text{subject to} \quad & f_i(x) - d_i^+ + d_i^- = \hat{f}_i, \quad i = 1, \ldots, k && (4.74) \\ & d_i^+ \cdot d_i^- = 0, \; i = 1, \ldots, k && (4.75) \\ & d_i^+ \geq 0, \; d_i^- \geq 0, \; i = 1, \ldots, k && (4.76) \end{aligned}$$

where $I_l \neq \emptyset$ is the index set of objective functions in the lth priority class. When there are k distinct ordinal ranking classes (i.e., $L = k$) with $f_i(x)$ belonging to the ith priority class, then the objective function becomes simply

$$\sum_{i=1}^{k} P_i (w_i^+ d_i^+ + w_i^- d_i^-). \tag{4.77}$$

Obviously, the solution strategy of nonlinear goal programming is the same as that of linear goal programming. Starting from trying to achieve the goals of all objective functions in the highest priority class, a sequence of priority class subproblems is solved successively until either a unique solution is obtained at some stage or all priority classes are considered.

To be more specific mathematically, for $l = 1$, start by solving the lth priority class subproblem

$$f_l^* = \underset{x \in \tilde{X} \cap X_{l-1}}{\text{minimize}} \quad \sum_{i \in I_l} (w_i^+ d_i^+ + w_i^- d_i^-) \tag{4.78}$$

where

$$\tilde{X} = \{(x, d^+, d^-) \in R^{n+2k} \mid x \in X,\ f_i(x) - d_i^+ + d_i^- = \hat{f}_i,\quad i = 1, \ldots, k\} \tag{4.79}$$

$$X_{l-1} = \Big\{(x, d^+, d^-) \in R^{n+2k} \,\Big|\, \sum_{i \in I_l} (w_i^+ d_i^+ + w_i^- d_i^-) \le f_r^*,\ r = 1, \cdots, l-1\Big\}; X_0 = \tilde{X} \tag{4.80}$$

and stop if the optimal solution of the lth priority class problem is unique or $l = L$.

It should be noted here that any existing algorithms of nonlinear programming can be used to solve a sequence of priority class nonlinear subproblems.

Another obvious approach, as suggested by Griffith and Stewart (1961), is to approximate the nonlinear functions by linear functions and solve the resulting linear problems using the modified simplex method. Ignizio (1976), however, points out the drawbacks of this approach and suggests using a modified Hooke and Jeeves pattern search to solve the nonlinear goal programming problems directly.

Further details concerning the algorithm, extensions, and applications can be found in the text of Lee (1972) and Ignizio (1976, 1982).

4.2.4 Compromise programming

A well-known extension of the goal programming approach is obtained if the goal vector $\hat{f} = (\hat{f}_1, \ldots, \hat{f}_k)$ is replaced by the so-called ideal or utopia vector $f^{\min} = (f_1^{\min}, \ldots, f_k^{\min})$, where $f_i^{\min} = \min_{x \in X} f_i(x)$, $1, \ldots, k$. The resulting problem can be interpreted as an attempt to minimize deviation from the ideal or utopia vector (point). Realizing that the ideal vector is generally infeasible in most of the MONLP with conflicting objective functions, Yu (1973) and Zeleny (1973) introduced the concept of compromise solution. Geometrically, the compromise solution defined by Yu and Zeleny is the solution which is the minimum distance from the ideal point.

To be more specific mathematically, given a weight vector w, x_w^p is called a compromise solution of the MONLP with respect to p if and only if it solves

$$\underset{x \in X}{\text{minimize}} \quad d_p^w(f(x), f^{\min}) \triangleq \Big(\sum_{i=1}^{k} w_i |f_i(x) - f_i^{\min}|^p\Big)^{1/p} \tag{4.81}$$

or equivalently, for $1 \leq p < \infty$,

$$\underset{x \in X}{\text{minimize}} \quad \tilde{d_p^w}(f(x), f^{\min}) \triangleq \sum_{i=1}^{k} w_i (f_i(x) - f_i^{\min})^p, \tag{4.82}$$

and for $p = \infty$,

$$\underset{x \in X}{\text{minimize}} \quad \tilde{d_\infty^w}(f(x), f^{\min}) \triangleq \max_{i=1,\ldots,k} w_i (f_i(x) - f_i^{\min}). \tag{4.83}$$

Observe that for $p = 1$, all deviations from $f_i^{\min}$ are taken into account in direct proportion to their magnitudes, while for $2 \leq p < \infty$, the largest deviation has the greatest influence. Ultimately for $p = \infty$, the largest deviation is the only one taken into account.

It should be noted here that any solution of (4.82) for any $1 \leq p < \infty$ or a unique solution of (4.83) with $w_i > 0$ for all $i = 1, \ldots, k$ is a Pareto optimal solution of the MONLP.

The compromise set C_w, given the weight, is defined as the set of all compromise solutions x_w^p, $1 \leq p \leq \infty$. To be more explicit,

$$C_w = \{x \in X \mid x \text{ solves (4.82) or (4.83) given } w \text{ for some } 1 \leq p \leq \infty\}. \tag{4.84}$$

In the context of linear problems, Zeleny (1973) suggested that the compromise set C_w can be approximated by the Pareto optimal solutions of the following two-objective problem:

$$\underset{x \in X}{\text{minimize}} \quad (\tilde{d_1}(f(x), f^{\min}), \tilde{d_\infty}(f(x), f^{\min})). \tag{4.85}$$

Although it can be seen that the compromise solution set C_w is a subset of the set of Pareto optimal solutions, C_w may still be too large to select the final solution and, hence, should be reduced further.

Zeleny (1973, 1976) suggests several methods to reduce the compromise solution set C_w. One possible reduction method without the DM's aid is to generate a compromise solution set $\bar{C_w}$ similar to C_w by maximizing the distance from the so-called anti-ideal point $f^{\max} = (f_1^{\max}, \ldots, f_k^{\max})$, where $f^{\max} = \max_{x \in X} f_i(x)$. The problem to be solved thus becomes

$$\underset{x \in X}{\text{maximize}} \quad \left(\sum_{i=1}^{k} w_i |f_i^{\max} - f_i(x)|^p \right)^{1/p} \tag{4.86}$$

or equivalently, for $1 \leq p < \infty$,

$$\underset{x \in X}{\text{maximize}} \quad \sum_{i=1}^{k} w_i (f_i^{\max} - f_i(x))^p, \tag{4.87}$$

and for $p = \infty$,

$$\underset{x \in X}{\text{maximize}} \quad \min_{i=1,\ldots,k} w_i (f_i^{\max} - f_i(x)). \tag{4.88}$$

The compromise set $\bar{C}_w$ is defined by

$$\bar{C}_w = \{x \in X \mid x \text{ solves (4.87) or (4.88) given } w \text{ for some } 1 \leq p \leq \infty\}. \quad (4.89)$$

The compromise set C_w based on the ideal point is not identical with the compromise set $\bar{C}_w$ based on the anti-ideal point. Zeleny (1976) suggests using this fact to further reduce the compromise set by considering the intersection $C_w \cap \bar{C}_w$.

An interactive strategy for reducing the compromise solution set proposed by Zeleny (1976) is based on the concept of the so-called displaced ideal and thus is called the method of the displaced ideal. In this approach, the ideal point with respect to the new C_w displaces the previous ideal point and the (reduced) compromise solution set eventually encloses the new ideal point, terminating the process.

The basic steps of the method of displaced ideal are summarized as follows.

The algorithm of the method of displaced ideal

Step 1: Let $C_w^{(0)} = X$ and set the iteration index $r = 1$.

Step 2: Find the ideal point $f^{\min(r)}$ by solving

$$\underset{x \in C_w^{(r-1)}}{\text{minimize}} \quad f_i(x), \ i = 1, \ldots, k.$$

Step 3: Construct the compromise solution set $C_w^{(r)}$ by finding the Pareto optimal solution set of

$$\underset{x \in C_w^{(r)}}{\text{minimize}} \quad (\tilde{d}_1(f(x), f^{\min(r)}), \tilde{d}_\infty(f(x), f^{\min(r)})).$$

Step 4: If the DM can select the final solution from $C_w^{(r)}$, or if $C_w^{(r)}$ contains $f^{\min(r)}$, stop. Otherwise, set $r = r + 1$ and return to step 2.

It should be noted here that the method of displaced ideal can be viewed as the best ideal-seeking process, not the ideal itself.

Further refinements and details can be found in Zeleny (1976, 1982).

4.2.5 The surrogate worth trade-off method

In 1974, Y.Y. Haimes et al. proposed the surrogate worth trade-off (SWT) method by realizing that, given all the levels of the objective functions corresponding to any Pareto optimal solution, it is much easier for the decision maker (DM) to assess the relative value of the trade-off of marginal increases and decreases between any two objectives rather than their absolute values. In this subsection, the SWT method using the constraint problems is briefly introduced.

First, without loss of generality, take $f_1(x)$ as the standing objective function and consider the following constraint problem:

$$\left.\begin{array}{rl} \text{minimize} & f_1(x) \\ \text{subject to} & f_i(x) \leq \varepsilon_i, \; i = 2, \ldots, k \\ & x \in X. \end{array}\right\} \tag{4.90}$$

When a Pareto optimal solution is obtained by solving the constraint problem for some selected values of ε_i, $i = 2, \ldots, k$, if all the ε-constraints are active, i.e., $f_i(x) - \varepsilon_i = 0$, $i = 2, \ldots, k$, then, as discussed in the constraint method, the corresponding Lagrange multipliers λ_{1i}, $i = 2, \ldots, k$, give the trade-off rates as follows:

$$\lambda_{1i} = -\frac{\partial f_1(x)}{\partial f_i(x)}, \quad i = 2, \ldots, k \tag{4.91}$$

λ_{1i} represents the marginal decrease of $f_1(x)$ when $f_i(x)$ is increased by one unit.

Thus, by changing the values of ε_i, $i = 2, ..., k$, and solving the corresponding constraint problems, both Pareto optimal solutions and trade-off information for the Lagrange multipliers of the active constraints can be obtained. Then the analyst asks the DM the following question.

Question: How much are additional λ_{1i} units of f_1 worth in relation to one additional unit of f_i given the levels of all the objective functions $f_1, f_2, \ldots, f_k$ of the current Pareto optimal solution?

For this question, the DM responds by expressing an ordinal preference for $f_1, f_2, \ldots, f_k$ through the surrogate worth functions W_{1i}, $i = 2, \ldots, k$, using a scale from -10 to $+10$, where the values have the following meaning:

$$W_{1i} \begin{cases} > 0 \Leftrightarrow & \left\{\begin{array}{l}\text{when } \lambda_{1i} \text{ units of } f_1 \text{ are preferred over one marginal unit of} \\ f_i, \text{ given the levels of achievement of all objectives.}\end{array}\right. \\ = 0 \Leftrightarrow & \left\{\begin{array}{l}\text{when } \lambda_{1i} \text{ units of } f_1 \text{ are equivalent over one marginal unit of} \\ f_i, \text{ given the levels of achievement of all objectives.}\end{array}\right. \\ < 0 \Leftrightarrow & \left\{\begin{array}{l}\text{when } \lambda_{1i} \text{ units of } f_1 \text{ are not preferred over one marginal unit} \\ \text{of } f_i, \text{ given the levels of achievement of all objectives.}\end{array}\right. \end{cases}$$

It is significant to realize here that the DM is provided with not only all the levels of objectives but also the trade-off rate information between any two objective functions associated with the current Pareto optimal solution.

Geometrical interpretation of the surrogate worth function for the case of two objectives is as follows:

In $f_1 - f_2$ space shown in Figure 4.3, at point A, let the slope of the indifference curve of the DM be denoted by $-m^A$. Then at point A, the DM is willing to

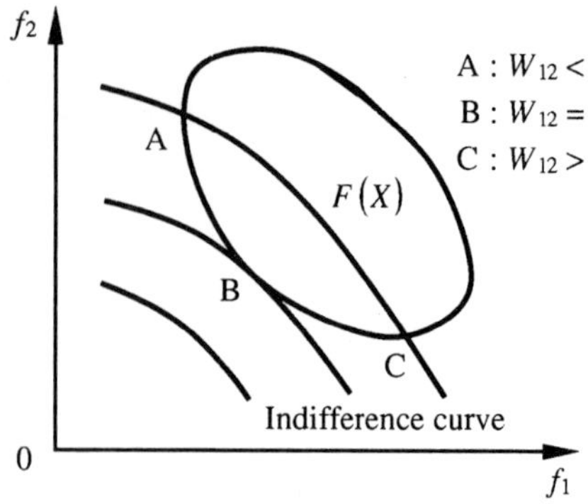

Figure 4.3 Geometrical interpretation of the SWT method

accept one additional unit of f_2 to get m^A less units of f_1. However, since $\lambda_{12}^A < m^A$, when asked if the DM is willing to take one additional unit of f_2 for λ_{12}^A less units of f_1, the response will be no, and thus, $W_{12} < 0$ at point A. Similarly, at point C, since $\lambda_{12}^C > m^C$, the response will be $W_{12} > 0$. At point B, since $\lambda_{12}^B = m^B$, the response will be $W_{12} = 0$.

Based on the above discussions, we can now introduce the algorithm of the SWT method using the constraint method as follows:

The algorithm of the SWT method

Step 1: Obtain a representative subset of Pareto optimal solutions together with the trade-off information between the standing objective and each of the other objectives by appropriately varying $\varepsilon_2, \ldots, \varepsilon_k$ and solve the corresponding constraint problems.

Step 2: If enough information has been generated, the DM is asked to assess the preference for each of the given Pareto optimal solutions through the surrogate worth functions W_{1i}, $i = 2, \ldots, k$, by taking into account all levels of objective functions together with the trade-off information.

Step 3: If there exists a Pareto optimal solution satisfying $W_{1i}(f_2, \ldots, f_k) = 0$, $i = 2, \ldots, k$, then it is the preferred solution of the DM. Otherwise, develop approximate relations for all surrogate worth functions and solve the corresponding simultaneous equations for obtaining the preferred solution of the DM.

While the original version of the SWT method is, in principle, noninteractive, Chankong and Haimes (1978, 1983a) proposed the interactive version of the SWT method, called the ISWT method. They realized that the values of the surrogate worth functions can be used to determine the direction in which the preference function of the DM, although unknown, increases most rapidly. Details of the ISWT method can be found in Chankong and Haimes (1978, 1983a). Other extensions and applications of the SWT method may also be found in Haimes, Hall, and Freedman (1975), Haimes (1977), Chankong and Haimes (1983a) and Haimes, Tarvainen, Shima, and Thadathil (1990).

4.3 Interactive multiobjective nonlinear programming

4.3.1 Interactive Frank-Wolfe method

An interactive nonlinear programming approach first developed by Geoffrion, Dyer, and Feinberg (1972) is designed for the solution of multiobjective decision-making problems. This approach assumes the existence of an underlying preference function of the decision maker (DM), but never actually requires this preference function to be identified explicitly. The basic idea behind their approach is that even if the DM cannot specify an overall preference function explicitly, local information regarding a preference at a particular solution can be provided.

Their interactive approach is described in the context of the well-known Frank-Wolfe algorithm, a method for solving nonlinear programming problems that moves from an initial feasible solution towards the optimal solution by finding the direction of steepest ascent and the optimal step size in that direction. It is particularly important to point out that during the solution process explicit knowledge of the overall preference function is not essential. Instead, only local information concerning the preferences of the DM is required and this, in turn, is sufficient to determine both the direction and the step size which will lead to an improvement over the current solution. This interactive approach which combines the Frank-Wolfe algorithm with local information on the DM's preference at each iteration for solving multiobjective decision-making problems is called the interactive Frank-Wolfe (IFW) method.

The multiobjective decision-making problem (MDMP) considered is of the following type:

$$\underset{x \in X}{\text{maximize}} \quad V(f_1(x), f_2(x), \ldots, f_k(x)), \tag{4.92}$$

where $f_1(x), \ldots, f_k(x)$ are k distinct objective functions of the decision vector x, $V(\cdot)$ is the DM's overall preference function defined on the values of the objective functions and X is the constrained set defined as

$$X = \{x \in R^n \mid g_j(x) \leq 0, \; j = 1, \ldots, m\}. \tag{4.93}$$

Observe that the functions $f_i(x)$, $i = 1, \ldots, k$, and the set X are assumed to be explicitly known, but the preference function $V(\cdot)$ is assumed to be known only implicitly to the DM. Furthermore, it is assumed that $V(\cdot)$ is concave, strictly decreasing, and continuously differentiable. All $f_i(x)$, $i = 1, \ldots, k$, are convex and continuously differentiable and the constrained set X is convex and compact.

Now, given a feasible solution $x^0 \in X$, a linear, or first-order approximation of the preference function $V(f(x)) = V(f_1(x), \ldots, f_k(x))$ can be made by neglecting the second- and higher-order terms in the Taylor series about x^0:

$$V(f(x)) \cong V(f(x^0)) + \nabla_x V(f(x^0))(x - x^0). \tag{4.94}$$

Using this linear approximation, the direction-finding problem to seek an increased value of $V(f(x))$ becomes

$$\underset{x \in X}{\text{maximize}} \quad \nabla_x V(f(x^0))x. \tag{4.95}$$

Let x^* be the solution to this problem, then $d = (x^* - x^0)$ determines the steepest ascent direction of movement from the current point x^0.

Although $V(f(x))$ is not known explicitly, the concept of marginal rates of substitution by the DM can be used to estimate the values of the gradient $\nabla_x V(f(x_0))$ of the preference function $V(f(x))$. The marginal rates of substitution or trade-off ratios among objective functions of the DM is defined as follows:

Definition 4.5 (Marginal rates of substitution (MRS))

At any point of $f = (f_1, \ldots, f_j, \ldots, f_i, \ldots, f_k)$, the amount of f_j that the DM is willing to acquire for sacrificing an additional unit of f_i is called the marginal rates of substitution (MRS) between f_j and f_i. Mathematically, the MRS $m_{ji}(f)$ is represented by the negative slope of the indifference curve of a preference at f:

$$\begin{aligned} m_{ji}(f) &= [\partial V(f)/\partial f_i]/[\partial V(f)/\partial f_j] \\ &= -df_j/df_i \mid_{dV=0,\, df_r=0,\, r \neq j,i} , \end{aligned} \tag{4.96}$$

where each indifference curve is a locus of points in an objective function space, for which the DM is indifferent.

Usually, the decision analyst assesses the MRS by presenting the following prospects to the DM:

$$\left.\begin{aligned} f &= (f_1, \ldots, f_j, \ldots, f_i, \ldots, f_k), \\ f' &= (f_1, \ldots, f_j + \Delta f_j, \ldots, f_i - \Delta f_i, \ldots, f_k). \end{aligned}\right\} \tag{4.97}$$

For a small fixed Δf_j, small enough so the indifference curve is approximately linear but large enough so the increment is meaningful, the analyst varies Δf_i until the DM is indifferent between f and f'. At this level, $m_{ji}(f) \cong \Delta f_j / \Delta f_i$: in Figure 4.4, $df_j = -\Delta f_j$ and $df_i = \Delta f_i$.

Using the concept of the MRS, we can formulate the gradient $\nabla_x V(f(x^0))$ of the preference function. Applying the chain rule, we have

$$\nabla_x V(f(x^0)) = \sum_{i=1}^{k} \frac{\partial V(f(x^0))}{\partial f_i} \nabla_x f_i(x^0). \tag{4.98}$$

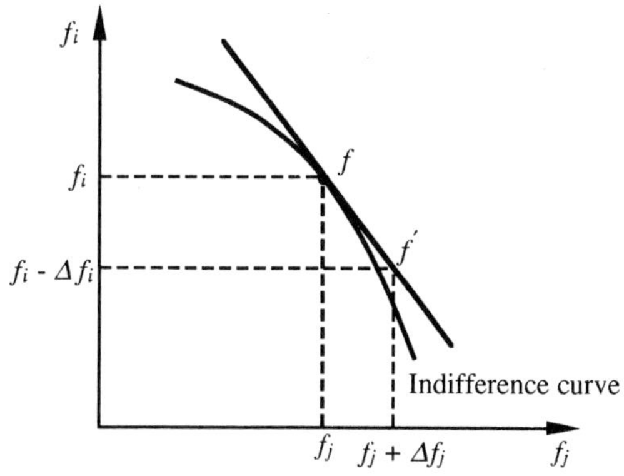

Figure 4.4 Assessing marginal rates of substitution

Assuming, without loss of generality, that the standing objective function is the first one and using the definition of MRS $m_{1i}(f(x^0))$, the objective function in the direction-finding problem can be expressed by

$$\frac{\partial V(f(x^0))}{\partial f_1}\left[\sum_{i=1}^{k} m_{1i}(f(x^0))\nabla_x f_i(x^0)\right] x, \tag{4.99}$$

where $m_{11}(f(x^0)) = 1$. From the strict monotonicity of $V(\cdot)$ with respect to f_1, $\partial V(\cdot)/\partial f_1$ is always negative. Hence, dividing the objective function by $\partial V(\cdot)/\partial f_1$, the direction-finding problem is equivalent to

$$\underset{x\in X}{\text{maximize}} \quad -\left[\sum_{i=1}^{k} m_{1i}(f(x^0))\nabla_x f_i(x^0)\right] x. \tag{4.100}$$

Once the steepest ascent direction $d = x^* - x^0$ is adopted, it is necessary to determine the step size α $(0 \leq \alpha \leq 1)$ which maximizes $V(f(x^0) + \alpha d)$ along this direction; the step size problem

$$\underset{0\leq\alpha\leq 1}{\text{maximize}} \quad V(f(x^0) + \alpha d) \tag{4.101}$$

must be solved to determine the optimal move along the direction d for obtaining the best step size α.

Since the preference function $V(\cdot)$ is not explicitly available, the step size problem must be solved through the interaction with the DM. One possible way to obtain the best step size α is to display the values of all k objective functions $f_i(x + \alpha d)$ as a function of α over the selected values of α between 0 and 1 in a tabular or graphic way. The DM then determines a value of α for the most preferred corresponding values of the objective functions.

Based on the above discussions, we can now describe the IFW method for solving the MDMP with an implicitly known preference function. The steps marked with an asterisk involve interaction with the DM.

The algorithm of the IFW method

Step 1: Select an initial point $x^{(0)} \in X$. Set the iteration index $r = 1$.

Step 2*: Assess the MRS $m_{1i}(f(x^{(r)}))$, $i = 2, \ldots, k$, of the DM at $f(x^{(r)})$.

Step 3: Determine the direction vector

$$d^{(r)} = x^{*(r)} - x^{(r)}$$

where $x^{*(r)}$ is the optimal solution to the direction-finding problem

$$\underset{x \in X}{\text{maximize}} \quad -\left[\sum_{i=1}^{k} m_{1i}(f(x^{(r)}))\nabla_x f_i(x^{(r)})\right] x.$$

Step 4*: Determine the step size $\alpha^{(r)}$ $(0 \leq \alpha \leq 1)$ which maximizes the preference function $V(f(x^{(r)} + \alpha^{(r)} d^{(r)}))$ as follows: Display the functions $f(x^{(r)} + \alpha d^{(r)})$ for the selected values of α $(0 \leq \alpha \leq 1)$ in a tabular or graphic way and have the DM subjectively locate the best step size according to preference.

Step 5*: If $V(f(x^{(r)} + \alpha^{(r)} d^{(r)})) \cong V(f(x^{(r)}))$, stop. Then $x^{(r)}$ is the preferred solution of the DM. Otherwise, set $x^{(r+1)} = x^{(r)} + \alpha d^{(r)}, r = r + 1$ and return to step 2.

It is important to note here that the IFW method relies mainly on the DM's estimation of the MRS, which requires the DM to indicate the degree of willingness to sacrifice the values of objective function $f_i(x)$ to acquire a specified increment in objective function $f_1(x)$ with all other objective functions held constant. In practice, however, the DM usually experiences difficulty in accomplishing such a task unaided. To decrease this difficulty, Dyer (1972, 1973) developed an algorithmic procedure for approximating the MRS through a series of ordinal paired comparisons, where the DM is required to indicate a preference for one of two possible choices, or to indicate an indifference between the two. For the fixed increment Δf_1, Δf_i is increased if the DM prefers f and Δf_i is decreased if f' is preferred. The process is continued until a value Δf_i is obtained for which the DM is indifferent between f and f'.

Formally, the MRS estimate algorithm of Dyer is summarized as follows:

Step 1: Let $f = (f_1, \ldots, f_i, \ldots, f_k)$ and $i = 2$.

Step 2: Select Δf_1 small enough so the indifference curve is approximately linear but large enough so the increment is meaningful.

Step 3: Set the iteration index $s = 1$ and select a convenient value of $\Delta f_i^{(s)} > 0$.

Step 4: Let $f' = (f_1 - \Delta f_1, \ldots, f_i + \Delta f_i^{(s)}, \ldots, f_k)$. Compare f and f'.

Step 5: If f' is preferred to f, set $\Delta f_i^{(s+1)} = 2(\Delta f_i^{(s)})$, $s = s + 1$, and return to step 4. If f is preferred to f', set $\theta^{(s)} = \Delta f_i^{(s)}$, and go to step 7. If the DM is indifferent between f and f', go to step 9.

Step 6: Compare f and f'. If f is preferred, go to step 7. If f' is preferred, go to step 8. If the DM is indifferent between f and f', go to step 9.

Step 7: Increase the desirability of f' by letting $\Delta f_i^{(s+1)} = \Delta f_i^{(s)} - \theta^{(s)}/2$. Set $\theta^{(s+1)} = \theta^{(s)}/2$, $s = s + 1$, and return to step 6.

Step 8: Decrease the desirability of f' by letting $\Delta f_i^{(s+1)} = \Delta f_i^{(s)} + \theta^{(s)}/2$. Set $\theta^{(s+1)} = \theta^{(s)}/2$, $s = s + 1$, and return to step 3. Otherwise, terminate.

Step 9: Compute $m_{1i} = \Delta f_1/\Delta f_i$. If $i \neq k$, set $i = i + 1$ and return to step 3. Otherwise, terminate.

The interactive goal programming method proposed by Dyer (1972, 1973) combines the goal programming formulation and the IFW method. In this approach, it is assumed that the DM's preference for each objective function is nondecreasing and that the DM is only concerned about minimizing the under-achievement. Consequently, the goal programming formulation under consideration can be written as the following so-called one-sided goal programming problem:

$$\begin{aligned} &\underset{x \in X}{\text{minimize}} \ \sum_{i=1}^{k} w_i^- d_i^- && (4.102)\\ &\text{subject to } f_i(x) - d_i^+ + d_i^- = \hat{f}_i, \ i = 1, \dots, k && (4.103)\\ &\qquad d_i^+ \cdot d_i^- = 0, \ i = 1, \dots, k && (4.104)\\ &\qquad d_i^+ \geq 0, d_i^- \geq 0, \ i = 1, \dots, k && (4.105) \end{aligned}$$

where $\hat{f}_i$ is the goal specified by the DM and w_i^- is the under-achievement weight for the ith objective function, respectively.

The relationship between the MDMP and the one-sided goal programming problem can be clarified by noting that

$$d_i^- = \hat{f}_i - f_i(x) + d_i^+, \ i = 1, \dots, k. \tag{4.106}$$

Observe that the objective function in the one-sided goal programming formulation for the MDMP can be interpreted as the piecewise linear approximation of an implicit additive separable preference function of the DM.

By considering that the one-sided goal programming formulation can be viewed as another mathematical expression of the direction-finding problem in the IFW method, Dyer (1972, 1973) proposed the interactive goal programming procedure, similar in spirit to the IFW method, with the modification that the direction-finding problem is replaced by the one-sided goal programming problem. In the interactive goal programming method of Dyer, all the assumptions postulated in the IFW method are assumed to be satisfied and each $\hat{f}_i$ should be chosen so that the marginal increase in preference associated with additional units of $f_i(x)$ beyond $\hat{f}_i$ is equal to zero; namely, $\partial V(f(x))/\partial f_i(x) = 0$ for $f_i(x) \geq \hat{f}_i$, $i = 1, \ldots, k$. Observe that the interactive goal programming method can be viewed as another mathematical expression of the concept of the IFW method and thus the computational procedures are essentially the same.

Oppenheimer (1978) introduced the local proxy preference functions in the IFW method. In his procedure, instead of solving the direction-finding problem and the step size problem, at each iteration, the local proxy preference function is updated by assessing a new MRS vector. Then the proxy is maximized to find a better point. Unfortunately, this method, like the IFW method, does not guarantee the generated solution in each iteration to be Pareto optimal.

As another modification of the IFW method, Musselman and Talavage (1980) suggested using the MRS assessed by the DM to establish the cutting plane in the objective space, called the trade-off cut, which removes a portion of the feasible region having lower values of the preference function than the current point. However, having obtained the usable direction by solving the linear programing problem, their trial and error step size determination scheme requires the DM to assess the MRS at each trial solution, which seems to add a considerable assessment burden.

To decrease the drawbacks of the multiobjective programming methods discussed thus far, Sakawa (1980, 1982a) proposed a new interactive multiobjective decision-making technique, which was called the sequential proxy optimization technique (SPOT), by incorporating the desirable features of the above methods. In his interactive on-line scheme, after solving the constraint problem, the values of the MRS assessed by the DM are used to determine the direction in which the preference function increases most rapidly. The local proxy preference function is updated to determine the optimal step size and Pareto optimality of the generated solution is guaranteed. SPOT, however, requires a great number of precise MRS estimates by the DM. In practice, it is a question of whether the DM can respond with precise and consistent values of MRS through the whole searching process because the DM's actions are often erratic, inconsistent due to the imprecise nature of human decision processes. Although by performing the MRS consistency test some of this erratic behavior is usually reduced, it is required to cope with imprecise nature of the DM's judgments. By considering the imprecise nature of the DM's judgments, techniques based on fuzzy set theory (Zadeh 1965) which are similar to the first method of Baptistella and Ollero (1980) have been incorporated into the algorithm of SPOT. On the basis of the DM's MRS presented in a fuzzy form, which can be interpreted as *L-R* fuzzy numbers, the revised version of SPOT, called the fuzzy sequential proxy optimization technique (FSPOT), has been presented (Sakawa 1984a; Sakawa and Seo 1983). Also, the SPOT algorithm has been extended to include the nonconvex problems by introducing the concept of local Pareto optimality (Sakawa and Mori 1983a, 1983b; Sakawa and Yano 1984b). For further details of the SPOT,

FSPOT and their extensions together with some applications, the readers might refer to the corresponding papers.

4.3.2 Reference point method and its extension

The reference point method (RPM), developed by Wierzbicki (1979), is a relatively practical interactive approach to multiobjective optimization problems that introduces the concept of a reference point suggested by the decision maker (DM) which reflects in some sense the desired values of the objective functions. The basic idea behind the RPM is that the DM can specify reference values for the objective functions, called a reference point, and can change the reference objective levels interactively due to learning or improved understanding during the solution process. Thus, the RPM can be seen as a successful combination of the advantages of the goal programming method (e.g. Charnes and Cooper 1963) and the method of the displaced ideal (Zeleny 1976), and hence, yields a satisficing solution in the same sense as March and Simon (1958).

Wierzbicki introduced the penalty scalarizing functions which are able to generate all the Pareto optimal solutions regardless of the convexity assumptions. In this procedure, when the DM specifies a reference point, optimization of the corresponding penalty scalarizing function provides the Pareto optimal solution close to the DM's reference point or better than that if the reference point is attainable. Then the DM either chooses the current Pareto optimal solution or modifies the reference point in order to find a satisficing solution.

Consider the following multiobjective nonlinear programming (MONLP) problem:

$$\underset{x \in X}{\text{minimize}} \quad f(x) \triangleq (f_1(x), f_2(x), \ldots, f_k(x)) \tag{4.107}$$

where $f_1(x), \ldots, f_k(x)$ are k distinct objective functions and X is the constrained set defined by

$$X = \{x \in R^n \mid g_j(x) \leq 0, \quad j = 1, \ldots, m\}. \tag{4.108}$$

As was discussed in subsection 4.2.1, there are several techniques for generating Pareto optimal solutions depending on the different methods for scalarizing the MONLP. For the RPM, the utilization of a penalty scalarizing function for obtaining Pareto optimal solutions to the MONLP with or without convexity assumptions is suggested (Wierzbicki 1977, 1979, 1980, 1982).

The typical penalty scalarizing function which might be convenient for nonlinear problems has the following form:

$$s_c(f(x) - \bar{f}) = \sum_{i=1}^{k}(f_i(x) - \bar{f}_i)^2 - c\sum_{i=1}^{k}(\max[0, f_i(x) - \bar{f}_i])^2. \tag{4.109}$$

For linear problems, however, the penalty scalarizing function based on the sum of the absolute values norm

$$s_c(f(x) - \bar{f}) = \sum_{i=1}^{k} \mid f_i(x) - \bar{f}_i \mid - c\sum_{i=1}^{k} \max[0, f_i(x) - \bar{f}_i] \tag{4.110}$$

or the Chebyshev norm

$$s_c(f(x) - \bar{f}) = \max_{i=1,\ldots,k} \mid f_i(x) - \bar{f}_i \mid - c \max_{i=1,\ldots,k} [0, f_i(x) - \bar{f}_i] \tag{4.111}$$

would be appropriate.

Here $c > 1$ is a scalar penalty coefficient and $\bar{f} = (\bar{f}_1, \ldots, \bar{f}_k)$ is the so-called reference point or reference objective level suggested by the DM reflecting a desired level for the objective functions.

It should be emphasized here that the purpose of the penalty scalarizing function is to generate a Pareto optimal point which is in some sense close to the DM's reference point or better than that if the reference point is attainable.

In the RPM, the Pareto optimal solutions to the MONLP are obtained by solving the penalty scalarizing problem defined by

$$\underset{x \in X}{\text{minimize}} \quad s_c(f(x) - \bar{f}) \tag{4.112}$$

or equivalently, in objective function space,

$$\underset{f \in F}{\text{minimize}} \quad s_c(f - \bar{f}) \tag{4.113}$$

where

$$F = \{f(x) \mid x \in X\}. \tag{4.114}$$

The following theorems, owed to Wierzbicki, show that the Pareto optimal solution set of the MONLP coincides with the optimal solution set of the penalty scalarizing problem under suitable assumptions.

Theorem 4.15

If $x^* \in X$ is an optimal solution of the penalty scalarizing problem for any $\bar{f}$, then x^* is a Pareto optimal solution to the MONLP.

Theorem 4.16

Let $x^* \in X$ be a Pareto optimal solution to the MONLP.

(1) *If $\bar{f} = f(x^*)$, then x^* is an optimal solution of the penalty scalarizing problem for a large enough c.*

(2) *If $\bar{f} \in \text{int}\, F$, then x^* is an optimal solution of the penalty scalarizing problem for some $\bar{f}$.*

(3) *If $\bar{f} \notin F$, then x^* is an optimal solution of the penalty scalarizing problem for some $\bar{f}$.*

Using the penalty scalarizing problem as a means for generating Pareto optimal solutions, a relatively practical procedure can be constructed where the DM interactively modifies the reference point on the basis of the knowledge gained by looking at the intermediary solutions obtained. The basic steps of the RPM can be sketched as follows. The steps marked with an asterisk involve interaction with the DM.

The algorithm of the RPM

Step 1*: Ask the DM to select the initial reference point. If the DM finds it difficult or impossible to identify such a point, the ideal point $f^{\min} = (f_1^{\min}, \ldots, f_k^{\min})$, where $f_i^{\min} = \min_{x \in X} f_i(x)$, can be used as an initial reference point.

Step 2: Solve the corresponding penalty scalarizing problem for obtaining a Pareto optimal solution.

Step 3*: If the DM is satisfied with the current levels of the objective functions, stop. Then the current Pareto optimal solution is the satisficing solution of the DM. Otherwise, ask the DM to modify the current reference point to a new reference point by considering the current levels of the objective functions and return to step 2.

Further refinements and details can be found in Wierzbicki (1979, 1981). Computer software implementing the RPM, called DIDASS (dynamic interactive decision analysis supporting system), has been developed to deal with both linear and nonlinear multiobjective optimization problems. (Lewandowski and Grauer 1982; Grauer 1983a, 1983b). An edited volume by Lewandowski and Wierzbicki (1989) covers the major theoretical and practical advances of the so-called DIDAS family.

Before leaving the RPM, it is significant to point out a few extensions of the RPM to deal with the imprecise nature of human decision processes. Recently, Sakawa and Yano (1984a) developed an interactive fuzzy multiobjective programming method which can be viewed as the fuzzy version of the RPM. Sakawa, Yumine, and Yano (1984) and Sakawa and Yano (1985b, 1985d, 1986a, 1986b, 1986f) also developed several interactive fuzzy multiobjective programming methods which can be interpreted as the extended fuzzy versions of the RPM in the sense that the DM is supplied with the trade-off information. Although the details of some of these methods and extensions will be discussed in section 4.5, it would certainly be appropriate to discuss their extended nonfuzzy version of the RPM.

In their nonfuzzy version, when the DM specifies the reference point $\bar{f} = (\bar{f}_1, \ldots, \bar{f}_k)$, the corresponding Pareto optimal solution, which is, in the minimax sense, nearest to the reference point or better than that if the reference point is attainable, is obtained by solving the following minimax problem instead of the penalty scalarizing problem:

$$\left.\begin{array}{ll} \text{minimize} & \max_{i=1,\ldots,k} \{f_i(x) - \bar{f}_i\} \\ \text{subject to} & x \in X, \end{array}\right\} \quad (4.115)$$

or equivalently,

$$\left.\begin{array}{ll} \text{minimize} & v \\ \text{subject to} & f_i(x) - \bar{f}_i \leq v, \ i = 1, ..., k \\ & x \in X. \end{array}\right\} \qquad (4.116)$$

The relationships between the optimal solutions of the minimax problem and the Pareto optimal concept of the MONLP can be characterized by the following theorems.

Theorem 4.17

If $x^* \in X$ is a unique optimal solution to the minimax problem for any reference point $\bar{f} = (\bar{f}_1, \ldots, \bar{f}_k)$, then x^* is a Pareto optimal solution to the MONLP.

Theorem 4.18

If $x^* \in X$ is a Pareto optimal solution to the MONLP, then there exists $\bar{f} = (\bar{f}_1, \ldots, \bar{f}_k)$ such that x^* is an optimal solution to the minimax problem.

The proofs of these theorems follow immediately from the definitions of unique optimality and Pareto optimality as in the proofs of Theorems 3.8 and 3.9.

If x^*, an optimal solution to the minimax problem, is not unique, then as discussed in subsection 4.2.1, the Pareto optimality test for x^* can be performed by solving the following problem:

$$\left.\begin{array}{ll} \text{maximize} & \sum_{i=1}^{k} \varepsilon_i \\ \text{subject to} & f_i(x) + \varepsilon_i = f_i(x^*), \ i = 1, \ldots, k \\ & x \in X, \ \varepsilon = (\varepsilon_1, ..., \varepsilon_k)^T \geq 0. \end{array}\right\} \qquad (4.117)$$

Let $\bar{x}$ be an optimal solution to this problem. If all $\varepsilon_i = 0$, then x^* is a Pareto optimal solution. If at least one $\varepsilon_i > 0$, not x^* but $\bar{x}$ is a Pareto optimal solution.

Now, given the Pareto optimal solution for the reference point specified by the DM by solving the corresponding minimax problem, the DM must either be satisfied with the current Pareto optimal solution, or modify the reference point. To help the DM express a degree of preference, trade-off information between a standing objective function $f_1(x)$ and each of the other objective functions is very useful. Such a trade-off between $f_1(x)$ and $f_i(x)$ for each $i = 2, \ldots, k$ is easily obtainable since it is closely related to the strict positive Lagrange multipliers of the minimax problem. Let the Lagrange multipliers associated with the constraints of the minimax problem be denoted by λ_i, $i = 1, \ldots, k$. If all $\lambda_i > 0$ for each i, then by extending the results in Haimes and Chankong

(1979), it can be proved that the following expression holds (see the Appendix for details):

$$-\frac{\partial f_i(x)}{\partial f_1(x)} = \frac{\lambda_1}{\lambda_i}, \quad i = 2, \ldots, k. \tag{4.118}$$

Here, mathematically, it is assumed that

(1) (v^*, x^*) is a regular point of the constraints of the minimax problem;
(2) the second-order sufficiency conditions are satisfied at (v^*, x^*); and
(3) there are no degenerate constraints at (v^*, x^*)

where (v^*, x^*) is an optimal solution to the minimax problem.

It should be noted here that to obtain the trade-off rate information from this relation, all the constraints of the minimax problem must be active. Therefore, if there are inactive constraints, it is necessary to replace $\bar{f}_i$ for inactive constraints by $f_i(x^*)$ and solve the corresponding minimax problem to obtain the Lagrange multipliers.

From the above discussions, we can now describe the interactive algorithm to obtain the desired satisficing solution for the DM from the Pareto optimal solution set. The steps marked with an asterisk involve interaction with the DM.

The algorithm of the RPM with trade-off information

Step 0: Calculate the individual minimum $f_i^{\min}$ and maximum $f_i^{\max}$ of each objective function $f_i(x)$ under the given constraints.

Step 1*: Ask the DM to select the initial reference point. If the DM finds it difficult or impossible to identify such a point, the ideal point $f^{\min} = (f_1^{\min}, \ldots, f_k^{\min})$ can be used as an initial reference point.

Step 2: Solve the corresponding minimax problem for obtaining a Pareto optimal solution together with the trade-off rate information between the objective functions.

Step 3*: If the DM is satisfied with the current levels of the objective functions, stop. Then the current Pareto optimal solution is the satisficing solution of the DM. Otherwise, ask the DM to modify the current reference point to the new reference point by considering the current levels of the objective functions together with the trade-off rates between the objective functions and return to step 2.

Here it should be stressed for the DM that any improvement of one objective function can be achieved only at the expense of at least one of the other objective functions.

An obvious advantage of the above method over the original RPM is that the DM is supplied with the trade-off information as well as the levels of all objective functions in trying to change the reference point. It should be emphasized here that the Pareto optimality test can easily be dropped, if one replaces the minimax problem by the augmented minimax problem in a similar manner to Steuer and Choo (1983) or Choo and Atkins (1983). This fact will be discussed further in section 4.5.

4.4 Fuzzy multiobjective nonlinear programming

There seems to be no fuzzy programming approach to cope with multiobjective nonlinear programming problems, except the theoretical papers of Feng (1983) and Buckley (1983), until Sakawa (1984b) first introduced fuzzy multiobjective nonlinear programming as a natural extension of his fuzzy linear programming (Sakawa 1983a, 1983b) discussed in the last part of the section 3.4. In his method, assuming that the decision maker (DM) has fuzzy goals for each of the objective functions in multiobjective nonlinear programming (MONLP) problems, the fuzzy goals of the DM are quantified using the five types of membership functions; linear, exponential, hyperbolic, hyperbolic inverse, and piecewise linear functions as proposed in the linear case. Then by selecting one of the three possible fuzzy decisions, the compromise or satisficing solution of the DM can be derived from among the Pareto optimal solution set. Furthermore, to cope with the fuzzy goals of the DM, such as "the objective function should be in the vicinity of some value," generalized multiobjective nonlinear programming problems are introduced together with extended Pareto optimality. Although this method has been further extended to more powerful and attractive interactive programming versions, as will be discussed in the remainder of this book, it may be significant to overview it in the historic sense.

Consider the following multiobjective nonlinear programming (MONLP) problem:

$$\left.\begin{array}{ll} \text{minimize} & f(x) \triangleq (f_1(x), f_2(x), \ldots, f_k(x))^T \\ \text{subject to} & x \in X \triangleq \{x \in R^n \mid g_j(x) \leq 0, \ j = 1, \ldots, m\} \end{array}\right\} \tag{4.119}$$

where $f_1(x), ..., f_k(x)$ are k distinct objective functions of the decision vector x, $g_1(x), ..., g_m(x)$ are m inequality constraints and X is the feasible set of constrained decisions.

Here, for simplicity, assume that all $f_i(x),\ i = 1, ..., k$, are convex and differentiable and the constrained set X is convex and compact.

Assuming that the DM has fuzzy goals for each of the objective functions in the MONLP, similar to fuzzy multiobjective linear programming proposed by Zimmermann (1978), it is possible to soften the rigid requirements of the MONLP to strictly minimize the k objective functions under the given constraints. In such a situation, the MONLP may be softened into the following fuzzy version:

$$\left.\begin{array}{ll} \widetilde{\text{minimize}} & f(x) \triangleq (f_1(x), f_2(x), \ldots, f_k(x))^T \\ \text{subject to} & x \in X \triangleq \{x \in R^n \mid g_i(x) \leq 0, \ j = 1, \ldots, m\} \end{array}\right\} \tag{4.120}$$

where the symbol "$\widetilde{\text{minimize}}$" denotes a relaxed or fuzzy version of "minimize" with the interpretation that "the k objective functions should be minimized as much as possible" under the given constraints.

Such fuzzy requirements of the DM can be quantified by eliciting membership functions $\mu_i(f_i(x))$, $i = 1, \ldots, k$, from the DM for all the objective functions $f_i(x)$, $i = 1, \ldots, k$. In general in a minimization problem, a fuzzy goal stated by the DM may be to achieve "substantially less than or equal to some value p_i," and the DM is asked to determine the subjective membership function $\mu_i(f_i(x))$ which is a strictly monotone decreasing function with respect to f_i in the following way:

$$\mu_i(f_i(x)) = \begin{cases} 1 & ; \quad f_i(x) \leq f_i^1 \\ d_i(x) & ; \quad f_i^1 \leq f_i(x) \leq f_i^0 \\ 0 & ; \quad f_i(x) \geq f_i^0 \end{cases} \tag{4.121}$$

where, as in Figure 4.5, f_i^1 or f_i^0 represents the value of f_i such that the grade of the membership function $\mu_i(f_i(x))$ is 1 or 0 and the grades of membership for the intermediate function values are expressed by a strictly monotone decreasing function $d_i(x)$ with respect to f_i.

To elicit a membership function $\mu_i(f_i(x))$ from the DM for each of the objective functions $f_i(x)$ of the MONLP, Sakawa (1984b) suggested the following approach. First, calculate the individual minimum $f_i^{\min}$ and maximum $f_i^{\max}$ of each objective function $f_i(x)$ under the given constraints. Then by taking account of the calculated individual minimum and maximum of each objective function, together with the rate of increase of membership of satisfaction, the DM is asked to select a membership function in a subjective manner from among the following five types of functions; linear, exponential, hyperbolic, hyperbolic inverse, and piecewise linear functions. The parameter values are determined through the interaction with the DM. Here, except for hyperbolic membership functions, it is assumed that $\mu_i(f_i(x)) = 0$ if $f_i(x) \geq f_i^0$ and $\mu_i(f_i(x)) = 1$ if $f_i(x) \leq f_i^1$.

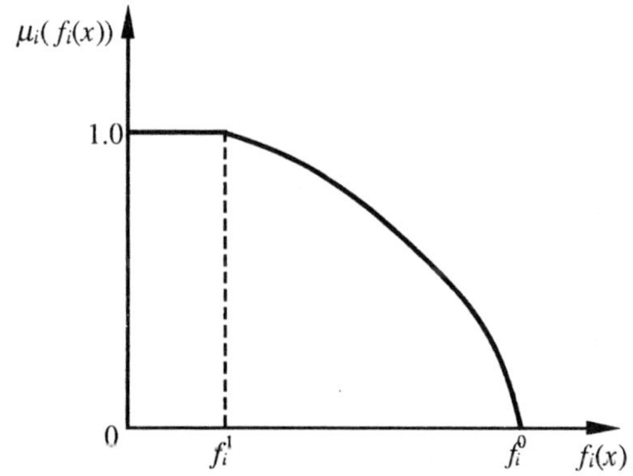

Figure 4.5 Strictly monotone decreasing membership function

(1) Linear membership function

$$\mu_i(f_i(x)) = [f_i(x) - f_i^0]/[f_i^1 - f_i^0]. \tag{4.122}$$

The linear membership function can be determined by asking the DM to specify the two points f_i^0 and f_i^1 within $f_i^{\max}$ and $f_i^{\min}$.

(2) Exponential membership function

$$\mu_i(f_i(x)) = a_i[1 - \exp\{-\alpha_i(f_i(x) - f_i^0)/(f_i^1 - f_i^0)\}] \tag{4.123}$$

where $a_i > 1, \alpha_i > 0$ or $a_i < 0, \alpha_i < 0$, and α_i is a shape parameter. The exponential membership function can be determined by asking the DM to specify the three points $f_i^0, f_i^{0.5}$, and f_i^1 within $f_i^{\max}$ and $f_i^{\min}$, where f_i^a represents the value of $f_i(x)$ such that the degree of membership function $\mu_{f_i}(x)$ is a.

(3) Hyperbolic membership function

$$\mu_i(f_i(x)) = \frac{1}{2}\tanh((f_i(x) - b_i)\alpha_i) + \frac{1}{2} \tag{4.124}$$

where $\alpha_i < 0$ and α_i is a shape parameter and b_i is associated with the point of inflection. The hyperbolic membership function can be determined by asking the DM to specify the two points $f_i^{0.25}$ and $f_i^{0.5}$ within $f_i^{\max}$ and $f_i^{\min}$.

(4) Hyperbolic inverse membership function

$$\mu_i(f_i(x)) = a_i \tanh^{-1}((f_i(x) - b_i)\alpha_i) + \frac{1}{2} \tag{4.125}$$

where $a_i > 0$ and $\alpha_i < 0$, and α_i is a shape parameter and b_i is associated with the point of inflection. The hyperbolic inverse membership function can be determined by asking the DM to specify the three points $f_i^0, f_i^{0.25}$, and $f_i^{0.5}$ within $f_i^{\max}$ and $f_i^{\min}$.

(5) Piecewise linear membership function

$$\mu_i(f_i(x)) = \sum_{j=1}^{N_i} \alpha_{ij}|f_i(x) - g_{ij}| + \beta_i f_i(x) + \gamma_i. \tag{4.126}$$

Here it is assumed that $\mu_i(f_i(x)) = t_{ir}f_i(x) + s_{ir}$ for each segment $g_{ir-1} \leq f_i(x) \leq g_{ir}$. The piecewise linear membership function can be determined by asking the DM to specify the degree of membership in each of several values of objective functions within $f_i^{\max}$ and $f_i^{\min}$.

Having determined the membership functions for each of the objective functions, Sakawa (1984b) proposed asking the DM to select one of the following three possible fuzzy decisions:

(1) a fuzzy decision (Bellman and Zadeh 1970) defined by

$$\min_{1\leq i\leq k} (\mu_i(f_i(x)), \mu_i(f_2(x)), ..., \mu_i(f_k(x))), \tag{4.127}$$

(2) a convex-fuzzy decision (Bellman and Zadeh 1970) defined by

$$\sum_{i=1}^{k} \alpha_i \mu_i(f_i(x)), \qquad \sum_{i=1}^{k} \alpha_i = 1, \ \alpha_i \geq 0, \tag{4.128}$$

(3) a product-fuzzy decision (Zimmermann 1978) defined by

$$\prod_{i=1}^{k} \mu_i(f_i(x)). \tag{4.129}$$

In each case, following the maximizing decision which maximizes one of the three fuzzy decisions, the compromise or satisficing solution of the DM can be derived as follows.

If the DM selects the fuzzy decision or minimum operator, like Zimmermann's approach (1978), the resulting problem to be solved is equivalent to solving the following problem:

$$\left.\begin{array}{ll} \text{maximize} & \lambda \\ \text{subject to} & \lambda \leq \mu_i(f_i(x)), \ i = 1, \ldots, k \\ & x \in X. \end{array}\right\} \tag{4.130}$$

If the DM selects the convex-fuzzy decision, like Sommer's approach (1978), the problem to be solved is

$$\underset{x \in X}{\text{maximize}} \ \sum_{i=1}^{k} \alpha_i \mu_i(f_i(x)). \tag{4.131}$$

If the DM selects the product-fuzzy decision or product operator, like Zimmermann's approach (1978), the problem to be solved becomes

$$\underset{x \in X}{\text{maximize}} \ \prod_{i=1}^{k} \mu_i(f_i(x)). \tag{4.132}$$

Fundamental to the MONLP is the Pareto optimal concept and thus the DM must select his compromise solution from among Pareto optimal solutions. The relationships between the optimal solutions of the above three types of problems and the Pareto optimal concept of the MONLP can be characterized by the following theorems.

Theorem 4.19

If x^* is a unique optimal solution to the problem (4.130), then x^* is a Pareto optimal solution to the MONLP.

Theorem 4.20

If x^* is an optimal solution to the problem (4.131) with $0 < \mu_i(f_i(x^*)) < 1$ holding for all i, then x^* is a Pareto optimal solution to the MONLP.

Theorem 4.21

If x^* is an optimal solution to the problem (4.132) with $0 < \mu_i(f_i(x^*)) < 1$ *holding for all* i*, then* x^* *is a Pareto optimal solution to the MONLP.*

The proofs of these theorems follow immediately from the definitions of (unique) optimality and Pareto optimality by contradictory arguments.

If x^* is an optimal solution to the problem (4.130), (4.131), or (4.132), and if none of the sufficiency conditions for Pareto optimality in Theorem 4.19, 4.20, or 4.21 is satisfied, then we can test the Pareto optimality for x^* by solving the following problem:

$$\left.\begin{array}{ll} \text{maximize} & \sum_{i=1}^{k} \varepsilon_i \\ \text{subject to} & f_i(x) + \varepsilon_i = f_i(x^*) \\ & x \in X, \ \varepsilon = (\varepsilon_1, ..., \varepsilon_k)^T \geq 0. \end{array}\right\} \quad (4.133)$$

Let $\bar{x}$ be an optimal solution to this problem. If all $\varepsilon_i = 0$, then x^* is a Pareto optimal solution. If at least one $\varepsilon_i > 0$, we adopt the solution $\bar{x}$ as the compromise or satisficing solution of the DM, because $\bar{x}$ not x^* is a Pareto optimal solution of the MONLP.

So far we have considered a minimization problem and consequently assumed that the DM has a fuzzy goal such as "$f_i(x)$ should be substantially less than or equal to p_i." In the fuzzy approaches, it is also possible to deal with the following general multiobjective nonlinear programming (GMONLP) problem first suggested by Sakawa (1984b):

$$\left.\begin{array}{lll} \text{fuzzy min} & f_i(x) & i \in I_1 \\ \text{fuzzy max} & f_i(x) & i \in I_2 \\ \text{fuzzy equal} & f_i(x) & i \in I_3 \\ \text{subject to} & x \in X & \end{array}\right\} \quad (4.134)$$

where $I_1 \cup I_2 \cup I_3 = \{1, 2, ..., k\}$, $I_i \cap I_j = \emptyset$, $i, j = 1, 2, 3$, $i \neq j$.

Here "fuzzy min $f_i(x)$" or "fuzzy max $f_i(x)$" represents the fuzzy goal of the DM such as "$f_i(x)$ should be substantially less than or equal to p_i or greater than or equal to q_i," and "fuzzy equal $f_i(x)$" represents the fuzzy goal of the DM such as "$f_i(x)$ should be in the vicinity of r_i."

Concerning the membership function for the fuzzy goal of the DM such as "$f_i(x)$ should be in the vicinity of r_i," it is obvious that a strictly monotone increasing function and a strictly monotone decreasing function corresponding to the left and right sides of r_i must be determined through interaction with the DM. In practice, the DM may select the left and right functions from among the five types of membership functions described previously (excluding hyperbolic functions).

As an example, Figure 4.6 illustrates the graph of the possible shape of the fuzzy equal membership functions where the left function is hyperbolic inverse and the right function is exponential.

When the fuzzy equal is included in the fuzzy goals of the DM, it is desirable that $f_i(x)$ should be as close as possible to r_i. As a result, the notion of Pareto optimal solutions defined in terms of objective functions is not applicable. With this in mind, Sakawa (1984b) first introduced the concept of fuzzy Pareto or M-Pareto optimal solutions. These are defined in terms of membership functions instead of objective functions. M refers to membership.

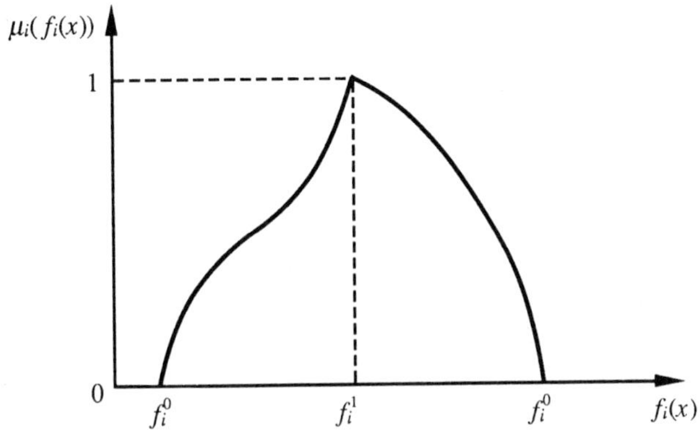

Figure 4.6 Fuzzy equal membership function

Definition 4.6 (M-Pareto optimal solution)

$x^* \in X$ is said to be an M-Pareto optimal solution to the GMONLP if and only if there does not exist another $x \in X$ such that $\mu_i(f_i(x)) \geq \mu_i(f_i(x^*))$ for all i and $\mu_j(f_j(x)) \neq \mu_j(f_j(x^*))$ for at least one j.

Note that the set of Pareto optimal solutions is a subset of the set of M-Pareto optimal solutions.

After determining the membership functions for all types of fuzzy goals of the DM, the compromise or satisficing solution of the DM can be derived by selecting and solving one of the three types of problems (4.130), (4.131), or (4.132) according to the DM's decision.

Using the concept of M-Pareto optimality, the following fuzzy versions of Theorems 4.19, 4.20, and 4.21 can be obtained under slightly different conditions.

Theorem 4.22

If x^* is a unique optimal solution to the problem (4.130), then x^* is an M-Pareto optimal solution to the GMONLP.

Theorem 4.23

If x^* is an optimal solution to the problem (4.131), then x^* is an M-Pareto optimal solution to the GMONLP.

Theorem 4.24

If x^* is an optimal solution to the problem (4.132) with $\mu_i(f_i(x^*)) \neq 0$ holding for all i, then x^* is an M-Pareto optimal solution to the GMONLP.

Similar to the minimization case, a numerical test of M-Pareto optimality for x^* can be performed by solving the following problem:

$$\left.\begin{array}{ll} \text{maximize} & \sum_{i=1}^{k} \varepsilon_i \\ \text{subject to} & \mu_i(f_i(x)) + \varepsilon_i = \mu_i(f_i(x^*)),\ i = 1, \ldots, k \\ & x \in X,\ \varepsilon = (\varepsilon_1, ..., \varepsilon_k)^T \geq 0. \end{array}\right\} \tag{4.135}$$

Let $\bar{x}$ be an optimal solution to this problem. If all $\varepsilon_i = 0$, then x^* is an M-Pareto optimal solution. If at least one $\varepsilon_i > 0$, we adopt the solution $\bar{x}$ as the compromise or satisficing solution of the DM because M-Pareto optimality of $\bar{x}$ can be established. Further details involving a numerical example can be found in Sakawa (1984b).

4.5 Interactive fuzzy multiobjective nonlinear programming

In the fuzzy multiobjective nonlinear programming discussed in the previous section, as well as the conventional fuzzy programming approaches proposed in the framework of Bellman and Zadeh (1970), it has been implicitly assumed that one of the possible three fuzzy decisions is the proper representation of the fuzzy preference of the decision maker (DM). Therefore, these approaches are preferable only when the DM feels that one of the three fuzzy decisions is appropriate when combining the fuzzy goals and/or constraints. However, such situations seem to happen infrequently, and consequently, it becomes evident that an interaction with the DM is necessary.

In this section, assuming that the DM has a fuzzy goal for each of the objective functions in multiobjective nonlinear programming problems, we present a few interactive fuzzy multiobjective nonlinear programming methods incorporating the desirable features of both goal programming and interactive programming approaches discussed thus far into the fuzzy approaches.

In general, the multiobjective nonlinear programming (MONLP) problem is represented as the following vector-minimization problem:

$$\left.\begin{array}{ll} \text{minimize} & f(x) \triangleq (f_1(x), f_2(x), ..., f_k(x))^T \\ \text{subject to} & x \in X \triangleq \{x \in R^n \mid g_j(x) \leq 0,\ j = 1, ..., m\} \end{array}\right\} \tag{4.136}$$

where x is an n-dimensional vector of decision variables, $f_1(x), ..., f_k(x)$ are k distinct objective functions of the decision vector x, $g_1(x), ..., g_m(x)$ are m inequality constraints, and X is the feasible set of constrained decisions.

As discussed before, fundamental to the MONLP is the concept of a Pareto optimal solution, which is also known as a noninferior solution. In practice, however, since only local solutions are guaranteed in solving a scalar optimization problem by any standard optimization technique, unless the problem is convex, instead of global Pareto optimal solutions, we deal with the local Pareto optimal solutions (Geoffrion 1968) as introduced in subsection 4.2.1.

Definition 4.7 (Local Pareto optimal solution)

$x^* \in X$ is said to be a local Pareto optimal solution to the MONLP if and only if there exists a real number $\delta > 0$ such that x^* is Pareto optimal in $X \cap N(x^*, \delta)$, i.e., there does not exist another $x \in X \cap N(x^*, \delta)$ such that $f_i(x) \leq f_i(x^*)$ for all i and $f_j(x) \neq f_j(x^*)$ for at least one j, where $N(x^*, \delta)$ denotes the δ neighborhood of x^* defined by $\{x \in R^n \mid \|x - x^*\| < \delta\}$.

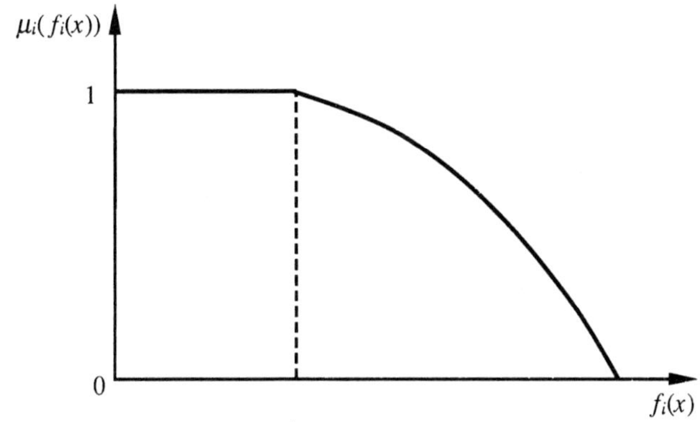

Figure 4.7 Membership function for minimization problem

Unfortunately, however, (local) Pareto optimal solutions consist of an infinite number of points, and thus, the DM must select a (local) final solution from among (local) Pareto optimal solutions as the satisficing solution.

In a minimization problem, a fuzzy goal stated by the DM may be to achieve "substantially less than or equal to some value p_i." This type of statement can be quantified by eliciting a corresponding membership function. Figure 4.7 illustrates the graph of the possible shape of the membership function representing the fuzzy goal to achieve substantially less than or equal to p_i.

To elicit a membership function $\mu_i(f_i(x))$ from the DM for each of the objective functions $f_i(x)$, $i = 1, \ldots, k$, we first calculate the individual minimum $f_i^{\min}$ and maximum $f_i^{\max}$ of each objective function $f_i(x)$ under the given constraints. By taking account of the calculated individual minimum and maximum of each objective function together with the rate of increase of membership of satisfaction, the DM must determine the subjective membership function $\mu_i(f_i(x))$, which is a strictly monotone decreasing function with respect to $f_i(x)$. Here, it is assumed that $\mu_i(f_i(x)) = 0$ or $\to 0$ if $f_i(x) \geq f_i^0$ and $\mu_i(f_i(x)) = 1$ or $\to 1$ if $f_i(x) \leq f_i^1$, where f_i^a represents the value of $f_i(x)$ such that the value of the membership function $\mu_i(f_i(x))$ is a within $f_i^{\min}$ and $f_i^{\max}$.

Up until now, we have considered only a minimization problem and as a consequence, assumed that the DM has a fuzzy goal such that "$f_i(x)$ ought to be substantially less than or equal to p_i." As discussed in a previous section, a fuzzy approach would be to treat a more general multiobjective nonlinear programming problem, in which the DM has two types of fuzzy goals, namely fuzzy goals expressed as "$f_i(x)$ should be in the vicinity of r_i" (called fuzzy equal) as well as "$f_i(x)$ should be substantially less than or equal to p_i or greater than or equal to q_i" (called fuzzy min or fuzzy max).

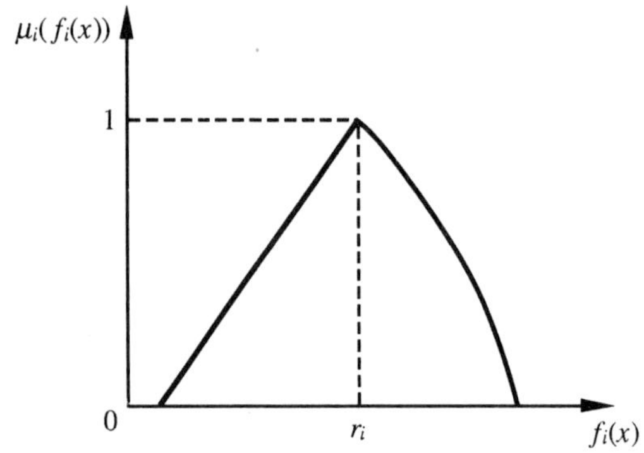

Figure 4.8 Fuzzy equal membership function

A generalized multiobjective nonlinear programming problem (GMONLP) can be expressed as

$$\left.\begin{array}{lll} \text{fuzzy min} & f_i(x) & i \in I_1 \\ \text{fuzzy max} & f_i(x) & i \in I_2 \\ \text{fuzzy equal} & f_i(x) & i \in I_3 \\ \text{subject to} & x \in X & \end{array}\right\} \tag{4.137}$$

where $I_1 \cup I_2 \cup I_3 = \{1, 2, ..., k\}$, $I_i \cap I_j = \emptyset$, $i, j = 1, 2, 3$, $i \neq j$.

To elicit a membership function from the DM for a fuzzy goal like "$f_i(x)$ should be in the vicinity of r_i," it is obvious that we can use different functions to the left and right sides of r_i. As an example, Figure 4.8 illustrates the graph of the possible shape of the fuzzy equal membership functions where the left function is linear and the right function is exponential.

When the fuzzy equal is included in the fuzzy goals of the DM, it is desirable that $f_i(x)$ should be as close to r_i as possible. Consequently, the notion of Pareto optimal solutions defined in terms of objective functions cannot be applied. For this reason, as discussed in the previous section, the concept of M-Pareto optimal solutions which is defined in terms of membership functions instead of objective functions is introduced, where M refers to membership.

Definition 4.8 ((local) M-Pareto optimal solution)

$x^* \in X$ is said to be a (local) M-Pareto optimal solution to the GMONLP if and only if there does not exist another $x \in X\ (\cap N(x^*, \delta)\ )$ such that $\mu_i(f_i(x)) \geq \mu_i(f_i(x^*))$ for all i and $\mu_j(f_j(x)) \neq \mu_j(f_j(x^*))$ for at least one j, where $N(x^*, \delta)$ denotes the δ neighborhood of x^*.

Observe that the set of (local) Pareto optimal solutions is a subset of the set of (local) M-Pareto optimal solutions.

Having elicited the membership functions $\mu_i(f_i(x)), i = 1, ..., k$, from the DM for each of the objective functions $f_i(x), i = 1, ..., k$, the MONLP and/or the GMONLP can be converted into the fuzzy multiobjective optimization problem (FMOP) defined by

$$\underset{x \in X}{\text{minimize}} \quad (\mu_1(f_1(x)), \mu_2(f_2(x)), ..., \mu_k(f_k(x))). \tag{4.138}$$

By introducing a general aggregation function

$$\mu_D(\mu(f(x))) = \mu_D(\mu_1(f_1(x)), \mu_2(f_2(x)), ..., \mu_k(f_k(x))), \tag{4.139}$$

a general fuzzy multiobjective decision-making problem (FMDMP) can be defined by

$$\underset{x \in X}{\text{maximize}} \quad \mu_D(\mu(f(x))). \tag{4.140}$$

Observe that the value of $\mu_D(\mu(f(x)))$ can be interpreted as representing an overall degree of satisfaction with the DM's multiple fuzzy goals.

The fuzzy decision or the minimum operator of Bellman and Zadeh (1970)

$$\min_{i=1,...,k} (\mu_1(f_1(x)), \mu_2(f_2(x)), ..., \mu_k(f_k(x))) \tag{4.141}$$

can be viewed only as one special example of $\mu_D(\mu(f(x)))$.

In the conventional fuzzy approaches, it has been implicitly assumed that the minimum operator is the proper representation of decision makers' fuzzy preferences, and hence, the FMDMP has been interpreted as follows:

$$\underset{x \in X}{\text{maximize}} \quad \min_{i=1,...,k} \quad (\mu_1(f_1(x)), \mu_2(f_2(x)), ..., \mu_k(f_k(x))) \tag{4.142}$$

or equivalently

$$\left.\begin{array}{ll} \text{maximize} & v \\ \text{subject to} & v \leq \mu_i(f_i(x)), \ i = 1, \ldots, k \\ & x \in X. \end{array}\right\} \tag{4.143}$$

There can be little doubt that the emphasis on this approach is only preferable when the DM feels that the minimum operator is the most appropriate situation. Needless to say, the DM does not always use the minimum operator in combinations of fuzzy goals and/or constraints. The most critical problem in the FMDMP is identifying an appropriate aggregation function representing the DM's fuzzy preferences. In circumstances where $\mu_D(\cdot)$ can be explicitly identifiable, then the FMDMP is reduced, as a result, to a standard mathematical programming problem. This, however, is a rare occurence and the alternative is an interaction with the DM to find a satisficing solution for the FMDMP.

Sakawa (1984c, 1986), Sakawa and Yano (1984a, 1985b, 1986a, 1986b, 1986f) and Sakawa, Yumine, and Yano (1984, 1987) have suggested several interactive fuzzy multiobjective nonlinear programming methods by adopting

(1) penalty scalarizing problems
(2) constraint problems
(3) goal programming problems
(4) minimax problems, and
(5) augmented minimax problems

as scalarizing methods for generating a candidate for the satisficing solution of the DM which is also (local) (M-) Pareto optimal. The DM then acts on this solution until satisfied with the current (local) (M-) Pareto optimal solution. Among them, in the following subsections, we will discuss the interactive methods based on the last three scalarizing problems because of their desirable features.

4.5.1 Interactive fuzzy multiobjective nonlinear goal programming

In the interactive fuzzy multiobjective nonlinear goal programming method proposed by Sakawa (1984c), after determining the membership functions $\mu_i(f_i(x))$, $i = 1, \dots, k$, for each of the objective functions $f_i(x)$, $i = 1, \dots, k$, the DM is asked to specify aspiration levels of achievement of goals in terms of membership values, called the goal membership levels, for all the membership functions. For the DM's goal membership levels $\overline{\mu}_i$, $i = 1, \dots, k$, the corresponding (local) (M-) Pareto optimal solution is obtained by solving the following goal programming problem in the membership functions space:

$$\left.\begin{array}{ll} \text{maximize} & \sum_{i=1}^{k} d_i^- \\ \text{subject to} & \mu_i(f_i(x)) + d_i^- - d_i^+ = \overline{\mu}_i, \ i = 1, \dots, k \\ & d_i^- \cdot d_i^+ = 0, \ i = 1, \dots, k \\ & d_i^-, \ d_i^+ \geq 0, \ i = 1, \dots, k \\ & x \in X \end{array}\right\} \tag{4.144}$$

where d_i^- and d_i^+ are, respectively, the negative and positive deviation variables, which provide us with a way to measure nonachievement of goal membership levels. This particular formulation may be called one-sided fuzzy nonlinear goal programming.

The relationships between the (local) optimal solutions of the fuzzy nonlinear goal programming problem and the (local) Pareto optimal concept of the MONLP can be characterized by the following theorems.

Theorem 4.25

If x^* is a (local) optimal solution to the fuzzy nonlinear goal programming problem with $0 < \mu_{f_i}(x^*) < 1$ and $d_i^+ = 0$ holding for all i, then x^* is a (local) Pareto optimal solution to the MONLP.

Theorem 4.26

If x^* is a (local) Pareto optimal solution to the MONLP with $0 < \mu_{f_i}(x^*) < 1$ holding for all i, then x^* is a (local) optimal solution to the fuzzy nonlinear goal programming with $d_i^+ = 0$ holding for all i.

The proofs of these theorems follow directly from the definitions of (local) optimality and (local) Pareto optimality by contradictory arguments.

If x^* is a (local) optimal solution to the fuzzy nonlinear goal programming problem, and if none of the sufficiency conditions for (local) Pareto optimality in Theorem 4.25 are satisfied (i.e., $^\exists i,\ \mu_{f_i}(x^*) = 0, 1$ or $d_i^+ = 0$), then, as discussed before, we can test the (local) Pareto optimality for x^* by solving the following Pareto optimality test problem :

$$\left.\begin{array}{ll} \text{maximize} & \sum_{i=1}^{k} \varepsilon_i \\ \text{subject to} & f_i(x) + \varepsilon_i = f_i(x^*),\ i = 1, \ldots, k \\ & x \in X,\ \varepsilon = (\varepsilon_1, \ldots, \varepsilon_k)^T \geq 0. \end{array}\right\} \tag{4.145}$$

Let $\overline{x}$ be a (local) optimal solution to this problem. If all $\varepsilon_i = 0$, then x^* is a (local) Pareto optimal solution. If at least one $\varepsilon_i > 0$, not x^* but $\overline{x}$ is a (local) Pareto optimal solution.

Using the concept of (local) M-Pareto optimality, the following (local) M-Pareto versions of Theorems 4.25 and 4.26 can be obtained immediately under slightly different conditions.

Theorem 4.27

If x^* is a (local) optimal solution to the fuzzy nonlinear goal programming problem with $d_i^+ = 0$ holding for all i, then x^* is a (local) M-Pareto optimal solution to the GMONLP.

Theorem 4.28

If x^* is a (local) M-Pareto optimal solution to the GMONLP, then x^* is a (local) optimal solution to the fuzzy nonlinear goal programming problem with $d_i^+ = 0$ holding for all i.

A numerical test of (local) M-Pareto optimality for x^*, similar to the minimization case, can be performed by solving the following M-Pareto optimality test problem:

$$\left.\begin{array}{ll} \text{maximize} & \sum_{i=1}^{k} \varepsilon_i \\ \text{subject to} & \mu_i(f_i(x)) - \varepsilon_i = \mu_i(f_i(x^*)),\ i = 1, \ldots, k \\ & x \in X,\ \varepsilon = (\varepsilon_1, \ldots, \varepsilon_k)^T \geq 0. \end{array}\right\} \quad (4.146)$$

Let $\overline{x}$ be a (local) optimal solution to this problem. If all $\varepsilon_i = 0$, then x^* is a (local) M-Pareto optimal solution. If at least one $\varepsilon_i > 0, \overline{x}$ becomes a (local) M-Pareto optimal solution.

We can now present the interactive fuzzy nonlinear goal programming algorithm to derive the (local) satisficing solution for the DM from the (local) (M-) Pareto optimal solution set. The steps marked with an asterisk involve interaction with the DM.

Interactive fuzzy multiobjective nonlinear goal programming

Step 0: Calculate the individual minimum $f_i^{\min}$ and maximum $f_i^{\max}$ of each objective function $f_i(x)$ under the given constraints.

Step 1*: Elicit a membership function from the DM for each of the objective functions.

Step 2: Set all the initial goal membership levels to one.

Step 3: Solve the corresponding fuzzy nonlinear goal programming problem to obtain the (local) (M-) Pareto optimal solution and membership function values.

Step 4*: If the DM is satisfied with the current values of membership functions, stop. Then the current (local) (M-) Pareto optimal solution is the (local) satisficing solution of the DM. Otherwise, ask the DM to update the current goal membership levels to the new goal membership levels by considering the current values of the membership functions together with the current values of the negative deviation variables and return to step 3.

Here it should be stressed for the DM that any improvement of one membership function can be achieved only at the expense of at least one of the other membership functions.

Further details, together with an application to water quality management, can be found in Sakawa (1984c).

4.5.2 Interactive fuzzy multiobjective nonlinear programming

An interactive fuzzy multiobjective nonlinear programming method using minimax problems as a means of generating (local) (M-) Pareto optimal solutions was proposed by Sakawa, Yumine, and Yano (1984, 1987), realizing that the minimax problems are much more favorable than the constraint problems especially in the sense that each objective function is treated equally.

In their method, after determining the membership functions $\mu_i(f_i(x))$, $i = 1, ..., k$, for each of the objective functions $f_i(x)$, $i = 1, ..., k$, the DM is asked to specify the reference membership levels for all the membership functions. For the DM's reference membership levels $\bar{\mu}_i$, $i = 1, 2, \cdots, k$, the corresponding (local) (M-) Pareto optimal solution which is, in the minimax sense, nearest to the requirement or better than that if the reference membership levels are attainable is obtained by solving the minimax problem

$$\underset{x \in X}{\text{minimize}} \quad \max_{i=1,\dots,k} \ \{\bar{\mu}_i - \mu_i(f_i(x))\} \tag{4.147}$$

or equivalently

$$\left.\begin{array}{ll} \text{minimize} & v \\ \text{subject to} & \bar{\mu}_i - \mu_i(f_i(x)) \leq v, \ i = 1, \dots, k \\ & x \in X. \end{array}\right\} \tag{4.148}$$

The relationships between the (local) optimal solutions of the minimax problem and the (local) Pareto optimal concept of the MONLP can be characterized by the following theorems.

Theorem 4.29

If x^* is a unique (local) optimal solution to the minimax problem for some $\bar{\mu}_i$, $i = 1, ..., k$, then x^* is a (local) Pareto optimal solution to the MONLP.

Theorem 4.30

If x^* is a (local) Pareto optimal solution to the MONLP with $0 < \mu_i(f_i(x^*)) < 1$ holding for all i, then there exists $\bar{\mu}_i$, $i = 1, ..., k$, such that x^* is a (local) optimal solution to the minimax problem.

The proofs of these theorems follow directly from the definitions of (local) optimality and (local) Pareto optimality by making use of contradiction arguments.

If x^*, a (local) optimal solution to the minimax, is not unique, then we can test the (local) Pareto optimality for x^* by solving the Pareto optimality test problem discussed in 4.2.1.

Using the concept of (local) M-Pareto optimality, the following theorem, which is similar to Theorems 4.29 and 4.30, can immediately be obtained under slightly different conditions.

Theorem 4.31

x^* is a (local) M-Pareto optimal solution to the GMONLP if and only if there exists a $\bar{\mu}$ such that x^* is a unique (local) optimal solution to the minimax problem.

Similar to the minimization case, a numerical test of (local) M-Pareto optimality for x^* can be performed by solving the M-Pareto optimality test problem as mentioned in 4.5.1.

The DM must now either be satisfied with the current (local) (M-) Pareto optimal solution or act on this solution by updating the reference membership levels. To help the DM express a degree of preference, as discussed before, trade-off information between a standing membership function $\mu_1(f_1(x))$ and each of the other membership functions is very useful. Such trade-off information between $\mu_1(f_1(x))$ and $\mu_i(f_i(x))$ for each $i = 2, ..., k$ is easily obtainable since it is related to the strict positive Lagrange multipliers of the minimax problem. Let the Lagrange multipliers associated with the constraints of the minimax problem be denoted by λ_i, $i = 1, 2, ..., k$. If all $\lambda_i > 0$ for each i, then by extending the results in Haimes and Chankong (1979), it can be proved that the following expression holds:

$$-\frac{\partial \mu_i(f_i(x))}{\partial \mu_1(f_1(x))} = \frac{\lambda_1}{\lambda_i}, \quad i = 2, ..., k. \tag{4.149}$$

In the proof of this relation, it is assumed that

(1) (v^*, x^*) is a regular point of the constraints of the minimax problem,

(2) the second-order sufficiency conditions of optimality are satisfied at (v^*, x^*), and

(3) there are no degenerate constraints at (v^*, x^*),

where (v^*, x^*) is an optimal solution to the minimax problem.

Although the formal proof of this relation can be found in the Appendix, geometrically, similar to the linear case, we can understand it as follows.

In the $(\mu_1, \mu_2, \cdots, \mu_k, v)$ space, the tangent hyperplane at some point on the Pareto surface can be described by

$$H(\mu_1, \mu_2, ..., \mu_k, v) = a_1\mu_1 + a_2\mu_2 + \cdots + a_k\mu_k + bv = c. \tag{4.150}$$

The necessary and sufficient condition for the small displacement from this point belonging to this tangent hyperplane is

$$\Delta H = 0 \tag{4.151}$$

or equivalently

$$a_1\Delta\mu_1 + a_2\Delta\mu_2 + \cdots + a_k\Delta\mu_k + b\Delta v = 0. \tag{4.152}$$

For fixed values of $\Delta\mu_j = 0$ $(j = 2, ..., k,\ j \neq i)$ and $\Delta v = 0$, except $\Delta\mu_1$ and $\Delta\mu_i$, we have

$$a_1\Delta\mu_1 + a_i\Delta\mu_i = 0. \tag{4.153}$$

Similarly, for fixed values of $\Delta\mu_i = 0$ $(i = 1, \ldots, k, i \neq j)$ except $\Delta\mu_j$ and v, we have

$$a_j\Delta\mu_j + b\Delta v = 0. \tag{4.154}$$

It follows from the last two relations that

$$-\frac{\Delta\mu_i}{\Delta\mu_1} = \frac{a_1}{a_i} = \frac{(-a_1/b)}{(-a_i/b)} = \frac{(\Delta v/\Delta\mu_1)}{(\Delta v/\Delta\mu_i)}. \tag{4.155}$$

Consequently, it holds that

$$-\frac{\partial\mu_i}{\partial\mu_1} = \frac{(\partial v/\partial\mu_1)}{(\partial v/\partial\mu_i)}. \tag{4.156}$$

Now, using the Lagrange multipliers λ_i, $i = 1, ..., k$, associated with all the active constraints of the minimax problem, we observe that

$$-\frac{\partial v}{\partial\mu_i} = -\lambda_i. \tag{4.157}$$

Hence, we have the result (4.149).

It should be stressed here that to obtain the trade-off rate information from (4.149), all the constraints of the minimax problem must be active. Therefore, if there are inactive constraints, it is necessary to replace $\bar{\mu}_i$ for inactive constraints by $\mu_i(f_i(x^*))$ and solve the corresponding minimax problem for obtaining the Lagrange multipliers.

We can now construct the interactive algorithm to derive the (local) satisficing solution for the DM from the (local) (M-) Pareto optimal solution set which can be interpreted as the fuzzy version of the reference point method (RPM) with trade-off information discussed in subsection 4.3.2.

For the multiobjective nonlinear programming problem, to circumvent the necessity to perform the (local) (M-) Pareto optimality tests in the minimax problems, an interactive fuzzy multiobjective nonlinear programming method using augmented minimax problems has been proposed by Sakawa and Yano (1985d, 1986f). Therefore, in the following, we shall detail their method using augmented minimax problems instead of minimax problems. In their method, after determining the membership functions $\mu_i(f_i(x))$, $i = 1, ..., k$, for each of the objective functions $f_i(x)$, $i = 1, ..., k$, the DM is asked to specify the reference membership levels for all the membership functions. For the DM's reference membership levels $\bar{\mu}_i$, $i = 1, ..., k$, the corresponding (local) (M-) Pareto optimal solution, which is, in the minimax sense, nearest to the requirement or better than that if the reference membership values are attainable, is obtained by solving the augmented minimax problem

$$\underset{x \in X}{\text{minimize}} \ \Big(\max_{i=1,\ldots,k} (\bar{\mu}_i - \mu_i(f_i(x))) + \rho \sum_{i=1}^{k} (\bar{\mu}_i - \mu_i(f_i(x))\Big), \tag{4.158}$$

or equivalently

$$\left.\begin{array}{rl} \text{minimize} & v+\rho\sum_{i=1}^{k}(\bar{\mu}_i-\mu_i(f_i(x))) \\ \text{subject to} & \bar{\mu}_i-\mu_i(z_i(x))\leq v,\ i=1,\ldots,k \\ & x\in X, \end{array}\right\} \tag{4.159}$$

or

$$\left.\begin{array}{rl} \text{minimize} & w \\ \text{subject to} & \bar{\mu}_i-\mu_i(f_i(x))\leq w-\rho\sum_{i=1}^{k}(\bar{\mu}_i-\mu_i(f_i(x))),\ i=1,...,k \\ & x\in X. \end{array}\right\} \tag{4.160}$$

The term augmented is adopted because the term $\rho\sum_{i=1}^{k}(\bar{\mu}_i-\mu_i(f_i(x)))$ is added to the standard minimax problem, where ρ is a sufficiently small positive scalar. Thus, the augmented minimax problem is a natural extension of the standard minimax problem and can be regarded as a modified fuzzy version of the augmented weighted Chebyshev norm problem of Steuer and Choo (1983) or Choo and Atkins (1983).

The relationships between the (local) optimal solutions of the augmented minimax problem and the (local) Pareto optimal concept of the MONLP can be characterized by the following theorems.

Theorem 4.32

If x^* is a (local) optimal solution to the augmented minimax problem for some $\bar{\mu}$, then x^* is a (local) Pareto optimal solution to the MONLP.

Proof

Assume that x^* is not a (local) Pareto optimal solution to the MONLP. Then there exists $x\in X$ $(\cap N(x^*,r))$ such that $f_i(x)\leq f_i(x^*)$, or equivalently, $\mu_i(f_i(x))\geq\mu_i(f_i(x^*))$, or $\bar{\mu}_i-\mu_i(f_i(x))\leq\bar{\mu}_i-\mu_i(f_i(x^*))$ with strict inequality holding at least one i. Hence, it follows that

$$\max_{i=1,\ldots,k}(\bar{\mu}_i-\mu_i(f_i(x)))\leq\max_{i=1,\ldots,k}(\bar{\mu}_i-\mu_i(f_i(x^*))),$$

$$\rho\sum_{i=1}^{k}(\bar{\mu}_i-\mu_i(f_i(x)))<\rho\sum_{i=1}^{k}(\bar{\mu}_i-\mu_i(f_i(x^*))).$$

This means

$$\max_{i=1,\ldots,k}(\bar{\mu}_i-\mu_i(f_i(x)))+\rho\sum_{i=1}^{k}(\bar{\mu}_i-\mu_i(f_i(x)))$$

$$<\max_{i=1,\ldots,k}(\bar{\mu}_i-\mu_i(f_i(x^*)))+\rho\sum_{i=1}^{k}(\bar{\mu}_i-\mu_i(f_i(x^*))),$$

which contradicts the fact that x^* is a (local) optimal solution to the augmented minimax problem. Hence, x^* is a (local) Pareto optimal solution to the MONLP. □

Theorem 4.33

If x^* is a (local) Pareto optimal solution to the MONLP with $0 < \mu_i(f_i(x^*)) < 1$ holding for all i, then there exists a $\bar{\mu}$ such that x^* is a (local) optimal solution to the augmented minimax problem.

Proof

Assume that x^* is not a (local) optimal solution to the augmented minimax problem for any $\bar{\mu}$ satisfying

$$\bar{\mu}_1 - \mu_1(f_1(x^*)) = \cdots = \bar{\mu}_k - \mu_k(f_k(x^*)).$$

Then there exists $x \in X$ $(\cap N(x^*, \delta))$ such that

$$\max_{i=1,\dots,k} (\bar{\mu}_i - \mu_i(f_i(x^*))) + \rho \sum_{i=1}^{k} (\bar{\mu}_i - \mu_i(f_i(x^*)))$$

$$> \max_{i=1,\dots,k} (\bar{\mu}_i - \mu_i(f_i(x))) + \rho \sum_{i=1}^{k} (\bar{\mu}_i - \mu_i(f_i(x))).$$

This implies

$$\max_{i=1,\dots,k} (\mu_i(f_i(x^*)) - \mu_i(f_i(x))) + \rho \sum_{i=1}^{k} (\mu_i(f_i(x^*)) - \mu_i(f_i(x))) < 0.$$

Now if either any $\mu_i(f_i(x^*)) - \mu_i(f_i(x))$ is positive or all $\mu_i(f_i(x^*)) - \mu_i(f_i(x))$, $i = 1, ..., k$, are zero, this inequality would be violated for sufficiently small positive ρ. Hence, it follows that

$$\mu_i(f_i(x^*)) - \mu_i(f_i(x)) \leq 0, \ i = 1, ..., k$$

with strict inequaliy holding at least one i. Because of the assumption $0 < \mu_i(f_i(x^*)) < 1$, we have $f_i(x^*) \geq f_i(x)$ with $f(x^*) \neq f(x)$. This contradicts the fact that x^* is a (local) Pareto optimal solution to the MONLP and the theorem is proved. □

Using the concept of (local) M-Pareto optimality, the following (local) M-Pareto version of Theorems 4.32 and 4.33 can also be obtained.

Theorem 4.34

x^* is a (local) M-Pareto optimal solution to the GMONLP if and only if there exists a $\bar{\mu}$ such that x^* is a (local) optimal solution to the augmented minimax problem.

The proof of this theorem follows in the same way as the Theorem 4.32 and 4.33.

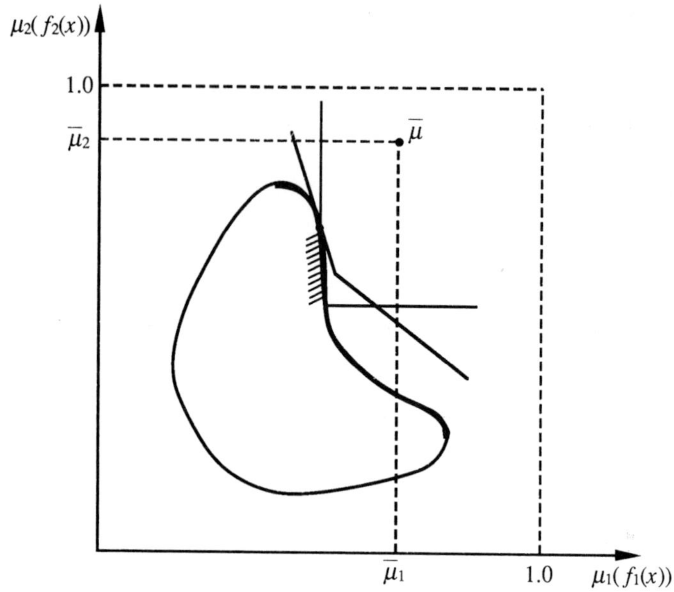

Figure 4.9 Isoquants of minimax and augmented minimax problems

As can be seen from the proofs, an obvious advantage of the augmented minimax problem over the standard minimax problem is that, because of the presence of the augmented terms, Pareto optimality is guaranteed without the uniqueness assumption for the solution.

Now it is significant to compare the augmented minimax problem with the standard minimax problem, which is the special case of the augmented minimax problem when $\rho = 0$.

Added insight can be obtained by comparing the isoquant of the augmented minimax problem

$$\bar{\mu}_i - \mu_i(f_i(x)) + \rho \sum_{i=1}^{k} (\bar{\mu}_i - \mu_i(f_i(x))) = \text{constant}, \ i = 1, ..., k \tag{4.161}$$

with the isoquant of the minimax problem

$$\bar{\mu}_i - \mu_i(f_i(x)) = \text{constant}, \ i = 1, ..., k \tag{4.162}$$

in the membership function space as depicted in Figure 4.9.

Observing that, in Figure 4.9, the normal vectors of the isoquant of the augmented minimax problem and the minimax problem become $(-\rho, ..., -\rho, -1-\rho, -\rho, ..., -\rho)$ and $(0, ..., 0, -1-\rho, 0, ..., 0)$, respectively, it easily follows that the cosine of angle θ between these two normal vectors is given by

$$\cos\theta = (1+\rho)/\sqrt{1+2\rho+k\rho^2}. \tag{4.163}$$

Hence, we have

$$\theta = \tan^{-1}(\sqrt{k-1}\ \rho/(1+\rho)). \tag{4.164}$$

This relation shows that θ is monotonically increasing with respect to ρ. Thus, for a sufficiently small positive scalar, the augmented minimax problems overcome the possibility of generating weak Pareto optimal solutions as shown in Theorems 4.31, 4.32, and 4.33. Hence, augmented minimax problems are attractive for generating Pareto optimal solutions even if appropriate convexity assumptions are absent.

Naturally, ρ should be a sufficiently small, but computationally significant, positive scalar. For practical purposes, however, a computationally significant lower bound of ρ may be

$$\rho = 10^{-(a-b)}$$

where a is the number of figures indicating the degree of precision of the computer and b is the number of figures in each membership value for which the DM can discriminate. In most cases, a computationally significant value of $\rho = 10^{-3} \sim 10^{-5}$ should suffice.

Now, given the (local) (M-) Pareto optimal solution for the reference membership levels specified by the DM by solving the corresponding augmented minimax problem, the DM must either be satisfied with the current (local) (M-) Pareto optimal solution or act on this solution by updating the reference membership levels. To help the DM express a degree of preference, as discussed in the previous two subsections, trade-off information between a standing membership function $\mu_1(f_1(x))$ and each of the other membership functions is very useful. Such a trade-off between $\mu_1(f_1(x))$ and $\mu_i(f_i(x))$ for each $i = 2, ..., k$ is easily obtainable since it is related to the strict positive Lagrange multipliers of the augmented minimax problem.

Let the Lagrange multipliers associated with the constraints of the augmented minimax problem be denoted by λ_i, $i = 1, ..., k$. If all $\lambda_i > 0$ for each i, then, similar to the minimax problems, the following expression holds:

$$-\frac{\partial \mu_i(f_i(x))}{\partial \mu_1(f_1(x))} = \frac{\lambda_1 + \rho}{\lambda_i + \rho}, \quad i = 2, ..., k. \tag{4.165}$$

It should be pointed out here that, to obtain the trade-off rate information from this relation all the constraints of the augmented minimax problem must be active. Therefore, if there are inactive constraints, it is necessary to replace $\bar{\mu}_i$ for inactive constraints by $\mu_i(f_i(x^*))$ and solve the corresponding augmented minimax problem for obtaining the Lagrange multipliers.

We can now construct the interactive algorithm to derive the (local) satisficing solution for the DM from the (local) (M-) Pareto optimal solution set. The steps marked with an asterisk involve interaction with the DM.

Interactive fuzzy multiobjective nonlinear programming

Step 0: Individual minimum and maximum

Calculate the individual minimum $f_i^{\min}$ and maximum $f_i^{\max}$ of each objective function $f_i(x)$ under the given constraints.

Step 1*: Membership functions

Elicit a membership function $\mu_i(f_i(x))$ from the DM for each of the objective functions.

Step 2: Initialization

Set the initial reference membership levels $\bar{\mu}_i^{(1)} = 1$, $i = 1, ..., k$, and set the iteration index $r = 1$.

Step 3: (M-) Pareto optimal solution

Set $\bar{\mu}_i = \bar{\mu}_i^{(r)}$, solve the corresponding augmented minimax problem to obtain the (local) (M-) Pareto optimal solution $x^{(r)}$, $f(x^{(r)})$, and the membership function value $\mu(f(x^{(r)}))$ together with the trade-off rate information between the membership functions.

Step 4*: Termination or New reference membership values

If the DM is satisfied with the current levels of $\mu_i(f_i(x^{(r)}))$, $i = 1, ..., k$, of the (local) (M-) Pareto optimal solution, stop. Then the current (local) (M-) Pareto optimal solution $f(x^{(r)}) = (f_1(x^{(r)}), ..., f_k(x^{(r)}))$ is the (local) satisficing solution for the DM. Otherwise, ask the DM to update the current reference membership levels $\bar{\mu}_1^{(r)}, ..., \bar{\mu}_k^{(r)}$ to the new reference membership levels $\bar{\mu}_1^{(r+1)}, ..., \bar{\mu}_k^{(r+1)}$ by considering the current levels of the membership functions together with the trade-off rates between the membership functions. Set $r = r + 1$ and return to step 3.

Here it should be stressed to the DM that any improvement of one membership function can be achieved only at the expense of at least one of the other membership functions.

Finally, it should be noted that the DM often cannot specify the reference membership levels in an exact way. In such a situation, instead of trying to obtain point-valued estimates of the reference membership levels from the DM, it may be more appropriate to obtain fuzzy-valued assessments of the reference membership levels such as "it should be between $\underline{\mu}_i$ and $\bar{\mu}_i$ for $f_i(x)$." Details of appropriate modifications of the above algorithm using such fuzzy-valued assessments can be found in Sakawa (1986) and Sakawa and Yano (1985b). Further extensions of the above algorithm to general multiobjective nonlinear programming problems involving fuzzy parameters, characterized by fuzzy parameters, which would be viewed as the more realistic version of the conventional multiobjective nonlinear programming problems, will be discussed in the next two chapters.

CHAPTER 5

INTERACTIVE MULTIOBJECTIVE LINEAR PROGRAMMING WITH FUZZY PARAMETERS

In this chapter, in contrast to the multiobjective linear programming problems discussed thus far, by considering the experts' imprecise or fuzzy understanding of the nature of the parameters in the problem-formulation process, multiobjective linear programming problems involving fuzzy parameters are formulated. Through the introduction of extended Pareto optimality concepts, linear programming-based interactive decision-making methods, both without and with the fuzzy goals of the decision maker (DM), for deriving a satisficing solution efficiently from an extended Pareto optimal solution set are presented together with illustrative numerical examples.

5.1 Introduction

In Chapter 3, after reviewing the basic concepts and methods of linear programming and fuzzy linear programming, we focused mainly on a multiobjective linear programming problem and presented major approaches of interactive multiobjective programming, fuzzy multiobjective linear programming, and interactive fuzzy multiobjective linear programming to derive the satisficing solution of the decision maker (DM) efficiently from a (M-) Pareto optimal solution set.

However, when formulating a multiobjective linear programming problem which closely describes and represents the real-world decision situation, various factors of the real-world system should be reflected in the description of the objective functions and the constraints. Naturally, these objective functions and constraints involve many parameters whose possible values may be assigned by the experts. In the conventional approaches, such parameters are required to fix at some values in an experimental and/or subjective manner through the experts' understanding of the nature of the parameters in the problem-formulation process.

It must be observed here that, in most real-world situations, the possible values of these parameters are often only imprecisely or ambiguously known to the experts. With this observation, it would be certainly more appropriate to interpret the experts' understanding of the parameters as fuzzy numerical data which can be represented by means of fuzzy sets of the real line known as fuzzy numbers. The resulting multiobjective linear programming problem involving fuzzy parameters would be viewed as a more realistic version than the conventional one.

The first attempt to incorporate fuzzy parameters into a linear programming framework can be found in the so-called robust programming problems of Negoita et al. (1976). Assuming the limited number of values of the membership functions of the fuzzy parameters and using the corresponding α-level sets, they showed that the original robust programming problem with "fuzzy set-inclusive" constraints can be transformed, via a linear programming problem with set coefficients, into a conventional linear programming problem. To decrease the serious drawback of high dimensionality of the resulting problem, when the fuzzy parameters are given as L-R fuzzy numbers, Dubois and Prade (1980a) showed the robust programming problem can be reduced to a conventional linear programming problem having three times the constraints of the original fuzzy ones.

Since then, several approaches have appeared for handling and tackling not only single- but also multiple-objective linear programming problems involving fuzzy parameters. For example, Tanaka and Asai (1984a, 1984b) formulated multiobjective linear programming problems with triangular fuzzy numbers and considered two types of fuzzy multiobjective linear programming problems, one to decide the nonfuzzy solution and the other to decide the fuzzy solution. From a more practical viewpoint, Slowinski (1986) considered a multiobjective linear programming problem involving L-R fuzzy numbers in the objective functions and the constraints together with the fuzzy aspiration levels for particular objectives. By introducing the two types of comparison indices of fuzzy numbers, the original problem can be transformed into a multiobjective linear fractional programming problem solvable by an interactive method proposed by Choo and Atkins (1980). His paper includes a brief survey of major approaches to fuzzy mathematical programming proposed before 1984 together with a successful application to water supply planning problems. More comprehensive surveys can be found in Dubois (1987) and Kacprzyk and Orlovski (1987).

Under these circumstances, to deal with multiobjective linear programming problems with fuzzy parameters characterized by fuzzy numbers, Sakawa and Yano (1986d, 1986g, 1987c, 1988a, 1990b) recently introduced the extended Pareto optimality concepts. Then they presented two types of linear programming-based interactive decision-making methods which are applicable and promising for handling and tackling not only the experts's fuzzy understanding of the nature of parameters in the problem-formulation process but also the fuzzy goals of the DM. These methods can be viewed as a natural generalization of the previous results for multiobjective linear programming problems without fuzzy parameters by Sakawa et al. (1983a, 1983b, 1987). In this chapter, these two types of interactive decision-making methods for multiobjective linear programming problems with fuzzy numbers to derive the satisficing solution of the DM efficiently from among the extended Pareto optimal solution set are presented in detail with numerical examples.

5.2 Problem formulation and solution concepts

First, recall the multiobjective linear programming (MOLP) problems discussed in Chapter 3. For convenience in our subsequent discussion, consider the MOLP of the following form:

$$\left.\begin{array}{rl} \text{minimize} & (c_1x, c_2x, ..., c_kx)^T \\ \text{subject to} & x \in X \triangleq \{x \in R^n \mid a_jx \leq b_j,\ j = 1, ..., m;\ x \geq 0\} \end{array}\right\} \tag{5.1}$$

where x is an n-dimensional column vector of decision variable, $c_1, c_2, ..., c_k$ are n-dimensional cost factor row vectors, $a_1, a_2, ..., a_m$ are n-dimensional constraint row vectors, and $b_1, b_2, ..., b_m$ are constants.

As discussed in Chapter 3, fundamental to the MOLP is the Pareto optimal concept, also known as a noninferior solution. Qualitatively, a Pareto optimal solution of the MOLP is one where any improvement of one objective function can be achieved only at the expense of another.

Definition 5.1 (Pareto optimal solution)

$x^* \in X$ is said to be a Pareto optimal solution to the MOLP if and only if there does not exist another $x \in X$ such that $c_ix \leq c_ix^*$, $i = 1, ..., k$, with strict inequality holding for at least one i.

In practice, however, it would certainly be more appropriate to consider that the possible values of the parameters in the description of the objective functions and the constraints usually involve the ambiguity of the experts' understanding of the real system. For this reason, in this chapter, we consider the following multiobjective linear programming problem involving fuzzy parameters (MOLP-FP):

$$\left.\begin{array}{rl} \text{minimize} & (C_1x, C_2x, ..., C_kx)^T \\ \text{subject to} & x \in X(A, B) \triangleq \{x \in R^n \mid A_jx \leq B_j,\ j = 1, ..., m;\ x \geq 0\}. \end{array}\right\} \tag{5.2}$$

$C_i = (C_{i1}, ..., C_{in})$, $A_j = (A_{j1}, ..., A_{jn})$, B_j represent, respectively, fuzzy parameters involved in the objective function C_ix and constraint $A_jx \leq B_j$.

These fuzzy parameters, reflecting the experts' ambiguous understanding of the nature of the parameters in the problem-formulation process, are assumed to be characterized as fuzzy numbers introduced by Dubois and Prade (1978, 1980b). In this chapter, we deal with a real fuzzy number M whose membership function $\mu_M(x)$ is defined as (Dubois and Prade 1978):

(1) A continuous mapping from R^1 to the closed interval $[0, 1]$
(2) $\mu_M(x) = 0$ for all $x \in (-\infty, c]$
(3) Strictly increasing and continuous on $[c, a]$
(4) $\mu_M(x) = 1$ for all $x \in [a, b]$

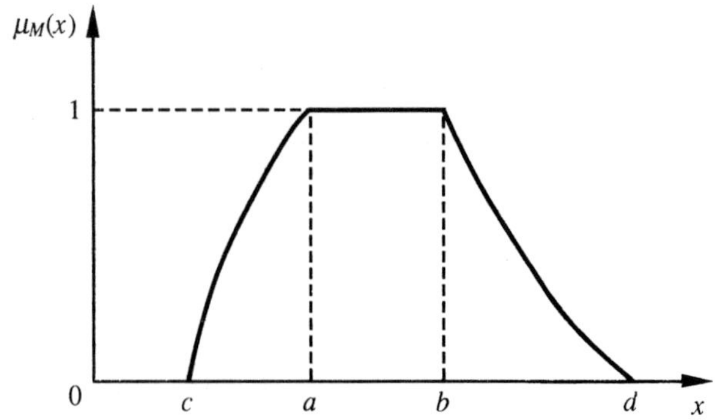

Figure 5.1 Fuzzy numbers

(5) Strictly decreasing and continuous on $[b, d]$

(6) $\mu_M(x) = 0$ for all $x \in [d, +\infty)$.

Figure 5.1 illustrates the graph of the possible shape of a fuzzy number M.

We now assume that all of the fuzzy parameters $C_{i1}, \ldots, C_{in}$, $A_{j1}, \ldots, A_{jn}$, and B_j in the MOLP-FP are fuzzy numbers whose membership functions are denoted by $\mu_{C_{i1}}(c_{i1})$, $\ldots$, $\mu_{C_{in}}(c_{in})$, $\mu_{A_{j1}}(a_{j1})$, $\ldots$, $\mu_{A_{jn}}(a_{jn})$, and $\mu_{B_j}(b_j)$, respectively. For simplicity in notation, define the following vectors:

$$c = (c_1, ..., c_k), \quad a = (a_1, ..., a_m), \quad b = (b_1, ..., b_m),$$

$$C = (C_1, ..., C_k), \quad A = (A_1, ..., A_m), \quad B = (B_1, ..., B_m).$$

Observing that the MOLP-FP involves fuzzy numbers both in the objective functions and the constraints, it is evident that the notion of Pareto optimality defined for the MOLP cannot be applied directly. Thus, it seems essential to extend the notion of usual Pareto optimality in some sense. For that purpose, we first introduce the α-level set of the fuzzy numbers A_{jr}, B_j, and C_{ir}.

Definition 5.2 (α-level set)

The α-level set of the fuzzy numbers A_{jr}, B_j, and C_{ir} is defined as the ordinary set $(A, B, C)_\alpha$ for which the degree of their membership functions exceeds the level α:

$$(A, B, C)_\alpha = \{\, (a, b, c) \mid \ \mu_{A_{jr}}(a_{jr}) \geq \alpha, \ \mu_{B_j}(b_j) \geq \alpha, \ \mu_{C_{ir}}(c_{ir}) \geq \alpha, \\ i = 1, ..., k, \ j = 1, ..., m, \ r = 1, ..., n\}. \quad (5.3)$$

The concept of the α-level set is illustrated in Figure 5.2.

From this figure, observe that the α-level sets have the following property:

$$\alpha_1 \leq \alpha_2 \Leftrightarrow (A, B, C)_{\alpha_1} \supseteq (A, B, C)_{\alpha_2}.$$

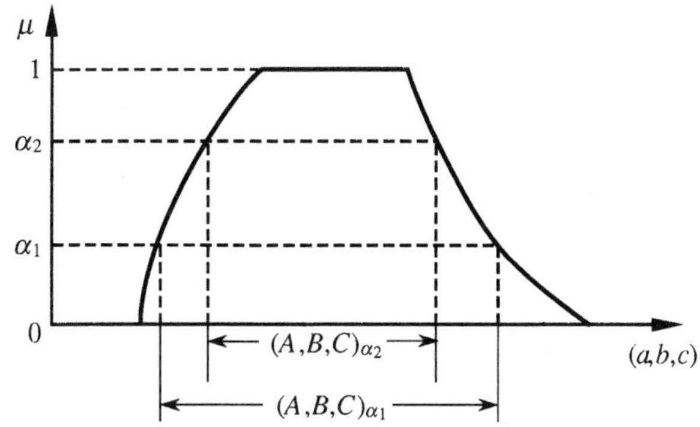

Figure 5.2 α-level set

Now suppose that the decision maker (DM) decides that the degree of all of the membership functions of the fuzzy numbers involved in the MOLP-FP should be greater than or equal to some value α. Then for such a degree α, the MOLP-FP can be interpreted as the following nonfuzzy multiobjective linear programming (MOLP-FP(a, b, c)) problem which depends on the coefficient vector $(a, b, c) \in (A, B, C)_\alpha$:

$$\left.\begin{array}{ll} \text{minimize} & (c_1x, c_2x, ..., c_kx)^T \\ \text{subject to} & x \in X(a,b) \triangleq \{x \in R^n \mid a_jx \leq b_j,\ j = 1, ..., m\ ;\ x \geq 0\}. \end{array}\right\} \quad (5.4)$$

Observe that there exists an infinite number of such MOLP-FP(a, b) depending on the coefficient vector $(a, b, c) \in (A, B, C)_\alpha$, and the values of (a, b, c) are arbitrary for any $(a, b, c) \in (A, B, C)_\alpha$ in the sense that the degree of all of the membership functions for the fuzzy numbers in the MOLP-FP exceeds the level α. However, if possible, it would be desirable for the DM to choose $(a, b, c) \in (A, B, C)_\alpha$ in the MOLP-FP(a, b) to minimize the objective functions under the constraints. From such a point of view, for a certain degree α, it seems to be quite natural to have the MOLP-FP as the following nonfuzzy α-multiobjective linear programming (α-MOLP) problem:

$$\left.\begin{array}{ll} \text{minimize} & (c_1x, c_2x, ..., c_kx)^T \\ \text{subject to} & x \in X(a,b) \triangleq \{x \in R^n \mid a_jx \leq b_j,\ j = 1, ..., m\ ;\ x \geq 0\} \\ & (a, b, c) \in (A, B, C)_\alpha. \end{array}\right\} \quad (5.5)$$

It should be emphasized here that, in the α-MOLP, the parameters (a, b, c) are treated as decision variables rather than constants.

On the basis of the α-level sets of the fuzzy numbers, we can introduce the concept of an α-Pareto optimal solution to the α-MOLP as a natural extension of the Pareto optimality concept for the MOLP.

Definition 5.3 (α-Pareto optimal solution)

$x^* \in X(a^*, b^*)$ is said to be an α-Pareto optimal solution to the α-MOLP if and only if there does not exist another $x \in X(a, b)$, $(a, b, c) \in (A, B, C)_\alpha$ such that $c_i x \leq c_i^* x^*$, $i = 1, ..., k$, with strict inequality holding for at least one i, where the corresponding values of parameters (a^*, b^*, c^*) are called α-level optimal parameters.

Observe that α-Pareto optimal solutions and α-level optimal parameters can be obtained through a direct application of the usual scalarizing methods for generating Pareto optimal solutions by regarding the decision variables in the α-MOLP as (x, a, b, c). However, as can be immediately understood from Definition 5.3, in general, α-Pareto optimal solutions to the α-MOLP consist of an infinite number of points and some kinds of subjective judgment should be added to the quantitative analyses by the DM. Namely, the DM must select a compromise or satisficing solution from an α-Pareto optimal solution set based on a subjective value judgment. Therefore, in the following, we first present an interactive programming approach to the MOLP-FP. Then we also present an interactive fuzzy programming approach to the MOLP-FP by incorporating the fuzzy goals of the DM for each of the objective functions of the α-MOLP.

5.3 Interactive programming

5.3.1 Minimax problems

To generate a candidate for the satisficing solution which is also α-Pareto optimal, in our interactive decision-making method, not considering the fuzzy goals of the DM for each of the objective functions of the α-MOLP, the DM is asked to specify the degree α of the α-level set and the reference levels of achievement of the objective functions, called reference levels. As discussed in Chapter 3, observe that the idea of the reference levels or the reference point first appeared in Wierzbicki (1979a). To be more explicit, for the DM's degree α and reference levels $\bar{z}_i$, $i = 1, ..., k$, the corresponding α-Pareto optimal solution, which is, in the minimax sense, nearest to the requirement or better than that if the reference levels are attainable, is obtained by solving the following minimax problem in an objective function space:

$$\left.\begin{array}{ll} \text{minimize} & \max\limits_{i=1,\ldots,k} (c_i x - \bar{z}_i) \\ \text{subject to} & x \in X(a, b) \\ & (a, b, c) \in (A, B, C)_\alpha \end{array}\right\} \quad (5.6)$$

or equivalently

$$\min_{x,v,a,b,c} \quad v \quad (5.7)$$

$$\text{subject to} \quad c_i x - \bar{z}_i \leq v, \ i = 1, ..., k, \quad (5.8)$$

$$a_j x \leq b_j, \ j = 1, ..., m, \ x \geq 0, \quad (5.9)$$

$$(a, b, c) \in (A, B, C)_\alpha. \tag{5.10}$$

A graphical description of the minimax problem in an objective space is depicted in Figure 5.3.

In this formulation, however, constraints (5.8) and (5.9) are nonlinear because the parameters a, b, and c are treated as decision variables. To deal with such nonlinearities, we introduce the following set-valued functions $S_i(\cdot)$ and $T_j(\cdot,\cdot)$:

$$S_i(c_i) = \{(x, v) \mid c_i x - \bar{z}_i \leq v, \ i = 1, ..., k\}, \tag{5.11}$$

$$T_j(a_j, b_j) = \{x \mid a_j x \leq b_j, \ j = 1, ..., m\}. \tag{5.12}$$

Then it can be easily verified that the following relations hold for $S_i(\cdot)$ and $T_j(\cdot,\cdot)$ when $x \geq 0$.

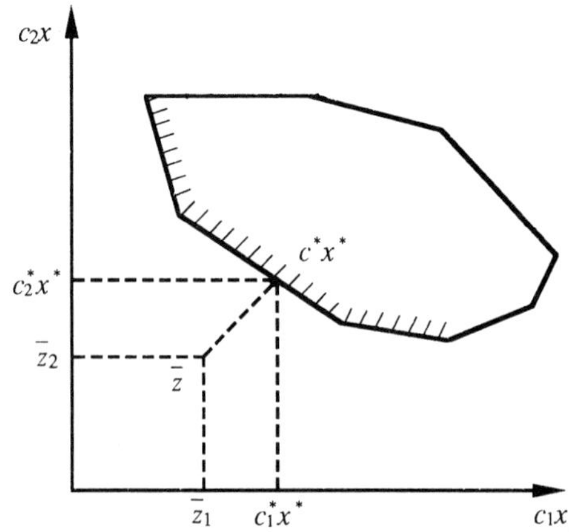

Figure 5.3 Minimax problem in objective space

Proposition 5.1

(1) *If* $c_i^1 \leq c_i^2$, *then* $S_i(c_i^1) \supseteq S_i(c_i^2)$.
(2) *If* $a_j^1 \leq a_j^2$, *then* $T_j(a_j^1, b_j) \supseteq T_j(a_j^2, b_j)$.
(3) *If* $b_j^1 \leq b_j^2$, *then* $T_j(a_j, b_j^1) \subseteq T_j(a_j, b_j^2)$.

Now from the properties of the α-level set for the vectors of fuzzy numbers C_i, A_j, and fuzzy numbers B_j, it should be noted here that the feasible regions for c_i, a_j, and b_j can be denoted respectively by the closed intervals $[c_{i\alpha}^L, c_{i\alpha}^R]$, $[a_{j\alpha}^L, a_{j\alpha}^R]$, and $[b_{j\alpha}^L, b_{j\alpha}^R]$ shown in Figure 5.4.

Therefore, through the use of Proposition 5.1, we can obtain an optimal solution of the problem (5.7) to (5.10) by solving the following linear programming problem:

$$\left.\begin{array}{ll} \text{minimize} & v \\ \text{subject to} & c_{i\alpha}^L x - z_i \leq v, \ i = 1, ..., k, \\ & a_{j\alpha}^L x \leq b_{j\alpha}^R, \ j = 1, ..., m, \ x \geq 0. \end{array}\right\} \tag{5.13}$$

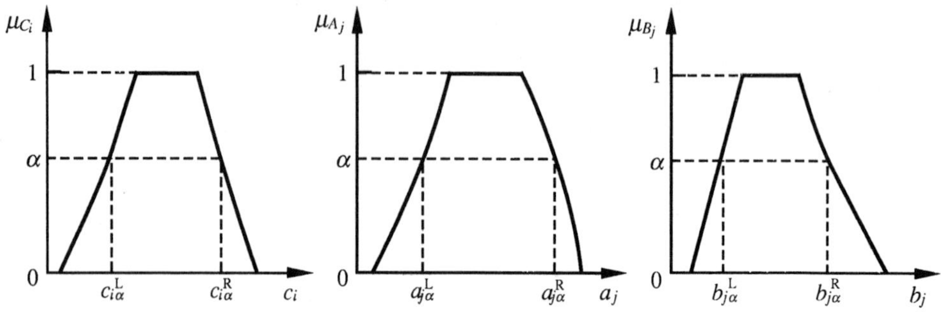

Figure 5.4 Feasible regions for c_i, a_j, and b_j

It is important to realize here that this problem is no more nonlinear but an ordinary linear programming problem, and consequently, it is easy to solve it by making use of the simplex method of linear programming.

Also, observe that the minimax problem is simply used as a means of generating an α-Pareto optimal solution, and if the DM is not satisfied with the current α-Pareto optimal solution, it is possible to improve the solution by updating the reference levels and/or the degree α.

The relationships between the optimal solutions to (5.13) and the α-Pareto optimal concept of the α-MOLP can be characterized by the following theorems.

Theorem 5.1

If x^* is a unique optimal solution to (5.13), then x^* is an α-Pareto optimal solution to the α-MOLP.

Proof

Assume that x^* is not an α-Pareto optimal solution to the α-MOLP, then there exist $x \in X(a,b)$ and $(a,b,c) \in (A,B,C)_\alpha$ such that $c_i x \leq c_{i\alpha}^L x^*, i = 1, ..., k$, with strict inequality holding for at least one i. Then it holds that

$$\max_{i=1,...,k}(c_{i\alpha}^L x - \bar{z}_i) \leq \max_{i=1,...,k}(c_i x - \bar{z}_i) \leq \max_{i=1,...,k}(c_{i\alpha}^L x^* - \bar{z}_i),$$

which contradicts the fact that x^* is a unique optimal solution to (5.13). □

Theorem 5.2

If x^* is an α-Pareto optimal solution and (a^*, b^*, c^*) is an α-level optimal parameter to the α-MOLP, then x^* is an optimal solution to (5.13) for some $\bar{z} = (\bar{z}_1, ..., \bar{z}_k)$.

Proof

Assume that x^* is not an optimal solution to (5.13) for any $\bar{z}$ satisfying

$$c_1^* x^* - \bar{z}_1 = \cdots = c_k^* x^* - \bar{z}_k = v^*.$$

Then there exist $x \in X(a,b)$ and $(a,b,c) \in (A,B,C)_\alpha$ such that

$$c_i^* x^* - \bar{z}_i > c_i x - \bar{z}_i \geq c_{i\alpha}^L x - \bar{z}_i.$$

This implies that $c_{i\alpha}^L x < c_i^* x^*$ which contradicts the fact that x^* is an α-Pareto optimal solution to the α-MOLP. □

It should be noted here that for generating α-Pareto optimal solutions using Theorem 5.1 , uniqueness of solution must be verified. In general, however, it is not easy to check numerically whether an optimal solution to (5.13) is unique or not.

Consequently, in order to perform a numerical test of α-Pareto optimality of a current optimal solution x^*, we formulate and solve the following linear programming problem:

$$\left.\begin{array}{ll} \text{maximize} & \sum_{i=1}^{k} \varepsilon_i \\ \text{subject to} & c_{i\alpha}^L x + \varepsilon_i = c_{i\alpha}^L x^*, \ \varepsilon_i \geq 0, \ i = 1, ..., k, \\ & a_{j\alpha}^L x \leq b_{j\alpha}^R, \ j = 1, ..., m, \ x \geq 0. \end{array}\right\} \tag{5.14}$$

Let $\bar{x}$ and $\bar{\varepsilon}$ be an optimal solution to (5.14). If all $\bar{\varepsilon}_i = 0$, then x^* is an α-Pareto optimal solution. If at least one $\varepsilon_i > 0$, as discussed in Chapter 3, it can be easily shown that $\bar{x}$ is an α-Pareto optimal solution.

5.3.2 Interactive algorithm

Now, given the α-Pareto optimal solution for the degree α and the reference levels specified by the DM by solving the corresponding minimax problem, the DM must either be satisfied with the current α-Pareto optimal solution or act on this solution by updating the reference levels and/or the degree α. To help the DM express a degree of preference, trade-off information between a standing objective function and each of the other objective functions as well as between the degree α and the objective functions is very useful. Such trade-off information is easily obtainable since it is closely related to the simplex multipliers of the problem (5.13).

To derive the trade-off information, we define the Lagrangian function L for the problem (5.13) as follows:

$$L = v + \sum_{i=1}^{k} \pi_i (c_{i\alpha}^L x - \bar{z}_i - v) + \sum_{j=1}^{m} \lambda_j (a_{j\alpha}^L x - b_{j\alpha}^R). \tag{5.15}$$

If all $\pi_i > 0$ for each i, then by extending the results in Haimes and Chankong (1979), it can be proved (for details, see the Appendix) that the following expression holds:

$$-\frac{\partial (c_{1\alpha}^L x)}{\partial (c_{i\alpha}^L x)} = \frac{\pi_i}{\pi_1}, \ i = 2, \ldots, k. \tag{5.16}$$

Regarding a trade-off rate between $c_i x$ and α, the following relation holds based on the sensitivity theorem (for details, see, e.g., Luenberger 1973 or Fiacco 1983):

$$\frac{\partial(c_{i\alpha}^L x)}{\partial \alpha} = \sum_{i=1}^{k} \pi_i \frac{\partial(c_{i\alpha}^L)}{\partial \alpha} x + \sum_{j=1}^{m} \lambda_j \left\{ \frac{\partial(a_{j\alpha}^L)}{\partial \alpha} x - \frac{\partial(b_{j\alpha}^R)}{\partial \alpha} \right\}, \ i = 1, ..., k. \tag{5.17}$$

It should be noted here that to obtain the trade-off information from (5.16) or (5.17), all the constraints $c_{i\alpha}^L x - \bar{z}_i \leq v$, $i = 1, ..., k$, of the problem (5.13) must be active for the current optimal solution. Therefore, if there are inactive constraints, it is necessary to replace $\bar{z}_i$ for inactive constraints by $c_{i\alpha}^L x^* - v^*$ and solve the corresponding minimax problem (5.13) for obtaining the simplex multipliers.

We can now construct the interactive algorithm to derive the satisficing solution of the DM from the α-Pareto optimal solution set. The steps marked with an asterisk involve interaction with the DM.

Interactive multiobjective linear programming with fuzzy parameters

Step 0: Individual minimum and maximum

Calculate the individual minimum and maximum of each objective function under the given constraints for $\alpha = 0$ and $\alpha = 1$.

Step 1*: Initialization

Ask the DM to select the initial value of $\alpha\,(0 \leq \alpha \leq 1)$ and the initial reference levels $\bar{z}_i$, $i = 1, ..., k$.

Step 2: α-Pareto optimal solution

For the degree α and the reference levels specified by the DM, solve the corresponding minimax problem and perform the α-Pareto optimality test to obtain the α-Pareto optimal solution together with the trade-off rates between the objective functions and the degree α.

Step 3*: Termination or updating

The DM is supplied with the corresponding α-Pareto optimal solution and the trade-off rates between the objective functions and the degree α. If the DM is satisfied with the current objective function values of the α-Pareto optimal solution, stop. Otherwise, the DM must update the reference levels and/or the degree α by considering the current values of the objective functions and α together with the trade-off rates between the objective functions and the degree α and return to step 2.

Here it should be stressed to the DM that (1) any improvement of one objective function can be achieved only at the expense of at least one of the other objective functions for some fixed degree α and (2) the greater value of the degree α gives the worse values of the objective functions for some fixed reference levels.

5.3.3 Illustrative numerical example

To clarify the concept of α-Pareto optimality as well as the proposed method, consider the following three-objective linear programming problem with fuzzy parameters.

Example 5.1

$$\begin{array}{lrcr} \text{minimize} & 2x_1 & + & C_{12}x_2 \\ \text{minimize} & -3x_1 & + & C_{22}x_2 \\ \text{minimize} & C_{31}x_1 & - & x_2 \\ \text{subject to} & x \in X & & \end{array}$$
$$\triangleq \{(x_1, x_2) \mid 3x_1 + x_2 \leq 12,\ x_1 + 2x_2 \leq 12, x_i \geq 0,\ i = 1, 2\},$$

where C_{12}, C_{22}, and C_{31} are fuzzy numbers whose membership functions are given by

$$\begin{aligned} \mu_{C_{12}}(c_{12}) &= \max(1 - 0.5\,|c_{12} - 4|,\ 0) \\ \mu_{C_{22}}(c_{22}) &= \max(1 - 2\,|c_{22} + 0.75|,\ 0) \\ \mu_{C_{31}}(c_{31}) &= \max(1 - |c_{31} - 2.5|,\ 0). \end{aligned}$$

Now, for illustrative purposes, assume that the hypothetical DM selects the initial value of the degree α to be 0.5 and the initial reference levels $(\bar{z}_1, \bar{z}_2, \bar{z}_3)$ to be $(9, -13, -3)$. Then the values of $(c_{12\alpha}^L, c_{22\alpha}^L, c_{31\alpha}^L)$ become $(3, -1, 2)$ as shown in Figure 5.5, and the corresponding α-Pareto optimal solution can be obtained by solving the following linear programming problem:

$$\begin{array}{lrcrcrl} \text{minimize} & v & & & & & \\ \text{subject to} & 2x_1 & + & 3x_2 & - & 11 & \leq v \\ & -3x_1 & - & x_2 & + & 11 & \leq v \\ & 2x_1 & - & x_2 & + & 1 & \leq v \\ & x \in X. & & & & & \end{array}$$

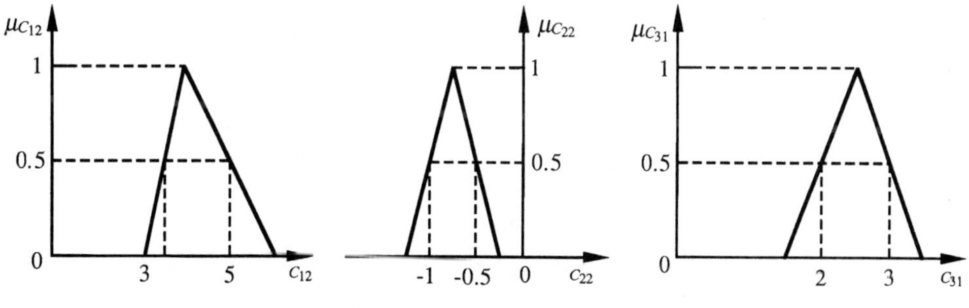

Figure 5.5 Fuzzy numbers C_{12}, C_{22}, and C_{31}

Solving this problem, we obtain an optimal solution $(x_1^*, x_2^*, v^*) = (2, 3, 4)$ and optimal values $(z_1^*, z_2^*, z_3^*) = (13, -9, 1)$. The simplex multipliers corresponding to the constraints for v of this problem become $(\pi_1^*, \pi_2^*, \pi_3^*) = (1/4, 2/5, 7/20)$. From (5.16), the trade-off rates among the objective functions become

$$-\left.\frac{\partial(c_{12}x)}{\partial(c_{22}x)}\right|_{\alpha=0.5} = \frac{\pi_2^*}{\pi_1^*} = \frac{8}{5}$$

$$-\left.\frac{\partial(c_{12}x)}{\partial(c_{31}x)}\right|_{\alpha=0.5} = \frac{\pi_3^*}{\pi_1^*} = \frac{7}{5}.$$

Concerning the trade-off rate between $c_i x$ and α, from (5.17), we have

$$\left.\frac{\partial(c_i x)}{\partial\alpha}\right|_{\alpha=0.5} = -(\pi_1^*, \pi_2^*, \pi_3^*)\,(2x_2^*, 0.5x_2^*, x_1^*)^T = \frac{14}{5}.$$

Observe that the DM can obtain a satisficing solution from an α-Pareto optimal solution set by updating the reference levels and/or the degree α on the basis of the current values of the objective functions and α together with the trade-off rates among the values of the objective functions and the degree α.

5.4 Interactive fuzzy programming

5.4.1 Fuzzy goals

As can be seen from the definition of α-Pareto optimality, in general, α-Pareto optimal solutions to the α-MOLP consist of an infinite number of points. The DM may be able to select a compromise or satisficing solution from an α-Pareto optimal solution set based on a subjective value judgment, making use of the interactive programming method presented in the previous section.

However, considering the imprecise nature of the DM's judgment, it is natural to assume that the DM may have imprecise or fuzzy goals for each of the objective functions in the α-MOLP. In a minimization problem, a goal stated by the DM may be to achieve "substantially less than or equal to some value p_i." This type of statement can be quantified by eliciting a corresponding membership function.

To elicit a membership function $\mu_i(c_i x)$ from the DM for each of the objective functions $c_i x, i = 1, \ldots, k$, in the α-MOLP, we first calculate the individual minimum and maximum of each objective function under the given constraints for $\alpha = 0$ and $\alpha = 1$. By taking account of the calculated individual minimum and maximum of each objective function for $\alpha = 0$ and $\alpha = 1$ together with the rate of increase of membership satisfaction, the DM may be able to determine a membership function $\mu_i(c_i x)$ in a subjective manner which is a strictly monotone decreasing function with respect to $c_i x$.

So far we have restricted ourselves to a minimization problem and consequently assumed that the DM has a fuzzy goal such as "$c_i x$ should be substantially less than or equal to p_i." In the fuzzy approaches, as discussed in Chapter

3, we can further treat a more general case where the DM has two types of fuzzy goals, namely, fuzzy goals expressed in words such as "$c_i x$ should be in the vicinity of r_i" (called fuzzy equal) as well as "$c_i x$ should be substantially less than or equal to p_i or greater than or equal to q_i" (called fuzzy min or fuzzy max). Such a generalized α-MOLP (Gα-MOLP) problem may now be expressed as

$$\left.\begin{array}{rll} \text{fuzzy min} & c_i x & i \in I_1 \\ \text{fuzzy max} & c_i x & i \in I_2 \\ \text{fuzzy equal} & c_i x & i \in I_3 \\ \text{subject to} & x \in X(a,b) & \\ & (a,b,c) \in (A,B,C)_\alpha & \end{array}\right\} \tag{5.18}$$

where $I_1 \cup I_2 \cup I_3 = \{1, 2, ..., k\}$, $I_i \cap I_j = \emptyset$, $i, j = 1, 2, 3$, $i \neq j$.

To elicit a membership function $\mu_i(c_i x)$ from the DM for a fuzzy goal like "$c_i x$ should be in the vicinity of r_i," it should be quite apparent that different functions can be utilized for both the left and right sides of r_i.

Concerning the membership functions of the Gα-MOLP, it is reasonable to assume that $\mu_i(c_i x)$, $i \in I_1$, and the right side functions of $\mu_i(c_i x)$, $i \in I_3$, are strictly monotone increasing and continuous functions with respect to $c_i x$. To be more explicit, each membership function $\mu_i(c_i x)$ of the Gα-MOLP for $i \in I_1$, $i \in I_2$, or $i \in I_3$ is defined, and its possible shape is depicted as follows:

(1) $i \in I_1$

$$\mu_i(c_i x) = \begin{cases} 1 \text{ or } \to 1 & \text{if } (c_i x)_R^1 \geq c_i x, \\ d_{iR}(c_i x) & \text{if } (c_i x)_R^1 \leq c_i x \leq (c_i x)_R^0, \\ 0 \text{ or } \to 0 & \text{if } c_i x \geq (c_i x)_R^0, \end{cases} \tag{5.19}$$

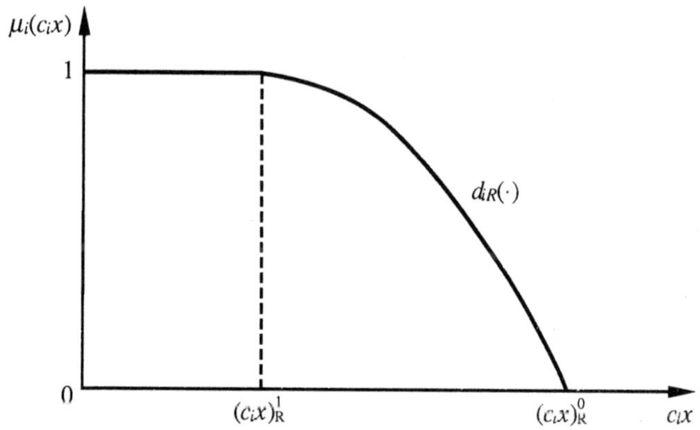

Figure 5.6 Fuzzy min membership function

(2) $i \in I_2$

$$\mu_i(c_ix) = \begin{cases} 0 \text{ or } \to 0 & \text{if } (c_ix)_L^0 \geq c_ix, \\ d_{iL}(c_ix) & \text{if } (c_ix)_L^0 \leq c_ix \leq (c_ix)_L^1, \\ 1 \text{ or } \to 1 & \text{if } c_ix \geq (c_ix)_L^1, \end{cases} \tag{5.20}$$

(3) $i \in I_3$

$$\mu_i(c_x) = \begin{cases} 0 \text{ or } \to 0 & \text{if } c_ix \leq (c_ix)_L^0, \\ d_{iL}(c_ix) & \text{if } (c_ix)_L^0 \leq c_ix \leq (c_ix)_L^1, \\ 1 & \text{if } (c_ix)_L^1 \leq c_ix \leq (c_ix)_R^1, \\ d_{iR}(c_x) & \text{if } (c_ix)_R^1 \leq c_ix \leq (c_ix)_R^0, \\ 0 \text{ or } \to 0 & \text{if } (c_ix)_R^0 \leq c_ix. \end{cases} \tag{5.21}$$

Here it is assumed that $d_{iR}(c_ix)$ is a strictly monotone decreasing continuous function with respect to c_ix and $d_{iL}(c_ix)$ is a strictly monotone increasing continuous function with respect to c_ix. Both may be linear or nonlinear. $(c_ix)_L^0$ and $(c_ix)_R^0$ are maximum values of unacceptable levels for c_ix, and $(c_ix)_L^1$ and $(c_ix)_R^1$ are minimum values of totally desirable levels for c_ix.

When a fuzzy equal is included in the fuzzy goals of the DM, it is desirable that c_ix should be as close to r_i as possible. Consequently, the notion of α-Pareto optimal solutions defined in terms of objective functions cannot be applied. For this reason, we introduce the concept of M-α-Pareto optimal solutions which is defined in terms of membership functions instead of objective functions, where M refers to membership.

Definition 5.4 (M-α-Pareto optimal solution)

$x^* \in X(a^*, b^*)$ is said to be an M-α-Pareto optimal solution to the Gα-MOLP if and only if there does not exist another $x \in X(a, b)$, $(a, b, c) \in (A, B, C)_\alpha$ such

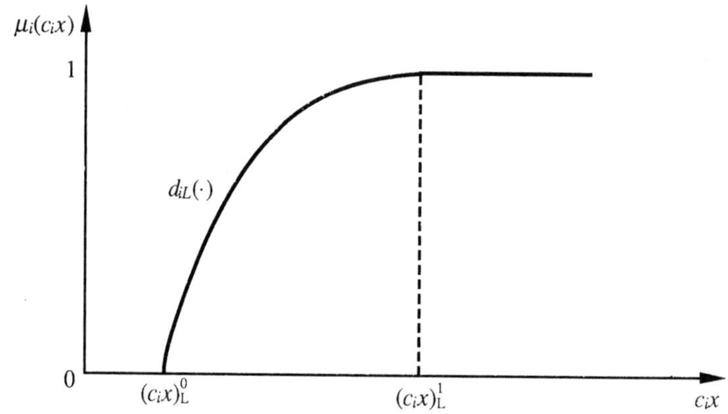

Figure 5.7 Fuzzy max membership function

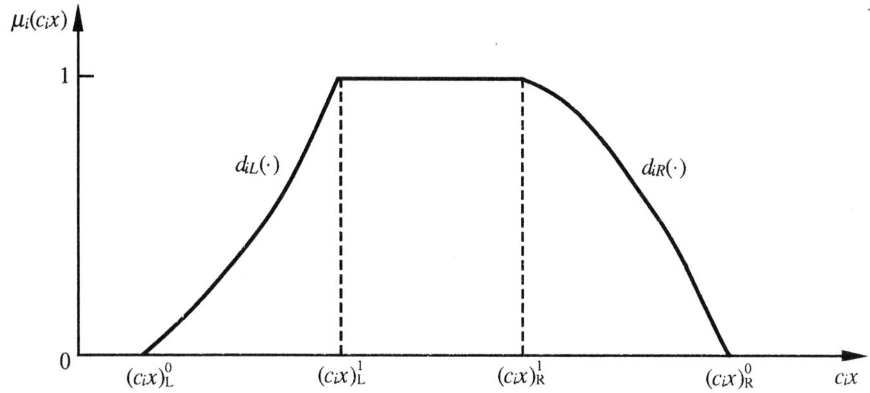

Figure 5.8 Fuzzy equal membership function

that $\mu_i(c_ix) \geq \mu_i(c_i^*x^*)$, $i = 1, \ldots, k$, *with strict inequality holding for at least one* i, *where the corresponding values of parameters* (a^*, b^*, c^*) *are called* α*-level optimal parameters.*

Observe that the concept of M-α-Pareto optimal solutions defined in terms of membership functions is a natural extension to that of α-Pareto optimal solutions defined in terms of objective functions when fuzzy equal is included in the fuzzy goals of the DM.

Having elicited the membership functions $\mu_i(c_ix)$, $i = 1, \ldots, k$, from the DM for each of the objective functions c_ix, $i = 1, \ldots, k$, if we introduce a general aggregation function

$$\mu_D(\mu_1(c_1x), \mu_2(c_2x), \ldots, \mu_k(c_kx), \alpha), \tag{5.22}$$

a general fuzzy α-multiobjective decision making problem (Fα-MODMP) can be defined by

$$\left.\begin{array}{ll} \text{maximize} & \mu_D(\mu_1(c_1x), \mu_2(c_2x), \ldots, \mu_k(c_kx), \alpha) \\ \text{subject to} & (x, a, b, c) \in P(\alpha), \;\; \alpha \in [0, 1], \end{array}\right\} \tag{5.23}$$

where $P(\alpha)$ is the set of M-α-Pareto optimal solutions and corresponding α-level optimal parameters to the Gα-MOLP.

Probably the most crucial problem in the Fα-MODMP is the identification of an appropriate aggregation function which well represents the human decision makers' fuzzy preferences. If $\mu_D(\cdot)$ can be explicitly identified, then the Fα-MODMP reduces to a standard mathematical programming problem. However, this happens rarely, and as an alternative approach, it becomes evident that an interaction with the DM is necessary.

To do so, assume that

(1) The fuzzy goals of the DM can be quantified by eliciting the corresponding membership functions through the interaction with the DM.

(2) $\mu_D(\cdot)$ exists and is known only implicitly to the DM, which means the DM cannot specify the entire form of $\mu_D(\cdot)$ but can provide local information concerning a preference. Moreover, it is strictly increasing and continuous with respect to $\mu_i(\cdot)$ and α.

5.4.2 Minimax problems

To generate a candidate for the satisficing solution, which is also M-α-Pareto optimal, in our decision making method, the DM is asked to specify the degree α of the α-level set and the reference membership values. Observe that the idea of the reference membership values, which first appeared in Sakawa, Yumine, and Yano (1984), can be viewed as an obvious extension of the idea of the reference point in Wierzbicki (1979a).

Once the DM's degree α and reference membership values $\bar{\mu}_i$, $i = 1, \ldots, k$, are specified, the corresponding M-α-Pareto optimal solution, which is, in the minimax sense, nearest to the requirement or better than that if the reference levels are attainable, is obtained by solving the following minimax problem:

$$\min_{\substack{x \in X(a,b) \\ (a,b,c) \in (A,B,C)_\alpha}} \max_{i=1,\ldots,k} (\bar{\mu}_i - \mu_i(c_i x)) \tag{5.24}$$

or equivalently

$$\left.\begin{array}{ll} \text{minimize} & v \\ \text{subject to} & \bar{\mu}_i - \mu_i(c_i x) \leq v, \; i = 1, \ldots, k, \\ & a_j x \leq b_j, \; j = 1, \ldots, m, \; x \geq 0, \\ & (a, b, c) \in (A, B, C)_\alpha. \end{array}\right\} \tag{5.25}$$

Figure 5.9 illustrates a graphical description of the minimax problem.

However, with the strictly monotone decreasing or increasing membership function given by (5.19) to (5.21), which may be nonlinear, the resulting problem becomes a nonlinear programming problem. To solve the formulated problem on the basis of the linear programming method, we first convert each constraint $\bar{\mu}_i - \mu_i(c_i x) \leq v$, $i = 1, \ldots, k$, of the minimax problem (5.25) into the following form using the strictly monotone property of $d_{iL}(\cdot)$ and $d_{iR}(\cdot)$:

$$c_i x \leq d_{iR}^{-1}(\bar{\mu}_i - v), \; i \in I_1 \cup I_3 \tag{5.26}$$

$$c_i x \geq d_{iL}^{-1}(\bar{\mu}_i - v), \; i \in I_2 \cup I_3 \tag{5.27}$$

Now we introduce the following set-valued functions $S_{iR}(\cdot)$, $S_{iL}(\cdot)$, and $T_j(\cdot, \cdot)$:

$$S_{iR}(c_i) = \{(x, v) \mid c_i x \leq d_{iR}^{-1}(\bar{\mu}_i - v)\}, \quad i \in I_1 \cup I_3 \tag{5.28}$$

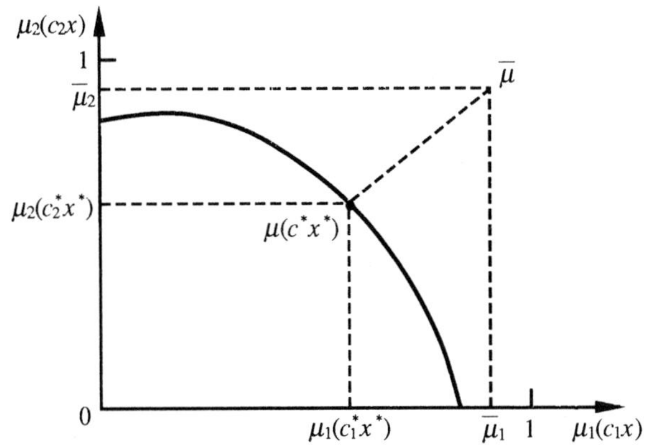

Figure 5.9 Minimax problem in membership space

$$S_{iL}(c_i) = \{(x, v) \mid c_i x \geq d_{iL}^{-1}(\bar{\mu}_i - v)\}, \quad i \in I_2 \cup I_3 \tag{5.29}$$

$$T_j(a_j, b_j) = \{x \mid a_j x \leq b_j\}, \quad j = 1, ..., k. \tag{5.30}$$

It can be verified that the following relations hold for $S_{iR}(\cdot), S_{iL}(\cdot)$, and $T_j(\cdot, \cdot)$ when $x \geq 0$.

Proposition 5.2

(1) *If* $c_i^1 \leq c_i^2$, *then* $S_{iR}(c_i^1) \supseteq S_{iR}(c_i^2)$ *and* $S_{iL}(c_i^1) \subseteq S_{iL}(c_i^2)$.
(2) *If* $a_j^1 \leq a_j^2$, *then* $T_j(a_j^1, b_j) \supseteq T_j(a_j^2, b_j)$.
(3) *If* $b_j^1 \leq b_j^2$, *then* $T_j(a_j, b_j^1) \subseteq T_j(a_j, b_j^2)$.

Using the properties of the α-level sets for the vectors of the fuzzy numbers C_i, A_j and the fuzzy numbers B_j, the feasible regions for c_i, a_j, and b_j can be denoted respectively by the closed intervals $[c_{i\alpha}^L, c_{i\alpha}^R]$, $[a_{j\alpha}^L, a_{j\alpha}^R]$, and $[b_{j\alpha}^L, b_{j\alpha}^R]$ as shown in Figure 5.10.

Consequently, using of the results in Proposition 5.2, we can obtain an optimal

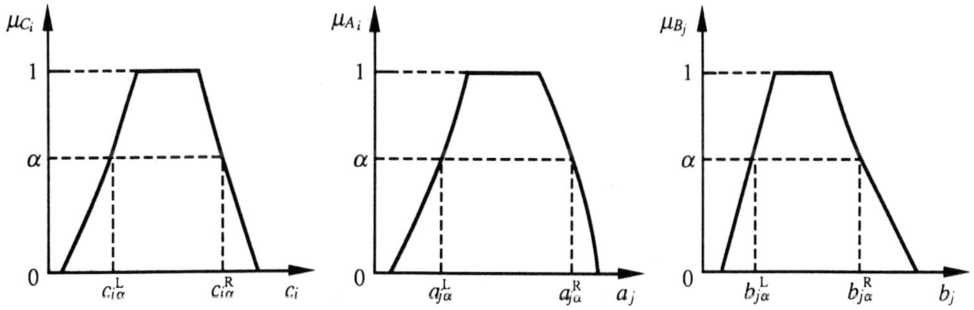

Figure 5.10 Feasible region for c_i, a_j, and b_j

solution to (5.25) by solving the following problem:

$$\left.\begin{array}{ll} \text{minimize} & v \\ \text{subject to} & c^L_{i\alpha}x \leq d^{-1}_{iR}(\bar{\mu}_i - v),\ i \in I_1 \cup I_3 \\ & c^R_{i\alpha}x \geq d^{-1}_{iL}(\bar{\mu}_i - v),\ i \in I_2 \cup I_3 \\ & a^L_{j\alpha} \leq b^R_{j\alpha},\ j = 1, ..., m,\ x \geq 0. \end{array}\right\} \quad (5.31)$$

It is important to note here that this formulation, if the value of v is fixed, can be reduced to a set of linear inequalities. Obtaining the optimal solution v^* to the above problem is equivalent to determining the minimum value of v so that there exists an admissible set satisfying the constraints of (5.31). Since v satisfies $\bar{\mu}_{\max} - 1 \leq v \leq \bar{\mu}_{\max}$, where $\bar{\mu}_{\max}$ denotes the maximum value of $\bar{\mu}_i$, $i = 1, ..., k$, we have the following method for solving this problem by combined use of the bisection method and the simplex method of linear programming.

Step 1: Set $v = \bar{\mu}_{\max}$ and test whether an admissible set satisfying the constraints of (5.31) exists or not by making use of phase one of the simplex method. If an admissible set exists, proceed. Otherwise, the DM must reassess the membership function.

Step 2: Set $v = \bar{\mu}_{\max} - 1$ and test whether an admissible set satisfying the constraints of (5.31) exists or not using phase one of the simplex method. If an admissible set exists, set $v^* = \bar{\mu}_{\max} - 1$. Otherwise, go to the next step since the minimum v which satisfies the constraints of (5.31) exists between $\bar{\mu}_{\max} - 1$ and $\bar{\mu}_{\max}$.

Step 3: For the initial value of $v = \bar{\mu}_{\max} - 0.5$, update the value of v using the bisection method as follows:

$$\left\{\begin{array}{ll} v_{n+1} = v_n - 1/2^{n+1} & \text{if an admissible set exists for } v_n, \\ v_{n+1} = v_n + 1/2^{n+1} & \text{if no admissible set exists for } v_n. \end{array}\right.$$

For each v_n, $n = 1, 2, ...$, test whether an admissible set of (5.31) exists or not using the sensitivity analysis technique for the changes in the right-hand side of the simplex method and determine the minimum value of v satisfying the constraints of (5.31).

In this way, we can determine the optimal solution v^*. Then the DM selects an appropriate standing objective from among the objectives $c_i x$, $i = 1, ..., k$. For notational convenience in the following without loss of generality, let it be $c_1 x$ and $1 \in I_1$. Then the following linear programming problem is solved for $v = v^*$:

$$\left.\begin{array}{ll} \text{minimize} & c^L_{1\alpha}x \\ \text{subject to} & c^L_{i\alpha}x \leq d^{-1}_{iR}(\bar{\mu}_i - v^*),\ i \in I_1 \cup I_3, \\ & c^R_{i\alpha}x \geq d^{-1}_{iL}(\bar{\mu}_i - v^*),\ i \in I_2 \cup I_3, \\ & a^L_{j\alpha}x \leq b^R_{j\alpha},\ j = 1, ..., m,\ x \geq 0. \end{array}\right\} \quad (5.32)$$

For convenience in our subsequent discussion, we assume that the optimal solution x^* to (5.32) satisfies the following conditions:

$$c_{i\alpha}^L x^* = d_{iR}^{-1}(\bar{\mu}_i - v^*),\ i \in I_1 \cup I_{3R}, \tag{5.33}$$

$$c_{i\alpha}^R x^* = d_{iL}^{-1}(\bar{\mu}_i - v^*),\ i \in I_2 \cup I_{3L}. \tag{5.34}$$

where $I_3 = I_{3L} \cup I_{3R}$ and $I_{3L} \cap I_{3R} = \emptyset$.

It is interesting to note that $c_{i\alpha}^L, i \in I_1 \cup I_{3R}$, $c_{i\alpha}^R, i \in I_2 \cup I_{3L}$, and $a_{j\alpha}^L$, $b_{j\alpha}^R$, $j = 1, ..., m$, are α-level optimal parameters for any M-α-Pareto optimal solution.

The relationships between the optimal solutions to (5.31) and the M-α-Pareto optimal concept of the Gα-MOLP can be characterized by the following theorems.

Theorem 5.3

If x^* is a unique optimal solution to (5.31), then x^* is an M-α-Pareto optimal solution to the Gα-MOLP.

Proof

Assume that x^* is not an M-α-Pareto optimal solution to the Gα-MOLP. Then, since $c_{i\alpha}^L, i \in I_1 \cup I_{3R}$, $c_{i\alpha}^R, i \in I_2 \cup I_{3L}$, and $a_{j\alpha}^L$, $b_{j\alpha}^R$, $j = 1, ..., m$, are α-level optimal parameters to the Gα-MOLP, there exist $x \in X(a,b)$ and $(a,b,c) \in (A,B,C)_\alpha$ such that $\mu_i(c_i x) \geq \mu_i(c_{i\alpha}^L x^*)$, $i \in I_1 \cup I_{3R}$, $\mu_i(c_i x) \geq \mu_i(c_{i\alpha}^R x^*)$, $i \in I_2 \cup I_{3L}$, with strict inequality holding for at least one i. Then it holds that

$$\begin{aligned}
\max_{i \in I_1 \cup I_{3R}} (\bar{\mu}_i - \mu_i(c_{i\alpha}^L x^*)) &\geq \max_{i \in I_1 \cup I_{3R}} (\bar{\mu}_i - \mu_i(c_i x)) \geq \max_{i \in I_1 \cup I_{3R}} (\bar{\mu}_i - \mu_i(c_{i\alpha}^L x)),\\
\max_{i \in I_2 \cup I_{3L}} (\bar{\mu}_i - \mu_i(c_{i\alpha}^R x^*)) &\geq \max_{i \in I_2 \cup I_{3L}} (\bar{\mu}_i - \mu_i(c_i x)) \geq \max_{i \in I_2 \cup I_{3L}} (\bar{\mu}_i - \mu_i(c_{i\alpha}^R x)),
\end{aligned}$$

which contradicts the fact that x^* is a unique optimal solution to (5.31). □

Theorem 5.4

If x^* is an M-α-Pareto optimal solution to the Gα-MOLP, then x^* is an optimal solution to (5.31) for some $\bar{\mu} = (\bar{\mu}_1, ..., \bar{\mu}_k)$.

Proof

Assume that (x^*, v^*) is not an optimal solution to (5.31) for any $\bar{\mu}$ satisfying

$$\bar{\mu}_i - \mu_i(c_{i\alpha}^L x^*) = v^*, \quad i \in I_1 \cup I_{3R}$$

$$\bar{\mu}_i - \mu_i(c_{i\alpha}^R x^*) = v^*, \quad i \in I_2 \cup I_{3L}$$

Then there exists $x \in X(a,b)$ such that

$$\bar{\mu}_i - \mu_i(c_{i\alpha}^L x) < \bar{\mu}_i - \mu_i(c_{i\alpha}^L x^*), \quad i \in I_1 \cup I_{3R}$$

$$\bar{\mu}_i - \mu_i(c_{i\alpha}^R x) < \bar{\mu}_i - \mu_i(c_{i\alpha}^R x^*), \quad i \in I_2 \cup I_{3L}$$

This implies that $\mu_i(c_{i\alpha}^L x) > \mu_i(c_{i\alpha}^L x^*)$, $i \in I_1 \cup I_{3R}$, $\mu_i(c_{i\alpha}^R x) > \mu_i(c_{i\alpha}^R x^*)$, $i \in I_2 \cup I_{3L}$, which contradicts the fact that x^* is an M-α-Pareto optimal solution to the Gα-MOLP. □

It must be observed here that for generating M-α-Pareto optimal solutions using Theorem 5.3, uniqueness of solution must be verified. In the ad hoc numeral approach, however, to test the M-α-Pareto optimality of a current optimal solution x^*, we formulate and solve the following linear programming problem:

$$\left.\begin{array}{ll} \text{maximize} & \sum_{i=1}^{k} \varepsilon_i \\ \text{subject to} & c_{i\alpha}^L x + \varepsilon_i = c_{i\alpha}^L x^*,\ \varepsilon_i \geq 0,\ i \in I_1 \cup I_{3R} \\ & c_{i\alpha}^R x - \varepsilon_i = c_{i\alpha}^R x^*,\ \varepsilon_i \geq 0,\ i \in I_2 \cup I_{3L} \\ & a_{j\alpha}^L x \leq b_{j\alpha}^R,\ j = 1, ..., m,\ x \geq 0. \end{array}\right\} \quad (5.35)$$

Let $\bar{x}$ and $\bar{\varepsilon}_i$ be an optimal solution to this problem. If all $\bar{\varepsilon}_i = 0$, then x^* is an M-α-Pareto optimal solution. If at least one $\bar{\varepsilon}_i > 0$, as discussed in Chapter 3, it can be easily shown that $\bar{x}$ is an M-α-Pareto optimal solution.

5.4.3 Interactive algorithm with fuzzy goals

Now given the M-α-Pareto optimal solution for the degree α and the reference membership values specified by the DM by solving the corresponding minimax problem, the DM must either be satisfied with the current M-α-Pareto optimal solution and α or update the reference membership values and/or the degree α. To help the DM express a degree of preference, trade-off information between a standing membership function and each of the other membership functions as well as between the degree α and the membership functions is very useful. Such trade-off information is easily obtainable since it is closely related to the simplex multipliers of the problem (5.32).

To derive the trade-off information, define the following Lagrangian function L corresponding to problem (5.32):

$$\begin{aligned} L = \ & c_{1\alpha}^L x + \sum_{i \in I_1 \cup I_{3R}} \pi_{iR}\{c_{i\alpha}^L x - d_{iR}^{-1}(\bar{\mu}_i - v^*)\} \\ & + \sum_{i \in I_2 \cup I_{3L}} \pi_{iL}\{d_{iL}^{-1}(\bar{\mu}_i - v^*) - c_{i\alpha}^R x\} + \sum_{j=1}^{m} \lambda_j (a_{j\alpha}^L x - b_{j\alpha}^R) \end{aligned} \quad (5.36)$$

where π_{iL}, π_{iR}, and λ_j are simplex multipliers corresponding to the constraints of (5.32).

Here we assume that problem (5.32) has a unique and nondegenerate optimal solution satisfying the following conditions:

(1) $\pi_{iR} > 0,\ i \in I_1 \cup I_{3R},\ i \neq 1$
(2) $\pi_{iL} > 0,\ i \in I_2 \cup I_{3L}$.

Then by using the results in Haimes and Chankong (1979), the following expression holds:

$$-\frac{\partial(c_{1\alpha}^L x)}{\partial(c_{i\alpha}^L x)} = \pi_{iR},\ i \in I_1 \cup I_{3R},\ i \neq 1 \quad (5.37)$$

$$-\frac{\partial(c_{1\alpha}^{L}x)}{\partial(c_{i\alpha}^{R}x)} = -\pi_{iL},\ i \in I_2 \cup I_{3L}. \tag{5.38}$$

Furthermore, using the strictly monotone decreasing or increasing property of $d_{iR}(\cdot)$ or $d_{iL}(\cdot)$ together with the chain rule, if $d_{iR}(\cdot)$ and $d_{iL}(\cdot)$ are differentiable at the optimal solution to (5.32), it holds that

$$-\frac{\partial\mu_1(c_{1\alpha}^{L}x)}{\partial\mu_i(c_{i\alpha}^{L}x)} = \frac{d'_{1R}(c_{1\alpha}^{L}x)}{d'_{iR}(c_{i\alpha}^{L}x)}\pi_{iR},\ i \in I_1 \cup I_{3R},\ i \neq 1 \tag{5.39}$$

$$-\frac{\partial\mu_1(c_{1\alpha}^{L}x)}{\partial\mu_i(c_{i\alpha}^{R}x)} = \frac{d'_{1R}(c_{1\alpha}^{L}x)}{d'_{iL}(c_{i\alpha}^{R}x)}\pi_{iL},\ i \in I_2 \cup I_{3L} \tag{5.40}$$

where $d'_{iR}(\cdot)$ and $d'_{iL}(\cdot)$ denote the differential coefficients of $d_{iR}(\cdot)$ and $d_{iL}(\cdot)$, respectively.

Regarding a trade-off rate between $\mu_1(c_{1\alpha}^{L}x)$ and α, the following relation holds based on the sensitivity theorem (for details, see, e.g., Luenberger 1973 or Fiacco 1983):

$$\begin{aligned}\frac{\partial\mu_1(c_{1\alpha}^{L}x)}{\partial\alpha} &= d'_{1R}(c_{1\alpha}^{L}x)\left\{\frac{\partial(c_{1\alpha}^{L})}{\partial\alpha}x + \sum_{i\in I_1\cup I_{3R}}\pi_{iR}\frac{\partial(c_{i\alpha}^{L})}{\partial\alpha}x\right.\\ &\left. - \sum_{i\in I_2\cup I_{3L}}\pi_{iL}\frac{\partial(c_{i\alpha}^{R})}{\partial\alpha}x + \sum_{j=1}^{m}\lambda_j\left\{\frac{\partial(a_{j\alpha}^{L})}{\partial\alpha}x - \frac{\partial(b_{j\alpha}^{R})}{\partial\alpha}\right\}\right\}.\end{aligned} \tag{5.41}$$

It should be noted that to obtain the trade-off rate information from (5.39) and (5.40), all the constraints of problem (5.32) must be active for the current optimal solution. Therefore, if there are inactive constraints, it is necessary to replace $\bar{\mu}_j$ for inactive constraints by $d_{iR}(c_{i\alpha}^{L}x^*) + v^*$ or $d_{iL}(c_{i\alpha}^{R}x^*) + v^*$ and solve the corresponding problem (5.32) for obtaining the simplex multipliers.

Now, following the above discussions, we can present the interactive algorithm to derive the satisficing solution for the DM from the M-α-Pareto optimal solution set. The steps marked with an asterisk involve interaction with the DM.

Interactive fuzzy multiobjective linear programming with fuzzy parameters

Step 0: Individual minimum and maximum

Calculate the individual minimum and maximum of each objective function under the given constraints for $\alpha = 0$ and $\alpha = 1$.

Step 1*: Membership functions

Elicit a membership function $\mu_i(c_i x)$ from the DM for each of the objective functions.

Step 2*: Initialization

Ask the DM to select the initial value of $\alpha\,(0 \leq \alpha \leq 1)$ and set the initial reference membership values $\bar{\mu}_i = 1,\ i = 1, \ldots, k$.

Step 3: M-α-Pareto optimal solution

For the degree α and the reference membership values specified by the DM, solve the minimax problem and perform the M-α-Pareto optimality test to obtain the M-α-Pareto optimal solution and the trade-off rates between the membership functions and the degree α.

Step 4*: Termination or updating

The DM is supplied with the corresponding M-α-Pareto optimal solution and the trade-off rates between the membership functions and the degree α. If the DM is satisfied with the current membership function values of the M-α-Pareto optimal solution and α, stop. Otherwise, the DM must update the reference membership values and/or the degree α by considering the current values of the membership functions and α together with the trade-off rates between the membership functions and the degree α and return to step 3.

Here it should be stressed to the DM that (1) any improvement of one membership function can be achieved only at the expense of at least one of the other membership functions for some fixed degree α and (2) the greater value of the degree α gives the worse values of the membership functions for some fixed reference membership values.

5.4.4 Illustrative numerical example

To clarify the concept of M-α-Pareto optimality as well as the proposed method, consider the following three-objective linear programming problem with fuzzy parameters.

Example 5.2

$$\begin{array}{rrrcr}
\text{fuzzy min} & z_1 = & 2x_1 & + & C_{12}x_2 \\
\text{fuzzy max} & z_2 = & 3x_1 & - & C_{22}x_2 \\
\text{fuzzy equal} & z_3 = & C_{31}x_1 & - & x_2 \\
\text{subject to} & x \in X & & &
\end{array}$$

$$\triangleq \{(x_1, x_2) \mid 3x_1 + x_2 \leq 12,\ x_1 + 2x_2 \leq 12,\quad x_i \geq 0,\ i = 1, 2\}$$

where C_{12}, C_{22}, and C_{31} are fuzzy numbers whose membership functions are the same as in Example 5.1 and given by

$$\begin{aligned}
\mu_{C_{12}}(c_{12}) &= \max(1 - 0.5\,|c_{12} - 4|,\ 0) \\
\mu_{C_{22}}(c_{22}) &= \max(1 - 2\,|c_{22} + 0.75|,\ 0) \\
\mu_{C_{31}}(c_{31}) &= \max(1 - |c_{31} - 2.5|,\ 0).
\end{aligned}$$

Now, for illustrative purposes, suppose that the interaction with the hypothetical DM establishes the following simple linear membership functions for the three objective functions:

$$\mu_1(z_1) = \begin{cases} 1 & ; & z_1 \leq 5 \\ d_{1R}(z_1) = (20 - z_1)/15 & ; & 5 \leq z_1 \leq 20 \\ 0 & ; & 20 \leq z_1 \end{cases}$$

$$\mu_2(z_2) = \begin{cases} 0 & ; & z_2 \leq 3 \\ d_{2L}(z_2) = (z_2 - 3)/9 & ; & 3 \leq z_2 \leq 12 \\ 1 & ; & 12 \leq z_2 \end{cases}$$

$$\mu_3(z_3) = \begin{cases} 0 & ; & z_3 \leq -3 \\ d_{3L}(z_3) = (z_3 + 3)/3 & ; & -3 \leq z_3 \leq 0 \\ 1 & ; & z_3 = 0 \\ d_{3R}(z_3) = (6 - z_3)/6 & ; & 0 \leq z_3 \leq 6 \\ 0 & ; & 6 \leq z_3. \end{cases}$$

Figure 5.11 illustrates a graphical description of the hypothetical DM's linear membership functions representing the fuzzy goals for each of the objectives of this problem.

Also assume that the hypothetical DM selects the initial value of the degree α to be 0.5 and the initial reference membership values $(\bar{\mu}_1, \bar{\mu}_2, \bar{\mu}_3)$ to be $(19/30, 5/6, 1)$. The corresponding M-α-Pareto optimal solution can be obtained

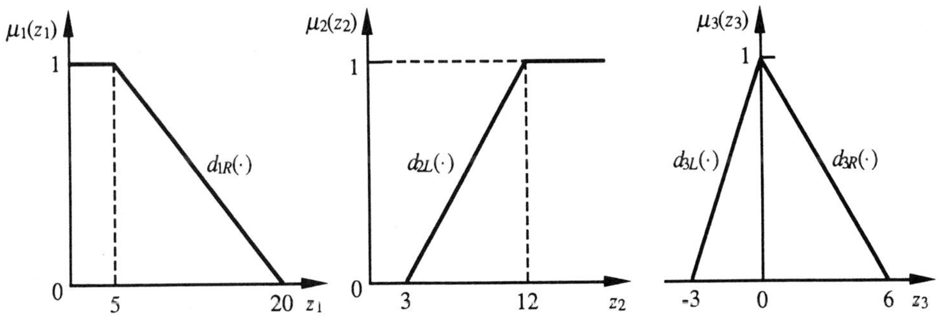

Figure 5.11 Linear membership functions representing fuzzy goals

by solving the following problem:

$$
\begin{array}{lllll}
\text{minimize} & v & & & \\
\text{subject to} & 2x_1 & + & 3x_2 & \leq d_{1R}^{-1}(\bar{\mu}_1 - v) \\
& 3x_1 & + & x_2 & \geq d_{2L}^{-1}(\bar{\mu}_2 - v) \\
& 2x_1 & - & x_2 & \leq d_{3R}^{-1}(\bar{\mu}_3 - v) \\
& 3x_1 & - & x_2 & \geq d_{3L}^{-1}(\bar{\mu}_3 - v) \\
& x \in X. & & &
\end{array}
$$

Solving this problem by combined use of the bisection method and the simplex method of linear programming, we obtain the optimal solution $v^* = 1/6$. To obtain the corresponding optimal values of the decision variable x^*, we solve the following linear programming problem for $v^* = 1/6$:

$$
\begin{array}{lllll}
\text{minimize} & 2x_1 & + & 3x_2 & \\
\text{subject to} & 3x_1 & + & x_2 & \geq d_{2L}^{-1}(\bar{\mu}_2 - v^*) \\
& 2x_1 & - & x_2 & \leq d_{3R}^{-1}(\bar{\mu}_3 - v^*) \\
& 3x_1 & - & x_2 & \geq d_{3L}^{-1}(\bar{\mu}_3 - v^*) \\
& x \in X. & & &
\end{array}
$$

As a result, we get the following optimal values for x^*, $z_1^* = z_1(x^*, c_{1\alpha}^L)$, $z_2^* = z_2(x^*, c_{2\alpha}^R)$, $z_3^* = z_3(x^*, c_{3\alpha}^L)$, $\mu_i^* = \mu_i(z_i^*)$, $i = 1, \ldots, k$, and the simplex multipliers $(\pi_{2L}^*, \pi_{3L}^*, \pi_{3R}^*)$:

$$(x_1^*, x_2^*) = (2, 3), \quad (z_1^*, z_2^*, z_3^*) = (13, 9, 1),$$

$$(\mu_1^*, \mu_2^*, \mu_3^*) = (7/15, 2/3, 5/6), \quad (\pi_{2L}^*, \pi_{3L}^*, \pi_{3R}^*) = (8/5, 0, 7/5).$$

From (5.39) and (5.40), the trade-off rates among the membership functions become

$$-\frac{\partial \mu_1(z_1^*)}{\partial \mu_2(z_2^*)} = -\frac{d'_{1R}(z_1^*)}{d'_{2L}(z_2^*)}\pi_{2L}^* = -\frac{(-1/15)}{(1/9)}\frac{8}{5} = \frac{72}{75},$$

$$-\frac{\partial \mu_1(z_1^*)}{\partial \mu_3(z_3^*)} = -\frac{d'_{1R}(z_1^*)}{d'_{3L}(z_3^*)}\pi_{3L}^* = -\frac{(-1/15)}{(-1/6)}\frac{7}{5} = \frac{42}{75}.$$

Concerning the trade-off rate between $\mu_1(z_1)$ and α, from (5.41), we have

$$\frac{\partial \mu_1(z_1^*)}{\partial \alpha} = d'_{1R}(z_1^*)\left\{\frac{\partial c_{1\alpha}^L}{\partial \alpha}x_2^* + \pi_{3R}^*\frac{\partial c_{3\alpha}^L}{\partial \alpha}x_1^* - \pi_{2L}^*\frac{\partial c_{2\alpha}^R}{\partial \alpha}x_2^*\right\}$$

$$= (-1/15)\{(1/2) \times 3 + (7/5) \times 1 \times 2 - (8/5) \times (-2) \times 3\} = -139/150.$$

Observe that the DM can obtain a satisficing solution from an M-α-Pareto optimal solution set by updating the reference membership values and/or the degree α on the basis of the current values of the membership functions and α

together with the trade-off rates among the values of the membership functions and the degree α.

5.5 Conclusions

In this chapter, we have introduced not only interactive programming but also interactive fuzzy programming methods for multiobjective linear programming problems with fuzzy parameters to cope with the imprecise nature of human judgments. To conclude this chapter, the desirable features of the proposed methods will be summarized as follows:

(1) The experts' ambiguous understanding of the nature of the parameters in the problem-formulation process can be incorporated.

(2) The fuzzy goals of the DM can also be quantified by eliciting the corresponding membership functions, which may be nonlinear.

(3) For the degree α and the reference objective (membership) values specified by the DM, the corresponding (M-) α-Pareto optimal solution can be easily obtained by solving the minimax problems based mainly on the well-known linear programming method.

(4) (M-) α-Pareto optimality of the generated solution in each iteration is guaranteed by performing the (M-) α-Pareto optimality test.

(5) The trade-off information between the objective and/or membership functions and the degree α is easily obtainable since it is closely related to the simplex multipliers of the minimax problems.

(6) The satisficing solution for the DM can be derived efficiently from among (M-) α-Pareto optimal solutions by updating the reference objective (membership) values and/or the degree α based on the current values of the (M-) α-Pareto optimal solution together with the trade-off information between the objective (membership) functions and the degree α.

It is significant to point out here that all the results presented in this chapter have already been extended by the authors to deal with multiobjective linear fractional programming problems with fuzzy parameters. A successful generalization along this line can be found in Sakawa and Yano (1985c, 1986h, 1987b), and the interested readers might refer to them for details. In the next chapter, we will proceed to the multiobjective nonlinear programming problems with fuzzy parameters as a generalized version of this chapter.

CHAPTER 6

INTERACTIVE MULTIOBJECTIVE NONLINEAR PROGRAMMING WITH FUZZY PARAMETERS

In this chapter, attention is focused on multiobjective nonlinear programming problems with fuzzy parameters, which reflect the experts' ambiguous or fuzzy understanding of the nature of the parameters in the problem-formulation process. By extending the usual Pareto optimality concepts, interactive decision-making methods, both without and with the fuzzy goals of the decision maker (DM), for deriving a satisficing solution efficiently from an extended Pareto optimal solution set are presented.

6.1 Introduction

The concept of fuzzy mathematical programming in the framework of the fuzzy decision of Bellman and Zadeh (1970) on a general level was first introduced by Tanaka et al. (1974). Under some assumptions for a fuzzy goal and fuzzy constraints for the decision maker (DM), they proved that the fuzzy mathematical programming problem can be reduced to a conventional mathematical programming problem. Since then, fuzzy mathematical programming has been developed on a more practical level, especially for dealing with linear models as discussed in Chapters 3 and 5. However, as discussed in Chapter 4, very few attempts have been made to deal with nonlinear programming problems until Sakawa and his collaborators (e.g., Sakawa 1984b, 1984c, Sakawa and Yano 1985b, 1985d, 1986a) developed several interactive approaches to multiobjective nonlinear programming problems beyond the framework of Bellman and Zadeh (1970).

However, in the conventional model building approaches, when formulating a multiobjective nonlinear programming problem, the possible values of the parameters involved in the objective functions and constraints are required to be fixed at a value in an experimental and/or subjective manner through the experts' understanding of the nature of the parameters. Realizing that, in most real-world situations, the possible values of these parameters are often only imprecisely or ambiguously known to the experts and should be interpreted as fuzzy numbers, Orlovski (1983, 1984) first formulated a multiobjective nonlinear programming problem involving fuzzy parameters which can be viewed as the more realistic version of the conventional one. He presented two approaches to the formulated problems by making systematic use of the extension principle of Zadeh (1965, 1975) and demonstrated that there exist, in some sense, equivalent nonfuzzy formulations. Unfortunately, however, no interactive decision-making methods have been proposed.

Under these circumstances, to deal with multiobjective nonlinear programming problems with fuzzy parameters characterized by fuzzy numbers, Sakawa and Yano (1986c, 1986e, 1987a, 1987d, 1989a, 1989b) recently presented several interactive decision-making methods for handling and tackling not only the experts' fuzzy understanding of the nature of parameters in the problem-formulation process but also the fuzzy goals of the DM by introducing the extended Pareto optimality concepts. These methods can be viewed as a natural generalization of the previous results for multiobjective nonlinear programming problems without fuzzy parameters by Sakawa, Yumine, and Yano (1984, 1987) and Sakawa and Yano (1985b, 1985d, 1986f).

In this chapter, these interactive decision-making methods for multiobjective nonlinear programming problems with fuzzy numbers for deriving the satisficing solution of the DM efficiently from the extended Pareto optimal solution set are presented as a nonlinear generalization of the previous chapter. However, it is interesting to point out here that, historically speaking, after the results of this chapter were presented, linear programming-based interactive decision-making methods for multiobjective linear programming problems with fuzzy parameters, discussed in the previous chapter, were developed.

6.2 Problem formulation and solution concepts

In general, the multiobjective nonlinear programming (MONLP) problem is represented as the following vector-minimization problem:

$$\left.\begin{array}{ll} \text{minimize} & f(x) \triangleq (f_1(x), f_2(x), ..., f_k(x))^T \\ \text{subject to} & x \in X \triangleq \{x \in R^n \mid g_j(x) \leq 0,\ j = 1, ..., m\} \end{array}\right\} \tag{6.1}$$

where x is an n-dimensional vector of decision variables, $f_1(x), ..., f_k(x)$ are k distinct objective functions of the decision vector x, $g_1(x), ..., g_m(x)$ are m inequality constraints, and X is the feasible set of constrained decisions.

As discussed in Chapter 4, fundamental to the MONLP is the Pareto optimal concept, also known as a noninferior solution. Qualitatively, a Pareto optimal solution of the MONLP is one where any improvement of one objective function can be achieved only at the expense of another. Mathematically, a formal definition of a Pareto optimal solution to the MONLP is given as follows.

Definition 6.1 (Pareto optimal solution)

$x^* \in X$ is said to be a Pareto optimal solution to the MONLP if and only if there does not exist another $x \in X$ such that $f_i(x) \leq f_i(x^*)$, $i = 1, ..., k$, with strict inequality holding for at least one i.

For practical purpose, as discussed in Chapter 4, since only local optimal solutions are guaranteed in solving a single-objective nonlinear programming problem by any available standard nonlinear programming technique, unless the problem is convex, the local Pareto optimality concept is also defined for the MONLP as follows.

Definition 6.2 (Local Pareto optimal solution)

$x^* \in X$ is said to be a local Pareto optimal solution to the MONLP if and only if there does not exist another $x \in X \cap N(x^*; r)$ such that $f_i(x) \leq f_i(x^*)$, $i = 1, \ldots, k$, with strict inequality holding for at least one i, where $N(x^*, r)$ denotes the r neighborhood of x^* defined by $\{x \in R^n \mid \| x - x^* \| < r\}$.

However, as was pointed out by Orlovski (1983, 1984), it would certainly be more appropriate to consider that the possible values of the parameters in the description of the objective functions and the constraints of the MONLP, although in the conventional approaches they are fixed at some values, usually involve the ambiguity of the experts' understanding of the real system in the problem-formulation process. For this reason, in this chapter, we consider the following multiobjective nonlinear programming problem with fuzzy parameters (MONLP-FP):

$$\left.\begin{array}{ll} \text{minimize} & f(x, A) \triangleq (f_1(x, A_1), f_2(x, A_2), \ldots, f_k(x, A_k))^T \\ \text{subject to} & x \in X(B) \triangleq \{x \in R^n \mid g_j(x, B_j) \leq 0, \; j = 1, \ldots, m\} \end{array}\right\} \quad (6.2)$$

where $A_i = (A_{i1}, \ldots, A_{ip_i})$, $B_j = (B_{j1}, \ldots, B_{jq_j})$ represent, respectively, a vector of fuzzy parameters involved in the objective function $f_i(x, A_i)$ and the constraint function $g_j(x, B_j)$. These fuzzy parameters, which reflect the experts' ambiguous understanding of the nature of the parameters in the problem-formulation process, are assumed to be characterized as fuzzy numbers. In this chapter, as mentioned in the previous chapter, we deal with a real fuzzy number M whose membership function $\mu_M(x)$ is defined as (Dubois and Prade 1978)

(1) A continuous mapping from R^1 to the closed interval $[0, 1]$
(2) $\mu_M(x) = 0$ for all $x \in (-\infty, c]$
(3) Strictly increasing and continuous on $[c, a]$
(4) $\mu_M(x) = 1$ for all $x \in [a, b]$
(5) Strictly decreasing and continuous on $[b, d]$
(6) $\mu_M(x) = 0$ for all $x \in [d, +\infty)$.

We now assume that A_{ir} and B_{js} in the MONLP-FP are fuzzy numbers whose membership functions are $\mu_{A_{ir}}(a_{ir})$ and $\mu_{B_{js}}(b_{js})$, respectively. For simplicity in the notation, define the following vectors:

$$\begin{array}{ll} a_i = (a_{i1}, \ldots, a_{ip_i}), & b_j = (b_{j1}, \ldots, b_{jq_j}), \\ a = (a_1, \ldots, a_k), & A = (A_1, \ldots, A_k), \\ b = (b_1, \ldots, b_m), & B = (B_1, \ldots, B_m). \end{array}$$

Since the MONLP-FP involves fuzzy numbers both in the objective functions and the constraints, it is necessary to extend the notion of usual Pareto optimality in some sense. For that purpose, as discussed in the previous chapter, we first introduce the following α-level set to the fuzzy numbers A_{ir} and B_{js}.

Definition 6.3 **(α-level set)**

The α-level set of the fuzzy numbers A_{ir}, $i = 1, ..., k$, $r = 1, ..., p_i$, and B_{js}, $j = 1, ..., m$, $s = 1, ..., q_j$, is defined as the ordinary set $(A, B)_\alpha$ for which the degree of their membership functions exceeds the level α:

$$(A, B)_\alpha = \{(a, b) \mid \mu_{A_{ir}}(a_{ir}) \geq \alpha,\ i = 1, ..., k,\ r = 1, ..., p_i\,; \\ \mu_{B_{js}}(b_{js}) \geq \alpha,\ j = 1, ..., m,\ s = 1, ..., q_j\,\}. \quad (6.3)$$

Now, as discussed in the previous chapter, suppose that the decision maker (DM) considers that the degree of all of the membership functions of the fuzzy numbers involved in the MONLP-FP should be greater than or equal to a certain value of α. Then, for such a degree α, the MONLP-FP can be interpreted as the following nonfuzzy multiobjective nonlinear programming (MONLP-FP(a, b)) problem which depends on the coefficient vector $(a, b) \in (A, B)_\alpha$:

$$\left.\begin{array}{ll} \text{minimize} & f(x, a) \triangleq (f_1(x, a_1), f_2(x, a_2), ..., f_k(x, a_k))^T \\ \text{subject to} & x \in X(b) \triangleq \{x \in R^n \mid g_j(x, b_j) \leq 0,\ j = 1, ..., m\}. \end{array}\right\} \quad (6.4)$$

It is significant to note here that there exists an infinite number of such MONLP-FP(a, b) depending on the coefficient vector $(a, b) \in (A, B)_\alpha$, and the values of (a, b) are arbitrary for any $(a, b) \in (A, B)_\alpha$ in the sense that the degree of all of the membership functions for the fuzzy numbers in the MONLP-FP exceeds the level α. However, from the viewpoint of the DM, if possible, it would be desirable to choose $(a, b) \in (A, B)_\alpha$ in the MONLP-FP(a, b) to minimize the objective functions under the constraints. For this reason, for a certain degree α, it seems to be quite natural for the DM to understand the MONLP-FP as the following nonfuzzy α-multiobjective nonlinear programming (α-MONLP) problem:

$$\left.\begin{array}{ll} \text{minimize} & f(x, a) \triangleq (f_1(x, a_1), f_2(x, a_2), ..., f_k(x, a_k))^T \\ \text{subject to} & x \in X(b) \triangleq \{x \in R^n \mid g_j(x, b_j) \leq 0,\ j = 1, ..., m\} \\ & (a, b) \in (A, B)_\alpha. \end{array}\right\} \quad (6.5)$$

It must be observed here that, in the α-MONLP, the parameters (a, b) are treated as decision variables rather than constants.

Through the introduction of the α-MONLP, we can now define the concepts of α-Pareto optimality and local α-Pareto optimality as follows.

Definition 6.4 **(α-Pareto optimal solution)**

$x^* \in X(b)$ is said to be an α-Pareto optimal solution to the α-MONLP if and only if there does not exist another $x \in X(b)$ and $(a, b) \in (A, B)_\alpha$ such that

$f_i(x, a_i) \leq f_i(x^*, a_i^*)$, $i = 1, \ldots, k$, with strict inequality holding for at least one i, where the corresponding values of parameters a^* and b^* are called α-level optimal parameters.

Definition 6.5 (Local α-Pareto optimal solution)

$x^* \in X(b)$ is said to be a local α-Pareto optimal solution to the α-MONLP if and only if there does not exist another $x \in X(b) \cap N(x^*; r)$ and $(a, b) \in (A, B)_\alpha \cap N(a^*, b^*; r')$ such that $f_i(x, a_i) \leq f_i(x^*, a_i^*)$, $i = 1, \ldots, k$, with strict inequality holding for at least one i, where the corresponding values of parameters a^* and b^* are called α-level local optimal parameters and $N(x^*; r)$ denotes the r neighborhood of x^* defined by $\{x \in R^n \mid \| x - x^* \| < r\}$.

Observe that (local) α-Pareto optimal solutions can be obtained through a direct application of the usual scalarizing methods for generating (local) Pareto optimal solutions just by regarding the decision variables in the α-MONLP as (x, a, b). However, as can be immediately understood from Definitions 6.4 and 6.5, in general, (local) α-Pareto optimal solutions to the α-MONLP consist of an infinite number of points, and the DM must select a (local) satisficing solution from an (local) α-Pareto optimal solution set based on a subjective value judgment. In the following, we first present interactive programming approaches to the MONLP-FP. We then present interactive fuzzy programming approaches to the MONLP-FP by incorporating the fuzzy goals of the DM for each of the objective functions of the α-MONLP.

6.3 Interactive nonlinear programming

6.3.1 Minimax and augmented minimax problems

To generate a candidate for the (local) satisficing solution which is also (local) α-Pareto optimal, the DM is asked to specify the degree α of the α-level set and the reference levels of achievement of the objective functions, called reference levels. Observe that the idea of the reference levels or the reference point first appeared in Wierzbicki (1979a). For the DM's degree α and reference levels $\bar{f}_i$, $i = 1, ..., k$, the corresponding (local) α-Pareto optimal solution, which is, in the minimax sense, nearest to the requirement or better than that if the reference levels are attainable, is obtained by solving the following minimax problem:

$$\left.\begin{array}{ll} \text{minimize} & \max\limits_{i=1,\ldots,k} (f_i(x, a_i) - \bar{f}_i) \\ \text{subject to} & (a, b) \in (A, B)_\alpha, \; x \in X(b) \end{array}\right\} \quad (6.6)$$

or equivalently

$$\left.\begin{array}{ll} \text{minimize} & v \\ \text{subject to} & f_i(x, a_i) - \bar{f}_i \leq v, \; i = 1, ..., k \\ & (a, b) \in (A, B)_\alpha, \; x \in X(b). \end{array}\right\} \quad (6.7)$$

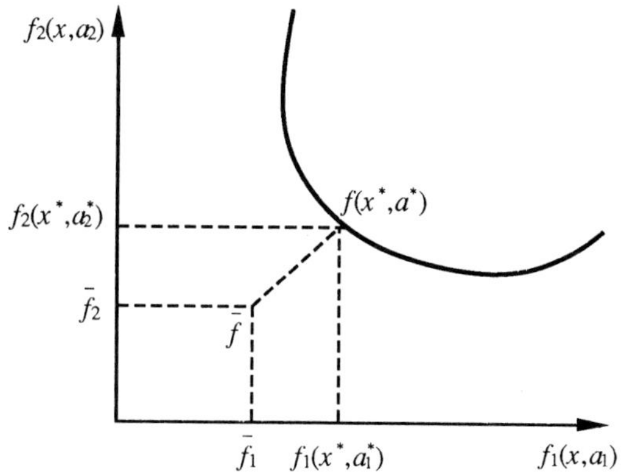

Figure 6.1 Minimax problem in objective space

A graphical description of the minimax problem in an objective space is depicted in Figure 6.1.

The relationships between the (local) optimal solutions of the minimax problem and the (local) α-Pareto optimal concept of the α-MONLP can be characterized by the following theorems.

Theorem 6.1

If (x^*, v^*, a^*, b^*) is a unique (local) optimal solution to the minimax problem for some $\bar{f} = (\bar{f}_1, ..., \bar{f}_k)$, then x^* is a (local) α-Pareto optimal solution to the α-MONLP.

Proof

Assume that x^* is not a (local) α-Pareto optimal solution to the α-MONLP, then there exist $x \in X(b)\,(\cap N(x^*; r))$ and $(a, b) \in (A, B)_\alpha\,(\cap N(a^*, b^*; r'))$ such that $f_i(x, a) \leq f_i(x^*, a^*)$ with strict inequality holding for at least one i. Hence, it follows that

$$\max_{i=1,...,k} (f_i(x, a_i) - \bar{f}_i) \leq \max_{i=1,...,k} (f_i(x^*, a_i^*) - \bar{f}_i),$$

which contradicts the fact that (x^*, v^*, a^*, b^*) is a unique (local) optimal solution to the minimax problem. □

Theorem 6.2

If x^* is a (local) α-Pareto optimal solution and (a^*, b^*) is a (local) α-level optimal parameter to the α-MONLP, then there exists $\bar{f} = (\bar{f}_1, ..., \bar{f}_k)$ such that (x^*, v^*, a^*, b^*) is a unique (local) optimal solution to the minimax problem.

Proof

Assume that (x^*, v^*, a^*, b^*) is not a unique (local) optimal solution to the minimax problem for any $\bar{f}$ satisfying

$$f_1(x^*, a_1^*) - \bar{f}_1 = \cdots = f_k(x^*, a_k^*) - \bar{f}_k.$$

Then there exist $x \in X(b)\,(\cap N(x^*;r))$ and $(a,b) \in (A,B)_\alpha\,(\cap N(a^*,b^*;r'))$ such that

$$\max_{i=1,\dots,k}(f_i(x^*,a_i^*) - \bar{f}_i) < \max_{i=1,\dots,k}(f_i(x,a_i) - \bar{f}_i).$$

This implies that

$$\max_{i=1,\dots,k}(f_i(x,a_i) - f_i(x^*,a_i^*)) < 0.$$

Now, if either any $f_i(x,a_i) - f_i(x^*,a_i^*)$ is positive or all $f_i(x,a_i) - f_i(x^*,a_i^*), i = 1,...,k$, are zero, this inequality would be violated, which contradicts the fact that x^* is a (local) α-Pareto optimal solution and (a^*,b^*) is a (local) α-level optimal parameter to the α-MONLP and the theorem is proved. □

It should be noted here that, for generating (local) α-Pareto optimal solutions using the result in Theorem 6.1, uniqueness of solution must be verified. In general, however, it is not easy to check numerically whether a (local) optimal solution to the minimax problem is unique or not.

Consequently, to test the (local) α-Pareto optimality of a current (local) optimal solution (x^*,v^*,a^*,b^*), we formulate and solve the following nonlinear programming problem:

$$\left.\begin{array}{ll} \text{maximize} & \sum_{i=1}^{k} \varepsilon_i \\ \text{subject to} & f_i(x,a_i) + \varepsilon_i = f_i(x^*,a_i^*),\ \varepsilon_i \geq 0,\ i = 1,...,k, \\ & x \in X(b),\ (a,b) \in (A,B)_\alpha. \end{array}\right\} \tag{6.8}$$

Let $(\bar{x},\bar{a},\bar{b})$, $\bar{\varepsilon}$ be a (local) optimal solution to this problem. If all $\bar{\varepsilon}_i = 0$, then x^* is a (local) α-Pareto optimal solution. If at least one $\bar{\varepsilon}_i > 0$, it can easily be shown that $\bar{x}$ is a (local) α-Pareto optimal solution.

However, in the framework of nonlinear programming, for circumventing the (local) α-Pareto optimality tests as in the minimax problems, it is recommended to use augmented minimax problems rather than minimax problems. To be more specific, for the DM's degree α and reference levels $\bar{f}_i, i = 1,...,k$, the corresponding (local) α-Pareto optimal solution, which is, in the minimax sense, nearest to the requirement or better than that if the reference levels are attainable, is obtained by solving the following augmented minimax problem:

$$\min_{\substack{x \in X(b) \\ (a,b)\in(A,B)_\alpha}} \left\{ \max_{i=1,\dots,k}(f_i(x,a_i) - \bar{f}_i) + \rho \sum_{i=1}^{k}(f_i(x,a_i) - \bar{f}_i) \right\} \tag{6.9}$$

or equivalently

$$\left.\begin{array}{ll} \text{minimize} & \left\{ v + \rho \sum_{i=1}^{k} (f_i(x,a_i) - \bar{f}_i) \right\} \\ \text{subject to} & f_i(x,a_i) - \bar{f}_i \leq v,\ i = 1,\dots,k \\ & (a,b) \in (A,B)_\alpha,\ x \in X(b) \end{array}\right\} \tag{6.10}$$

or

$$\left.\begin{array}{ll}\text{minimize} & v \\ \text{subject to} & f_i(x, a_i) - \bar{f}_i \leq v - \rho \sum_{i=1}^{k} (f_i(x, a_i) - \bar{f}_i), \ i = 1, \ldots, k \\ & (a, b) \in (A, B)_\alpha, \ x \in X(b).\end{array}\right\} \quad (6.11)$$

The term augmented is adopted because the term $\rho \sum_{i=1}^{k} (f_i(x, a_i) - \bar{f}_i)$ is added to the usual minimax problems, where ρ is a sufficiently small positive scalar. Such an augmented minimax problem can be viewed as a modified version of the augmented weighted Chebyshev norm problem of Steuer and Choo (1983) or Choo and Atkins (1983).

The relationships between the (local) optimal solutions of the augmented minimax problem and the (local) α-Pareto optimal concept of the α-MONLP can be characterized by the following theorems.

Theorem 6.3

If (x^*, a^*, b^*) is a (local) optimal solution to the augmented minimax problem for some $\bar{f} = (\bar{f}_1, \ldots, \bar{f}_k)$, then x^* is a (local) α-Pareto optimal solution to the α-MONLP.

Proof

Assume that x^* is not a (local) α-Pareto optimal solution to the α-MONLP, then there exists $x \in X(b) \ (\cap N(x^*; r))$, $(a, b) \in (A, B)_\alpha \ (\cap N(a^*, b^*; r'))$ such that $f_i(x, a) \leq f_i(x^*, a^*)$ with strict inequality holding for at least one i. Then it holds that

$$\max_{i=1,\ldots,k} (f_i(x, a_i) - \bar{f}_i) \leq \max_{i=1,\ldots,k} (f_i(x^*, a_i^*) - \bar{f}_i),$$

$$\rho \sum_{i=1}^{k} (f_i(x, a_i) - \bar{f}_i) < \rho \sum_{i=1}^{k} (f_i(x^*, a_i^*) - \bar{f}_i).$$

This means that

$$\max_{i=1,\ldots,k} (f_i(x, a_i) - \bar{f}_i) + \rho \sum_{i=1}^{k} (f_i(x, a_i) - \bar{f}_i)$$

$$< \max_{i=1,\ldots,k} (f_i(x^*, a_i^*) - \bar{f}_i) + \rho \sum_{i=1}^{k} (f_i(x^*, a_i^*) - \bar{f}_i),$$

which contradicts the fact that (x^*, a^*, b^*) is a (local) optimal solution to the augmented minimax problem. Hence, x^* is a (local) α-Pareto optimal solution to the α-MONLP. □

Theorem 6.4

If x^* is a (local) α-Pareto optimal solution and (a^*, b^*) is a (local) α-level optimal parameter to the α-MONLP, then there exists $\bar{f} = (\bar{f}_1, \ldots, \bar{f}_k)$ such that (x^*, a^*, b^*) is a (local) optimal solution to the augmented minimax problem.

Proof

Assume that (x^*, a^*, b^*) is not a (local) optimal solution to the augmented minimax problem for any $\bar{f} = (\bar{f}_1, \ldots, \bar{f}_k)$ satisfying

$$f_1(x^*, a_1^*) - \bar{f}_1 = \cdots = f_k(x^*, a_k^*) - \bar{f}_k.$$

Then there exist $x \in X(b)\,(\cap N(x^*; r))$ and $(a, b) \in (A, B)_\alpha\,(\cap N(a^*, b^*; r'))$ such that

$$\max_{i=1,\ldots,k} (f_i(x^*, a_i^*) - \bar{f}_i) + \rho \sum_{i=1}^{k} (f_i(x^*, a_i^*) - \bar{f}_i)$$

$$> \max_{i=1,\ldots,k} (f_i(x, a_i) - \bar{f}_i) + \rho \sum_{i=1}^{k} (f_i(x, a_i) - \bar{f}_i).$$

This implies that

$$\max_{i=1,\ldots,k} (f_i(x, a_i) - f_i(x^*, a_i^*)) + \rho \sum_{i=1}^{k} (f_i(x, a_i) - f_i(x^*, a_i^*)) < 0.$$

Now, if either any $f_i(x, a_i) - f_i(x^*, a_i^*)$ is positive or all $f_i(x, a_i) - f_i(x^*, a_i^*), i = 1, \ldots, k$, are zero, this inequality would be violated for sufficiently small positive ρ. Hence,

$$f_i(x, a_i) - f_i(x^*, a_i^*) \leq 0,\ i = 1, \ldots, k, \text{and } f_i(x, a_i) - f_i(x^*, a_i^*) < 0 \text{ for some } i$$

must hold, which contradicts the fact that x^* is a (local) α-Pareto optimal solution and (a^*, b^*) is a (local) α-level optimal parameter to the α-MONLP and the theorem is proved. □

As can be seen from the above proofs, it should be noted here that an obvious advantage of the augmented minimax problem over the usual minimax problem is that (local) α-Pareto optimality is guaranteed even if the uniqueness assumption for the solution is absent, because of the presence of the augmented term.

Added insight can be obtained by comparing the isoquants of the minimax problem and the augmented minimax problem as depicted in Figure 6.2, where, in general

$$\theta = \tan^{-1}(\sqrt{k-1}\rho/(1+\rho)). \tag{6.12}$$

This relation shows that θ is monotone increasing with respect to ρ. Thus, for a sufficiently small positive scalar, augmented minimax problems overcome the possibility of generating weak (local) α-Pareto optimal solutions as shown in Theorems 6.3 and 6.4. Hence, augmented minimax problems are attractive for generating (local) α-Pareto optimal solutions even if appropriate convexity assumptions are absent.

To verify this relation, we merely observe that the cosine of the angle θ between the normal vector

$$(-\rho, \ldots, -\rho, -1-\rho, \ldots, -\rho) \tag{6.13}$$

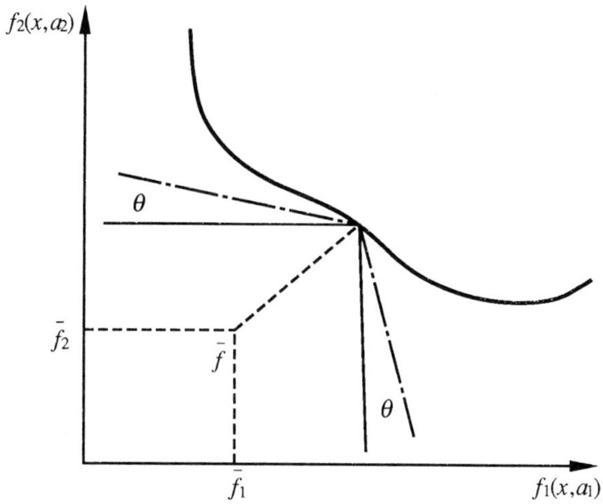

Figure 6.2 Isoquants of minimax and augmented minimax problems

of the isoquant of the augmented minimax problem and the normal vector

$$(0, \ldots, 0, -1, 0, \ldots, 0) \tag{6.14}$$

of the isoquant of the minimax problem is given by

$$\cos\theta = (1+\rho)/\sqrt{1+2\rho+k\rho^2}. \tag{6.15}$$

Naturally, ρ should be a sufficiently small, but computationally significant, positive scalar. However, for practical purposes, a computationally significant lower bound of ρ may be

$$\rho = 10^{(a-b)-c+1} \tag{6.16}$$

where a and b are the figures of $\max_{i=1,\ldots,k}(f_i(x,a_i) - \bar{f}_i)$ and $\sum_{i=1}^{k}(f_i(x,a_i) - \bar{f}_i)$ in the first and second terms in the augmented minimax problem, respectively, and c is the precision figure of the computer. Usually, since the values of a and b are unknown in advance, if we roughly estimate $a = b$, then we have $\rho = 10^{-c+1}$. In most cases, a computationally significant value of $\rho = 10^{-3} \sim 10^{-5}$ should suffice.

6.3.2 Algorithm using augmented minimax problems

Now given the (local) α-Pareto optimal solution for the degree α and the reference levels specified by the DM by solving the corresponding augmented minimax problem, the DM must either be satisfied with the current (local) α-Pareto optimal solution or update the reference levels and/or the degree α. In order to help the DM express a degree of preference, trade-off information between a standing objective function and each of the other objective functions as well as between the degree α and the objective functions is very useful. Fortunately, such trade-off information is easily obtainable since it is closely related to the strict positive Lagrange multipliers of the augmented minimax problem.

To derive the trade-off information, we first define the Lagrangian function L for the augmented minimax problem as follows:

$$L(x, v, a, b, \lambda^f, \lambda^a, \lambda^b, \lambda^g, \bar{f}, \alpha) = v + \rho \sum_{i=1}^{k} (f_i(x, a_i) - \bar{f}_i) + \sum_{i=1}^{k} \lambda_i^f (f_i(x, a_i) - \bar{f}_i - v)$$

$$+ \sum_{j=1}^{m} \lambda_j^g g_j(x, b_j) + \sum_{i=1}^{k} \sum_{r=1}^{p_i} \lambda_{ir}^a (\alpha - \mu_{A_{ir}}(a_{ir})) + \sum_{j=1}^{m} \sum_{s=1}^{q_j} \lambda_{js}^b (\alpha - \mu_{B_{js}}(b_{js})) \quad (6.17)$$

Then the following sensitivity theorem (recall Theorem 4.6), which is based on the implicit function theorem (Fiacco 1983), holds, where for notational convenience, the decision variables in the augmented minimax problem are denoted by $y = (x, v, a, b)$.

Theorem 6.5

Let $y^* = (x^*, v^*, a^*, b^*)$ be a unique local solution of the augmented minimax problem satisfying (1) y^* is a regular point of the constraints of the augmented minimax problem, (2) the second-order sufficiency conditions are satisfied at y^*, and (3) there are no degenerate constraints at y^*. Also, let $\lambda^* = (\lambda^{f^*}, \lambda^{a^*}, \lambda^{b^*}, \lambda^{g^*})$ denote the Lagrange multiplier vector corresponding to the constraints of the augmented minimax problem. Then there exist continuously differentiable vector valued functions $y(\cdot)$ and $\lambda(\cdot)$ defined on some neighborhood $N(\alpha^*)$ so that $y(\alpha^*) = y^*, \lambda(\alpha^*) = \lambda^*$, where $y(\alpha)$ is a unique local solution of the augmented minimax problem for any $\alpha \in N(\alpha^*)$ satisfying the three assumptions, and $\lambda(\alpha)$ is the Lagrange multiplier vector corresponding to the constraints of the augmented minimax problem.

Furthermore, the following relation holds on some neighborhood $N(\alpha^*)$ of α^*:

$$\frac{\partial \left\{ v + \rho \sum_{i=1}^{k} (f_i(x, a_i) - \bar{f}_i) \right\}}{\partial \alpha} = \frac{\partial L}{\partial \alpha} = \sum_{i=1}^{k} \sum_{r=1}^{p_i} \lambda_{ir}^a + \sum_{j=1}^{m} \sum_{s=1}^{q_j} \lambda_{js}^b. \quad (6.18)$$

If all the constraints for v of the augmented minimax problem are active, namely, if $v(\alpha^*) = f_i(x(\alpha^*), a_i(\alpha^*)) - \bar{f}_i$, $i = 1, \ldots, k$, then the following theorem holds.

Theorem 6.6

Let all the assumptions in Theorem 6.5 be satisfied. Also assume that all the constraints for v of the augmented minimax problem are active. Then it holds that

$$\left. \frac{\partial f_i(x, a_i)}{\partial \alpha} \right|_{\alpha = \alpha^*} = \frac{1}{1 + \rho k} \left(\sum_{i=1}^{k} \sum_{r=1}^{p_i} \lambda_{ir}^{a^*} + \sum_{j=1}^{m} \sum_{s=1}^{q_j} \lambda_{js}^{b^*} \right). \quad (6.19)$$

Regarding a trade-off rate between $f_1(x)$ and $f_i(x)$ for each $i = 2, \ldots, k$, by extending the results in Haimes and Chankong (1979), it can be proved that the following theorem holds (For details, see the Appendix).

Theorem 6.7

Let all the assumptions in Theorem 6.5 be satisfied. Also assume that all the constraints for v of the augmented minimax problem are active. Then it holds that

$$-\left.\frac{\partial f_i(x, a_i)}{\partial f_1(x, a_1)}\right|_{\alpha=\alpha^*} = \frac{\lambda_1^{f^*} + \rho}{\lambda_i^{f^*} + \rho}, \quad i = 2, \ldots, k. \tag{6.20}$$

It should be noted here that, to obtain the trade-off rate information from these relations, all the constraints for v of the augmented minimax problem must be active. Therefore, if there are inactive constraints, it is necessary to replace $\bar{f}_i$ for inactive constraints by $f_i(x^*, a_i^*)$ and solve the corresponding augmented minimax problem for obtaining the Lagrange multipliers.

We can now construct the interactive algorithm to derive the (local) satisficing solution for the DM from the (local) α-Pareto optimal solution set. The steps marked with an asterisk involve interaction with the DM.

Interactive multiobjective nonlinear programming with fuzzy parameters

Step 0: Individual minimum and maximum

Calculate the (local) individual minimum and maximum of each objective function under the given constraints for $\alpha = 0$ and $\alpha = 1$.

Step 1*: Initialization

Ask the DM to select the initial value of α $(0 < \alpha < 1)$ and the initial reference levels $\bar{f}_i$, $i = 1, \ldots, k$.

Step 2: α-Pareto optimal solution

For the degree α and the reference levels specified by the DM, solve the augmented minimax problem to obtain the α-Pareto optimal solution together with the trade-off rates between the objective functions and the degree α.

Step 3*: Termination or updating

The DM is supplied with the corresponding (local) α-Pareto optimal solution and the trade-off rates between the objective functions and the degree α. If the DM is satisfied with the current objective function values of the (local) α-Pareto optimal solution, stop. Otherwise, the DM must update the reference levels and/or the degree α by considering the current values of the objective functions and α together with the trade-off rates between the objective functions and the degree α and return to step 2.

Here it should be stressed to the DM that (1) any improvement of one objective function can be achieved only at the expense of at least one of the other objective functions for some fixed degree α and (2) the greater value of the degree α gives the worse values of the objective functions for some fixed reference levels.

6.4 Interactive fuzzy nonlinear programming

6.4.1 Fuzzy goals

As discussed above, (local) α-Pareto optimal solutions consist of an infinite number of points, and the DM may be able to select a (local) satisficing or compromise solution from (local) α-Pareto optimal solutions based on subjective value judgments by making use of the interactive programming method presented in the previous section.

However, considering the imprecise nature of the DM's judgments, it is quite natural to assume that the DM may have imprecise or fuzzy goals for each of the objective functions in the α-MONLP. In a minimization problem, a fuzzy goal stated by the DM may be to achieve "substantially less than or equal to some value p_i." This type of statement can be quantified by eliciting a corresponding membership function.

In order to elicit a membership function $\mu_i(f_i(x, a_i))$ from the DM for each of the objective functions $f_i(x, a_i)$ in the α-MONLP, we first calculate the individual (local) minimum $f_i^{\min}$ and maximum $f_i^{\max}$ of each objective function $f_i(x, a_i)$ under the given constraints for $\alpha = 0$ and $\alpha = 1$. By taking into account the calculated individual (local) minimum and maximum of each objective function for $\alpha = 0$ and $\alpha = 1$ together with the rate of increase of membership satisfaction, the DM may be able to determine a subjective membership function $\mu_i(f_i(x, a_i))$ which is a strictly monotone decreasing function with respect to $f_i(x, a_i)$.

Figure 6.3 illustrates the graph of the possible shape of the membership function representing the fuzzy goal to achieve substantially less than or equal to p_i.

If we restricted ourselves to a minimization problem, it is quite natural to assume that the DM has a fuzzy goal such as "$f_i(x, a_i)$ should be substantially less than or equal to p_i." In the fuzzy approaches, as discussed in Chapters 3, 4, and 5, we can further treat a more general case where the DM has two types of fuzzy goals, namely, fuzzy goals expressed in words such as "$c_i x$ should be in the vicinity of r_i" (called fuzzy equal) as well as "$c_i x$ should be substantially less than or equal to p_i or greater than or equal to q_i" (called fuzzy min or fuzzy max). Such a generalized α-MONLP (Gα-MONLP) problem can be expressed as

$$\left.\begin{array}{rll} \text{fuzzy min} & f_i(x, a_i) & i \in I_1 \\ \text{fuzzy max} & f_i(x, a_i) & i \in I_2 \\ \text{fuzzy equal} & f_i(x, a_i) & i \in I_3 \\ \text{subject to} & (a, b) \in (A, B)_\alpha, & x \in X(b) \end{array}\right\} \tag{6.21}$$

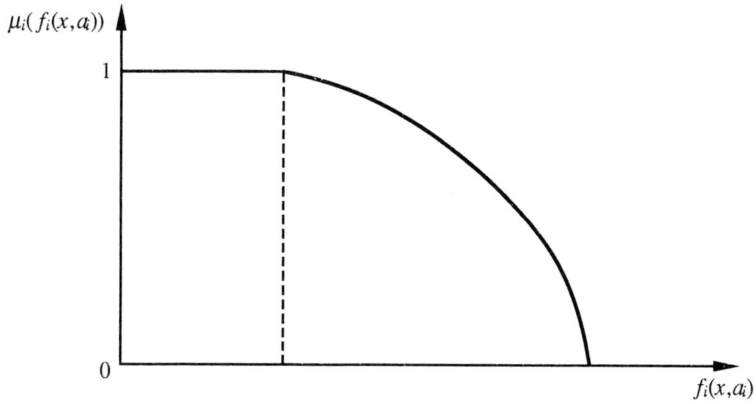

Figure 6.3 Monotone decreasing membership function

where $I_1 \cup I_2 \cup I_3 = \{1, 2, ..., k\}$, $I_i \cap I_j = \emptyset$, $i, j = 1, 2, 3$, $i \neq j$.

To elicit a membership function from the DM for a fuzzy goal like "$f_i(x, a_i)$ should be in the vicinity of r_i," it is obvious that we can use different functions for the left and right side of r_i.

As an example, Figure 6.4 illustrates the graph of the possible shape of the fuzzy equal membership function representing the fuzzy goal to be in the vicinity of r_i.

When the DM's fuzzy goals include a fuzzy equal, then preferably the $f_i(x, a_i)$ is as close to r_i as possible. However, as a consequence of this, the (local) α-Pareto optimal solution, defined in terms of objective functions, is not applicable. It is because of this, we introduce a concept of (local) M-α-Pareto optimal solutions which is defined in terms of membership functions, where M refers to membership.

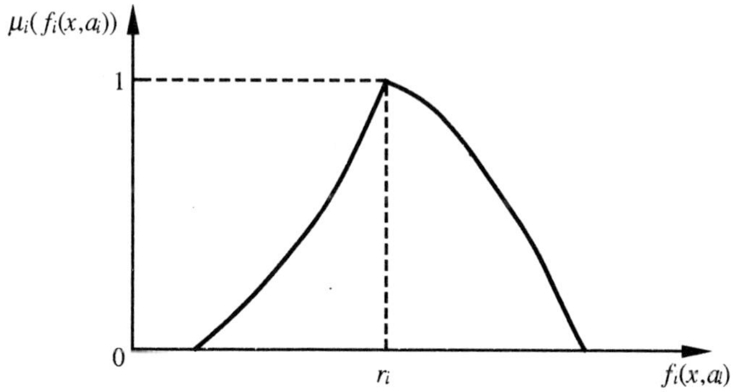

Figure 6.4 Fuzzy equal membership function

Definition 6.6 ((local) M-α-Pareto optimal solution)

$x^* \in X(b^*)$ is said to be a (local) M-α-Pareto optimal solution to the Gα-MONLP if and only if there does not exist another $x \in X(b)\,(\cap N(x^*; r))$, $(a, b) \in (A, B)_\alpha\ \ (\cap N(a^*, b^*; r'))$ such that $\mu_i(f_i(x, a_i)) \geq \mu_i(f_i(x^*, a_i^*)), i = 1, \ldots, k$, with strict inequality holding for at least one i, where the corresponding values of parameters a^* and b^* are called α-level (local) optimal parameters.

Having elicited the membership function $\mu_i(f_i(x, a_i)), i = 1, \ldots, k$, from the DM for each of the objective functions $f_i(x, a_i), i = 1, \ldots, k$, if we introduce a general aggregation function

$$\mu_D(\mu(f(x, a)), \alpha) = \mu_D(\mu_1(f_1(x, a_1)), \cdots, \mu_k(f_k(x, a_k)), \alpha) \tag{6.22}$$

a general fuzzy α-multiobjective decision-making problem (Fα-MODMP) can be defined by

$$\left.\begin{array}{ll} \text{maximize} & \mu_D(\mu_1(f_1(x, a_1)), \mu_2(f_2(x, a_2)), \ldots, \mu_k(f_k(x, a_k)), \alpha) \\ \text{subject to} & (x, a, b) \in P(\alpha), \ \ \alpha \in [0, 1] \end{array}\right\} \tag{6.23}$$

where $P(\alpha)$ is the set of M-α-Pareto optimal solutions and corresponding α-level optimal parameters to the Gα-MONLP.

Probably the most crucial problem in the Fα-MODMP is the identification of an appropriate aggregation function which well represents the human decision makers' fuzzy preferences. If $\mu_D(\cdot)$ can be explicitly identified, then the Fα-MODMP reduces to a standard mathematical programming problem. However, this seems to happen rarely, and as an alternative approach, interactive fuzzy programming methods under the following assumptions would be recommended.

(1) The fuzzy goals of the DM can be quantified by eliciting the corresponding membership functions through the interaction with the DM.

(2) $\mu_D(\cdot)$ exists and is known only implicitly to the DM, which means the DM cannot specify the entire form of $\mu_D(\cdot)$, but can provide local information concerning a preference. Moreover, it is strictly increasing and continuous with respect to $\mu_i(\cdot)$ and α.

(3) All $f_i(x, a_i)$, $i = 1, \ldots, k$, and all $g_j(x, b_j)$, $j = 1, \ldots, m$, are continuously differentiable in their respective domains.

6.4.2 Minimax and augmented minimax problems

Having determined the membership functions for each of the objective functions, to generate a candidate for the (local) satisficing solution which is also (local) M-α-Pareto optimal, the DM is asked to specify the degree α of the α-level set and the reference levels of achievement of the membership functions

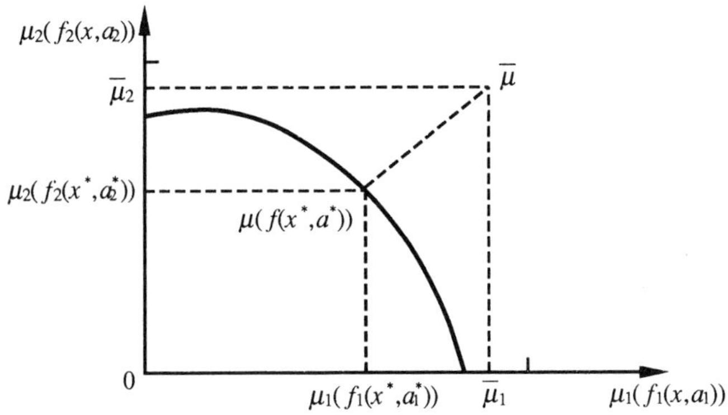

Figure 6.5 Minimax problem

called the reference membership values. Observe that the idea of the reference membership values (e.g. Sakawa, Yumine, and Yano 1984a, b; Sakawa and Yano 1984a, 1985f) can be viewed as an obvious extension of the idea of the reference point of Wierzbicki (1979a). For the DM's degree α and the reference membership values $\bar{\mu}_i$, $i = 1, \ldots, k$, the following minimax problem is solved to generate the (local) M-α-Pareto optimal solution, which is, in the minimax sense, nearest to the requirement or better than that if the reference membership values are attainable:

$$\min_{\substack{x \in X(b) \\ (a,b) \in (A,B)_\alpha}} \max_{i=1,\ldots,k} (\bar{\mu}_i - \mu_i(f_i(x, a_i))) \tag{6.24}$$

or equivalently

$$\left.\begin{array}{ll} \text{minimize} & v \\ \text{subject to} & \bar{\mu}_i - \mu_i(f_i(x, a_i)) \leq v, \ i = 1, \ldots, k \\ & (a, b) \in (A, B)_\alpha, \ x \in X(b). \end{array}\right\} \tag{6.25}$$

Figure 6.5 illustrates a graphical description of the minimax problem in a membership space.

The relationships between the (local) optimal solutions of the minimax problem and the (local) M-α-Pareto optimal concept of the Gα-MONLP can be characterized by the following theorems.

Theorem 6.8

If (x^*, a^*, b^*) is a unique (local) optimal solution to the minimax problem for some $\bar{\mu}_i, i = 1, \ldots, k$, then x^* is a (local) M-α-Pareto optimal solution and a^*, b^* are α-level (local) optimal parameters to the Gα-MONLP.

Proof

Assume that x^* is not a (local) M-α-Pareto optimal solution and a^*, b^* are not α-level (local) optimal parameters to the Gα-MONLP, then there exists $x \in X(b)$

$(\cap N(x^*;r))$ and $(a,b) \in (A,B)_\alpha$ $(\cap N(a^*,b^*;r'))$ such that $f_i(x,a) \leq f_i(x^*,a^*)$, $i=1,\ldots,k$, with strict inequality holding for at least one i. This implies that $\mu_i(f(x,a)) \geq \mu_i(f(x^*,a^*))$ or $\bar{\mu}_i - \mu_i(f(x,a)) \leq \bar{\mu}_i - \mu_i(f(x^*,a^*)$ with strict inequality holding for at least one i. Then it holds that

$$\max_{i=1,\ldots,k}(\bar{\mu}_i - \mu_i(f_i(x,a_i))) \leq \max_{i=1,\ldots,k}(\bar{\mu}_i - \mu_i(f_i(x^*,a_i^*)))$$

which contradicts the fact that (x^*,a^*,b^*) is a unique (local) optimal solution to the minimax problem. Hence x^* is a (local) M-α-Pareto optimal solution and a^*, b^* are α-level (local) optimal parameters to the Gα-MONLP. □

Theorem 6.9

If x^* is a (local) M-α-Pareto optimal solution and a^*,b^* are α-level (local) optimal parameters to the Gα-MONLP, then there exists $\bar{\mu}_i, i=1,\ldots,k$, such that (x^*,a^*,b^*) is a (local) optimal solution to the minimax problem.

Proof

Assume that (x^*,a^*,b^*) is not a (local) optimal solution to the minimax problem for any $\bar{\mu}_i$, $i=1,\ldots,k$, satisfying

$$\bar{\mu}_1 - \mu_1(f_1(x^*,a_1^*)) = \cdots = \bar{\mu}_k - \mu_k(f_k(x^*,a_k^*)).$$

Then there exist $x \in X$ $(\cap N(x^*;r))$ and $(a,b) \in (A,B)_\alpha$ such that

$$\max_{i=1,\ldots,k}(\bar{\mu}_i - \mu_i(f_i(x^*,a_i^*))) > \max_{i=1,\ldots,k}(\bar{\mu}_i - \mu_i(f_i(x,a_i))).$$

This implies that

$$\max_{i=1,\ldots,k}(\mu_i(f_i(x^*,a_i^*)) - \mu_i(f_i(x,a_i))) < 0.$$

Hence,

$$\mu_i(f_i(x^*,a_i^*)) - \mu_i(f_i(x,a_i)) < 0\ ,\ i=1,\ldots,k$$

must hold, which contradicts the fact that x^* is a (local) M-α-Pareto optimal solution and a^*,b^* are α-level (local) optimal parameters to the Gα-MONLP, and the theorem is proved. □

Quite similar to the minimization case, a numerical test of (local) M-α-Pareto optimality for x^* can be performed by solving the following M-α-Pareto optimality test problem:

$$\left.\begin{array}{ll} \text{maximize} & \sum_{i=1}^{k} \varepsilon_i \\ \text{subject to} & \mu_i(f_i(x,a_i)) - \varepsilon_i = \mu_i(f_i(x^*,a_i)),\ \varepsilon_i \geq 0,\ i=1,...,k, \\ & x \in X(b),\ (a,b) \in (A,B)_\alpha. \end{array}\right\} \quad (6.26)$$

Let $(\bar{x},\bar{a},\bar{b})$, $\bar{\varepsilon}$ be a (local) optimal solution to this problem. If all $\bar{\varepsilon}_i = 0$, then x^* is a (local) M-α-Pareto optimal solution. If at least one $\bar{\varepsilon}_i > 0$, it can easily be shown that $\bar{x}$ is a (local) M-α-Pareto optimal solution.

To circumvent the necessity to perform the (local) M-α-Pareto optimality test in the minimax problems, for the nonlinear case, it is reasonable to use augmented minimax problems instead of minimax problems. For the DM's degree α and the reference membership values $\bar{\mu}_i$, $i = 1, \ldots, k$, the following augmented minimax problem is solved for generating the (local) M-α-Pareto optimal solution, which is, in the minimax sense, nearest to the requirement or better than that if the reference membership values are attainable:

$$\min_{\substack{x \in X(b) \\ (a,b) \in (A,B)_\alpha}} \left\{ \max_{i=1,\ldots,k} (\bar{\mu}_i - \mu_i(f_i(x, a_i))) + \rho \sum_{i=1}^{k} \Big(\bar{\mu}_i - \mu_i(f_i(x, a_i)) \Big) \right\} \tag{6.27}$$

or equivalently

$$\left. \begin{array}{ll} \text{minimize} & v + \rho \sum_{i=1}^{k} \Big(\bar{\mu}_i - \mu_i(f_i(x, a_i)) \Big) \\ \text{subject to} & \bar{\mu}_i - \mu_i(f_i(x, a_i)) \leq v, \ i = 1, \ldots, k \\ & x \in X(b), \ (a, b) \in (A, B)_\alpha. \end{array} \right\} \tag{6.28}$$

Such an augmented minimax problem can be viewed as a modified fuzzy version of the augmented Chebyshev norm problem of Steuer and Choo(1983) or Choo and Atkins (1983).

The relationships between the (local) optimal solutions of the augmented minimax problem and the (local) M-α-Pareto optimal concept of the Gα-MONLP can be characterized by the following theorems.

Theorem 6.10

If (x^*, a^*, b^*) is a (local) M-α-Pareto optimal solution to the augmented minimax problem for some $\bar{\mu}_i, i = 1, \ldots, k$, then x^* is a (local) M-α-Pareto optimal solution and a^*, b^* are α-level (local) optimal parameters to the Gα-MONLP.

Proof

Assume that x^* is not a (local) M-α-Pareto optimal solution and a^*, b^* are not α-level (local) optimal parameters to the Gα-MONLP, then there exist $x \in X(b)\,(\cap N(x^*, r))$ and $(a, b) \in (A, B)_\alpha\,(\cap N(a^*; b^*; r'))$ such that $\mu_i(f(x, a)) \geq \mu_i(f(x^*, a^*))$ or $\bar{\mu}_i - \mu_i(f(x, a)) \leq \bar{\mu}_i - \mu_i(f(x^*, a^*))$ with strict inequality holding for at least one i. Hence, it follows that

$$\max_{i=1,\ldots,k} (\bar{\mu}_i - \mu_i(f_i(x, a_i))) \leq \max_{i=1,\ldots,k} (\bar{\mu}_i - \mu_i(f_i(x^*, a_i^*))).$$

This means that

$$\max_{i=1,\ldots,k} (\bar{\mu}_i - \mu_i(f_i(x, a_i))) + \rho \sum_{i=1}^{k} (\bar{\mu}_i - \mu_i(f_i(x, a_i)))$$

$$< \max_{i=1,\ldots,k} (\bar{\mu}_i - \mu_i(f_i(x^*, a_i^*))) + \rho \sum_{i=1}^{k} (\bar{\mu}_i - \mu_i(f_i(x^*, a_i^*))),$$

which contradicts that (x^*, a^*, b^*) is a (local) optimal solution to the augmented minimax problem. Hence, x^* is a (local) M-α-Pareto optimal solution and a^*, b^* are α-level (local) optimal parameters to the Gα-MONLP. □

Theorem 6.11

If x^* is a (local) M-α-Pareto optimal solution and a^*, b^* are α-level (local) optimal parameters to the Gα-MONLP, then there exists $\bar{\mu}_i, i = 1, \ldots, k$, such that (x^*, a^*, b^*) is a (local) optimal solution to the augmented minimax problem for a sufficiently small positive ρ.

Proof

Assume that (x^*, a^*, b^*) is not a (local) optimal solution to the augmented minimax problem for any $\bar{\mu}_i, i = 1, \ldots, k$, satisfying

$$\bar{\mu}_1 - \mu_1(f_1(x^*, a_1^*)) = \cdots = \bar{\mu}_k - \mu_k(f(x^*, a_k^*)).$$

Then there exist $x \in X$ $(\cap N(x^*; r))$ and $(a, b) \in (A, B)_\alpha$ $(\cap N(a^*, b^*; r'))$ such that

$$\max_{i=1,\ldots,k} (\bar{\mu}_i - \mu_i(f_i(x^*, a_i^*))) + \rho \sum_{i=1}^{k} \Big(\bar{\mu}_i - \mu_i(f_i(x^*, a_i^*))\Big)$$

$$> \max_{i=1,\ldots,k} (\bar{\mu}_i - \mu_i(f_i(x, a_i))) + \rho \sum_{i=1}^{k} \Big(\bar{\mu}_i - \mu_i(f_i(x, a_i))\Big).$$

This implies that

$$\max_{i=1,\ldots,k} \Big(\mu_i(f_i(x^*, a_i^*)) - \mu_i(f_i(x, a_i))\Big) + \rho \sum_{i=1}^{k} \Big(\mu_i(f_i(x^*, a_i^*)) - \mu_i(f_i(x, a_i))\Big) < 0$$

Now, if either any $\mu_i(f_i(x^*, a_i^*)) - \mu_i(f_i(x, a_i))$ is positive or all $\mu_i(f_i(x^*, a^*)) - \mu_i(f_i(x, a_i)), i = 1, \ldots, k$, are zero, this inequality would be violated for a sufficiently small positive ρ. Hence, it follows that

$$\mu_i(f_i(x^*, a_i^*)) - \mu_i(f_i(x, a_i)) \leq 0, \ i = 1, \ldots, k$$

with strict inequality holding for at least one i. This contradicts the fact that x^* is a (local) M-α-Pareto optimal solution and a^*, b^* are α-level (local) optimal parameters to the Gα-MONLP, and the theorem is proved. □

As can be seen from the above proofs, an obvious advantage of the augmented minimax problem over the usual minimax problem is that, because of the presence of the augmented term, (local) M-α-Pareto optimality is guaranteed without the uniqueness assumption for the solution.

Added insight can be obtained by comparing the isoquant of the augmented minimax problem

$$\bar{\mu}_i - \mu_i(f_i(x, a_i)) + \rho \sum_{i=1}^{k} \Big(\bar{\mu}_i - \mu_i(f_i(x, a_i))\Big) = \text{constant}, \ i = 1, \ldots, k \qquad (6.29)$$

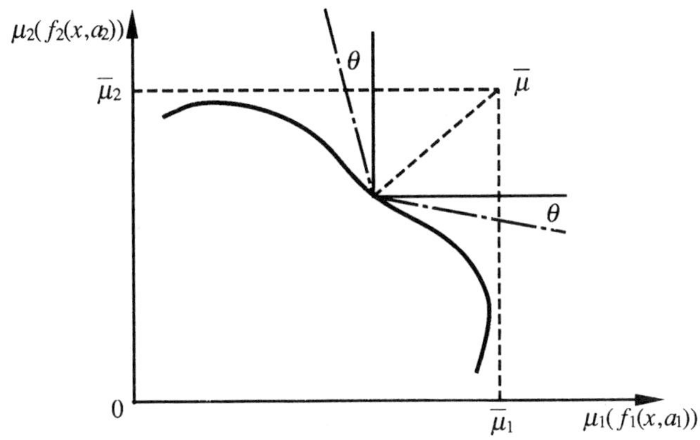

Figure 6.6 Isoquants of minimax and augmented minimax problems

with the isoquant of the minimax problem

$$\bar{\mu}_i - \mu_i(f_i(x, a_i)) = \text{constant}, \quad i = 1, \ldots, k \tag{6.30}$$

in the membership function space as depicted in Figure 6.6.

Observe that, in Figure 6.6, the normal vectors of the isoquant of the augmented minimax problem and the minimax problem become $(-\rho, \ldots, -\rho, -1 - \rho, -\rho, \ldots, -\rho)$ and $(0, \ldots, 0, -1, 0, \ldots, 0)$, respectively. It follows that the cosine of the angle θ between these two normal vectors is given by $\cos\theta = (1 + \rho)/\sqrt{1 + 2\rho + k\rho^2}$. Hence, we have

$$\theta = \tan^{-1}(\sqrt{k-1}\,\rho\,/\,(1+\rho)). \tag{6.31}$$

This relation shows that θ is monotone increasing with respect to ρ. Thus, for a sufficiently small positive scalar ρ, augmented minimax problems overcome the possibility of generating weak M-α-Pareto optimal solutions as shown in Theorems 6.10 and 6.11. Hence, augmented minimax problems are attractive for generating M-α-Pareto optimal solutions even if appropriate convexity assumptions are absent.

Naturally, ρ should be a sufficiently small but computationally significant positive scalar. However, for practical purposes, a computationally significant lower value of ρ may be

$$\rho = 10^{(a-b)-c+1} \tag{6.32}$$

where a and b are the figures of $\max_{i=1,\ldots,k}(\bar{\mu}_i - \mu_i(f_i(x, a_i)))$ and $\sum_{i=1}^{k}(\bar{\mu}_i - \mu_i(f_i(x, a_i)))$ in the first and second terms in the augmented minimax problem, respectively, and c is the precision figure of the computer. Usually, since the values of a and b are unknown in advance, if we roughly estimate $a = b$, then we have $\rho = 10^{-c+1}$. In most cases, a computationally significant value of $\rho = 10^{-3} \sim 10^{-5}$ should suffice.

6.4.3 Algorithm using augmented minimax problems

Now, given the (local) M-α-Pareto optimal solution for the degree α and the reference membership values specified by the DM for solving the corresponding augmented minimax problem, the DM must either be satisfied with the current (local) M-α-Pareto optimal solution and α or update the reference membership values and/or the degree α. To help the DM express à degree of preference, trade-off information between a standing membership function and each of the other membership functions as well as between the degree α and the membership functions is very useful. Fortunately, such trade-off information is easily obtainable since it is closely related to the strict positive Language multipliers of the augmented minimax problem.

To derive the trade-off information, we first define the Lagrangian function L for the augmented minimax problem as follows:

$$\begin{aligned} L &= v + \rho \sum_{i=1}^{k} \left(\bar{\mu}_i - \mu_i(f_i(x, a_i)) \right) \\ &+ \sum_{i=1}^{k} \lambda_i^{\mu} (\bar{\mu}_i - \mu_i(f_i(x, a_i)) - v) + \sum_{j=1}^{m} \lambda_j^{g} g_j(x, b_j) \\ &+ \sum_{i=1}^{k} \sum_{r=1}^{p_i} \lambda_{ir}^{a} (\alpha - \mu_{A_{ir}}(a_{ir})) + \sum_{j=1}^{m} \sum_{s=1}^{q_j} \lambda_{js}^{b} (\alpha - \mu_{B_{js}}(b_{js})). \end{aligned} \tag{6.33}$$

Then, similar to Theorems 6.5, 6.6, and 6.7, the following sensitivity theorem holds.

Theorem 6.12

Let $y^* = (x^*, v^*, a^*, b^*)$ be a unique local solution of the augmented minimax problem satisfying (1) y^* is a regular point of the constraints of the augmented minimax problem, (2) the second-order sufficiency conditions are satisfied at y^*, and (3) there are no degenerate constraints at y^*. Also, let $\lambda^* = (\lambda^{\mu^*}, \lambda^{a^*}, \lambda^{b^*}, \lambda^{g^*})$ denote the Lagrange multiplier vector corresponding to the constraints of the augmented minimax problem. Then there exist continuously differentiable vector-valued functions $y(\cdot)$ and $\lambda(\cdot)$ defined on some neighborhood $N(\alpha^*)$ so that $y(\alpha^*) = y^*, \lambda(\alpha^*) = \lambda^*$, where $y(\alpha)$ is a unique local solution of the augmented minimax problem for any $\alpha \in N(\alpha^*)$ satisfying the three assumptions and $\lambda(\alpha)$ is the Lagrange multiplier vector corresponding to the constraints of the augmented minimax problem.

Furthermore, the following relation holds on some neighborhood $N(\alpha^*)$ of α^*:

$$\frac{\partial \left(v + \rho \sum_{i=1}^{k} \left(\bar{\mu}_i - \mu_i(f_i(x, a_i)) \right) \right)}{\partial \alpha} = \frac{\partial L}{\partial \alpha} = \sum_{i=1}^{k} \sum_{r=1}^{p_i} \lambda_{ir}^{a} + \sum_{j=1}^{m} \sum_{s=1}^{q_j} \lambda_{js}^{b}. \tag{6.34}$$

If all the constraints for v of the augmented minimax problem are active, namely, if $v(\alpha^*) = \bar{\mu}_i - \mu_i(f_i(x(\alpha^*), a_i(\alpha^*)))$, $i = 1, \ldots, k$, then the following theorem holds.

Theorem 6.13

Let all the assumptions in Theorem 6.12 be satisfied. Also assume that the constraints for v are active. Then it holds that

$$-\frac{\partial \mu_i(f_i(x, a_i))}{\partial \alpha}\bigg|_{\alpha=\alpha^*} = \frac{1}{1+\rho k}\left(\sum_{i=1}^{k}\sum_{r=1}^{p_i}\lambda_{ir}^{a^*} + \sum_{j=1}^{m}\sum_{s=1}^{q_j}\lambda_{js}^{b^*}\right), \quad i = 1, \ldots, k. \tag{6.35}$$

For trade-off rates between $\mu_1(f_1(x, a_1))$ and $\mu_i(f_i(x, a_i))$ for each $i = 2, \ldots, k$, by extending the results in Haimes and Chankong (1979), it can be proved that the following theorem holds (see the Appendix, for details).

Theorem 6.14

Let all the assumptions in Theorem 6.12 be satisfied. Also assume that the constraints for v are active. Then it holds that

$$-\frac{\partial \mu_i(f_i(x, a_i))}{\partial \mu_1(f_1(x, a_1))}\bigg|_{y=y^*} = \frac{\lambda_1^{\mu^*} + \rho}{\lambda_i^{\mu^*} + \rho}, \quad i = 2, \ldots, k. \tag{6.36}$$

It should be noted here that to obtain the trade-off rate information from these relations, all the constraints for v of the augmented minimax problem must be active. Therefore, if there are inactive constraints, it is necessary to replace $\bar{\mu}_i$ for inactive constraints by $\mu_i(f_i(x^*, a_i^*)) + v^*$ and solve the corresponding augmented minimax problem for obtaining the Lagrange multipliers.

We can now construct the interactive algorithm to derive the (local) satisficing solution for the DM from the (local) M-α-Pareto optimal solution set. The steps marked with an asterisk involve interaction with the DM.

Interactive fuzzy multiobjective nonlinear programming with fuzzy parameters

Step 0: Individual minimum and maximum

Calculate the (local) individual minimum $f_i^{\min}$ and maximum $f_i^{\max}$ of each objective function $f_i(x)$ under the given constraints for $\alpha = 0$ and $\alpha = 1$.

Step 1*: Membership functions

Elicit a membership function from the DM for each of the objective functions.

Step 2*: Initialization

Ask the DM to select the initial values of α $(0 \le \alpha \le 1)$ and set the initial reference membership values $\bar{\mu}_i = 1$, $i = 1, \ldots, k$.

Step 3: M-α-Pareto optimal solution

For the degree α and the reference membership values specified by the DM, solve the corresponding augmented minimax problem to obtain the (local) M-α-Pareto optimal solution and the trade-off rates between the membership functions and the degree α.

Step 4*: Termination or updating

The DM is supplied with the corresponding M-α-Pareto optimal solution and the trade-off rates between the membership functions and the degree α. If the DM is satisfied with the current membership function values of the M-α-Pareto optimal solution and α, stop. Otherwise, the DM must update the reference membership values and/or the degree α by considering the current values of the membership functions and α together with the trade-off rates between the membership functions and the degree α and return to step 3.

Here it should be stressed to the DM that (1) any improvement of one membership function can be achieved only at the expense of at least one of the other membership function for some fixed degree α and (2) the greater value of the degree α gives the worse values of the membership functions for some fixed reference membership values.

6.5 Conclusions

As a nonlinear generalization of the previous chapter, for handling and tackling the imprecise nature of human judgments, not only interactive programming but also interactive fuzzy programming methods for multiobjective nonlinear programming problems with fuzzy parameters have been proposed in this chapter. Although the general conclusions of this chapter are essentially the same as in the previous chapter, the following is a brief summary of the desirable features of our proposed methods.

(1) The experts' ambiguous understanding of the nature of the parameters in the problem-formulation process can be incorporated.

(2) The fuzzy goals of the DM can also be quantified by eliciting the corresponding membership functions, which may be nonlinear.

(3) For the degree α and the reference membership values specified by the DM, the corresponding (local) (M-) α-Pareto optimal solution can be obtained by solving the augmented minimax problems based on the nonlinear program method.

(4) With the augmented minimax problems, (local) (M-) α-Pareto optimality of the generated solution in each iteration is guaranteed.

(5) The trade-off information between the objective and/or membership functions and the degree α is easily obtainable since it is closely related to the Lagrange multipliers of the augmented minimax problems.

(6) The (local) satisficing solution of the DM can be derived efficiently from among (local) (M-) α-Pareto optimal solutions by updating the reference objective (membership) values and/or the degree α based on the current values of the (local) (M-) α-Pareto optimal solution together with the trade-off information between the objective (membership) functions and the degree α.

Although no illustrative numerical examples are given in this chapter, interactive programming methods presented in this chapter will be demonstrated via interactive computer programs with numerical examples and the corresponding outputs in the next chapter.

CHAPTER 7

INTERACTIVE COMPUTER PROGRAMS

In this chapter, we present interactive computer programs developed by the author's group based on proposed methods discussed in Chapters 3, 4, 5, and 6 to facilitate the interaction processes for not only multiobjective programming problems but also multiobjective programming problems with fuzzy parameters. Moreover, to demonstrate the feasibility and efficiency of the proposed methods and the corresponding computer programs, interaction processes for several numerical examples of multiobjective linear and nonlinear programming problems, both without and with fuzzy parameters, are shown under the hypothetical decision maker along with the corresponding computer outputs.

7.1 Computer programs

Interactive fuzzy programming processes for multiobjective linear and nonlinear programming problems discussed in Chapters 3 and 4 include eliciting a membership function for each of the objective functions and reference membership levels from the decision maker (DM). Similarly, interactive programming or fuzzy interactive programming processes for multiobjective linear and nonlinear programming problems with fuzzy parameters discussed in Chapters 5 and 6 include eliciting reference levels and degree α from the DM or eliciting a membership function for each of the objective functions together with reference membership values and degree α from the DM. Thus, interactive utilization of computer facilities is indispensable for implementing the interactive decision making processes. Based on methods introduced in Chapters 3, 4, 5, and 6, we have developed interactive computer programs for multiobjective linear and nonlinear programming problems, both without and with fuzzy parameters. Our computer programs include graphical representations by which the DM can visualize the shapes of the membership functions, find incorrect assessments or inconsistent evaluations promptly, revise them immediately, and proceed to the next stage more easily.

In the original versions developed through the mid-1980s, the entire programs were written in FORTRAN, because at that time, FORTRAN language was very popular among many scientists and almost all mainframe computers were available for loading the programs through CRT terminals with some minor changes.

However, with electronic advances since the late 1980s, personal computers under the MS-DOS or PC-DOS operating system (OS), ranging from notebook size to desktop size, have made rapid progress and have spread remarkably in laboratories, factories, offices, universities, schools, banks, hospitals, and even homes. Roughly speaking, today's personal computers are nearly equivalent in basic computing power to the mainframes of the 1960s and the minicomputers of the 1970s. Hence, it becomes possible to solve relatively large linear programming problems with DOS-based 16-bit or 32-bit personal computers. Although some of the nonlinear models are too large to be solved on DOS machines, some commercial IBM PC Softwares for solving nonlinear programming have appeared (e.g. Liebman, Lasdon, Schrage, and Waren 1980, Brooke, Kendrick, and Meeraus 1988).

Under these circumstances, to develop user-friendly interactive programming or interactive fuzzy programming packages for personal computers (PCs) through the use of their monochrome or color graphics, C language is recommended rather than FORTRAN. While FORTRAN is one of the well-used high-level languages for mainframe computers, C has many of the characteristics of not only a higher-level language but also a lower-level language. The latter feature is very suitable for making full use of the graphics of PCs, and hence, in recent years, C has become the preferred software implementation language for MS-DOS or PC-DOS systems. From such a standpoint, the original programs for mainframe computers written in FORTRAN are converted into C language, and the well-known Microsoft C compiler for PCs is used to produce the executable code files that run on MS-DOS or PC-DOS systems.

7.1.1 Programs for multiobjective programming problems

Our programs for multiobjective linear and nonlinear programming problems are designed to derive the (local) satisficing solution for the DM from among the (local) (M-) Pareto optimal solution set by adopting one of the interactive fuzzy methods discussed in Chapters 3 and 4. The main menu of our computer program has several commands to be selected by the user (DM) according to specific requirements. When one of the commands is selected, each command has submenus with user-friendly prompt messages to facilitate the interactive decision-making processes. The functions of the commands prepared in the main menu are briefly summarized.

DATAIN: Inputs the data of the problem either in an interactive manner or by choosing the data file for the problem.

MINMAX: Displays the calculated individual minimum and maximum of each of the objective functions under the given constraints.

MF: Elicits a membership function from the DM for each of the objective functions.

GRAPH: Depicts graphically the shape of the membership function for each of the objective functions.

GO: Derives the (local) satisficing solution for the DM from the (local) (M-) Pareto optimal solution set.

LOOK: Displays the problem structure for review.

SUMMARY: Displays the summary of the whole interactive process.

STOP: Exits from the program.

In our computer software package for multiobjective linear programming problems, (M-) Pareto optimal solutions are calculated by solving the minimax problems by combined use of the bisection method and phase one of the simplex method of linear programming.

For multiobjective nonlinear programming problems, (local) (M-) Pareto optimal solutions are obtained by solving augmented minimax problems rather than minimax problems. For that purpose, the simplest way is to use one of the available effective nonlinear programming codes. For example, the FORTRAN code which employs augmented Lagrangians can be found in a text of Pierre and Lowe (1975). Other commercially available codes including MINOS (Murtagh and Saunders 1987) and the revised version of the generalized reduced gradient (GRG) program (Lasdon and co-workers 1974, 1976) called GRG2 (Lasdon and co-workers, 1980) would be recommended due to their robust optimizer ability for nonlinear programming problems. The PCs version of GRG2 called GINO (General Interactive Optimizer) (Liebman, Lasdon, Schrage, and Waren, 1980) and the PCs version of MINOS (Murtagh and Saunders 1987) called GAMS/MINOS (Brooke, Kendrick, and Meeraus 1988) can also be obtained through licensing arrangements with the publisher. In the early version of our computer programs for multiobjective nonlinear programming problems, written in FORTRAN, augmented Lagrangians or GRG2 code is employed. To make our software closed in its form in the sense of not depending on other software, a nonlinear programming code based on the recursive quadratic programming (RQP) or quasi-Newton method has been developed in FORTRAN and converted into C.

In our computer software packages, to elicit a membership function from the DM for each of the objective functions, the individual minimum and maximum of each objective function under the given constraints are calculated in advance. Taking into account the calculated individual minimum and maximum of each objective function together with the rate of increase of membership of satisfaction, the DM is asked to select a membership function in a subjective manner from among five types of functions: linear, exponential, hyperbolic, hyperbolic inverse, and piecewise linear functions. Then the parameter values are determined through the interaction with the DM. In the following explanations, for convenience, membership functions and objective functions are denoted by $\mu_i(z_i(x))$ and $z_i(x)$ instead of the notations $\mu_i(f_i(x))$ and $f_i(x)$ for the nonlinear case. Observe that, except for the hyperbolic functions, it is assumed that $\mu_i(z_i(x)) = 0$ if $z_i(x) \geq z_i^0$ and $\mu_i(z_i(x)) = 1$ if $z_i(x) \leq z_i^1$, where z_i^0 is a maximum value of an unacceptable level for $z_i(x)$ and z_i^1 is a minimum value of totally desirable level for $z_i(x)$. Also, z_i^a represents the value of $z_i(x)$ such that the degree

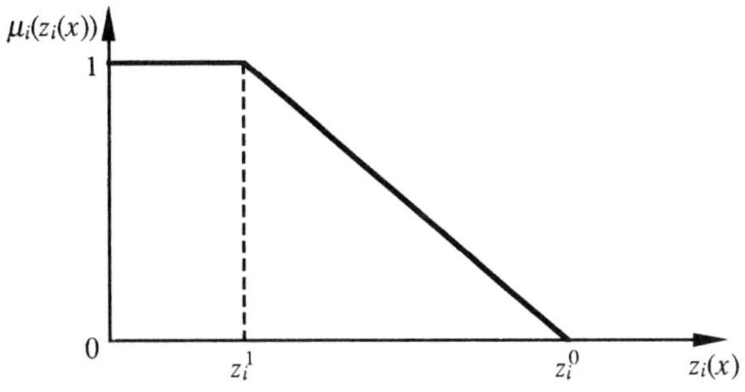

Figure 7.1 Linear membership function

of the membership function $\mu_i(z_i(x))$ is $a \in [0, 1]$ within the individual minimum $z_i^{\min}$ and individual maximum $z_i^{\max}$.

(1) Linear membership function

For each objective function, the corresponding linear membership function is defined by

$$\mu_i(z_i(x)) = [z_i(x) - z_i^0]/[z_i^1 - z_i^0]. \tag{7.1}$$

The linear membership function can be determined by asking the DM to specify the two points z_i^0 and z_i^1 within $z_i^{\max}$ and $z_i^{\min}$. Figure 7.1 illustrates the graphs of the linear membership function.

(2) Exponential membership function

For each objective function, the corresponding exponential membership function is defined by

$$\mu_i(z_i(x)) = a_i[1 - \exp\{-\alpha_i(z_i(x) - z_i^0)/(z_i^1 - z_i^0)\}], \tag{7.2}$$

where $a_i > 1, \alpha_i > 0$ or $a_i < 0, \alpha_i < 0$. The exponential membership function can be determined by asking the DM to specify the three points z_i^0, $z_i^{0.5}$, and z_i^1 within $z_i^{\max}$ and $z_i^{\min}$, where α_i is a shape parameter. Figure 7.2 illustrates the graph of the exponential membership function.

(3) Hyperbolic membership function

For each objective function, the corresponding hyperbolic membership function is defined by

$$\mu_i(z_i(x)) = \frac{1}{2}\tanh((z_i(x) - b_i)\alpha_i) + \frac{1}{2}, \tag{7.3}$$

where $\alpha_i < 0$. The hyperbolic membership function can be determined by asking the DM to specify the two points $z_i^{0.25}$ and $z_i^{0.5}$ within $z_i^{\max}$ and $z_i^{\min}$, where α_i is a shape parameter and b_i is associated with the point of inflection. Figure 7.3 illustrates the graph of the hyperbolic membership function.

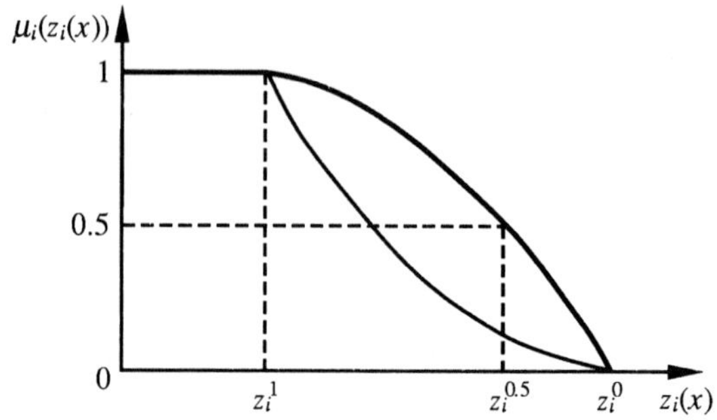

Figure 7.2 Exponential membership function

(4) Hyperbolic inverse membership function

For each objective function, the corresponding hyperbolic inverse membership function is defined by

$$\mu_i(z_i(x)) = a_i \tanh^{-1}((z_i(x) - b_i)\alpha_i) + \frac{1}{2}, \tag{7.4}$$

where $a_i > 0$ and $\alpha_i < 0$. The hyperbolic inverse membership function can be determined by asking the DM to specify the three points z_i^0, $z_i^{0.25}$, and $z_i^{0.5}$ within $z_i^{\max}$ and $z_i^{\min}$, where α_i is a shape parameter and b_i is associated with the point of inflection. Figure 7.4 illustrates the graph of the hyperbolic inverse membership function.

(5) Piecewise linear membership function

For each objective function, the corresponding piecewise linear membership function is defined by

$$\mu_i(z_i(x)) = \sum_{j=1}^{N_i} \alpha_{ij} \mid z_i(x) - g_{ij} \mid + \beta_i z_i(x) + \gamma_i \tag{7.5}$$

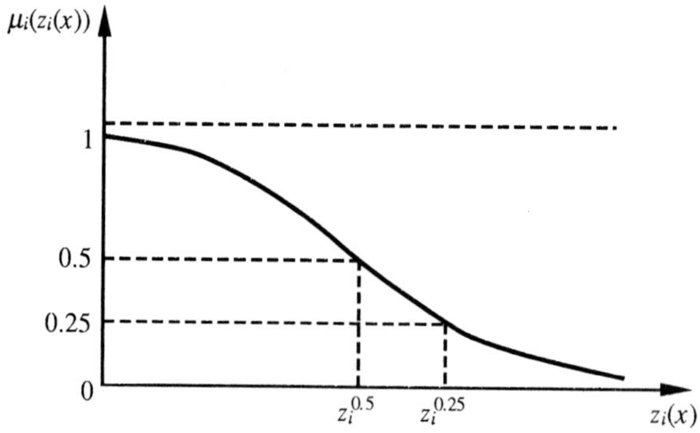

Figure 7.3 Hyperbolic membership function

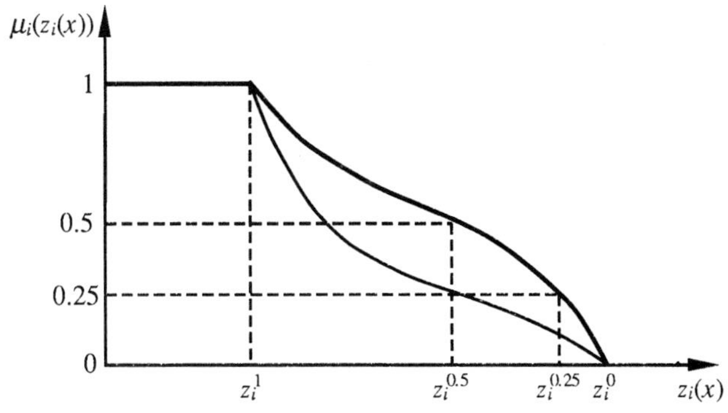

Figure 7.4 Hyperbolic inverse membership function

where

$$\alpha_{ij} = (t_{i,j+1} - t_{ij})/2, \quad \beta_i = (t_{i,N_i+1} + t_{i1})/2, \quad \gamma_i = (s_{i,N_i+1} + s_{i1})/2.$$

It is assumed that $\mu_i(z_i(x)) = t_{ir}z_i(x) + s_{ir}$ for each segment $g_{ir-1} \leq z(x) \leq g_{ir}$, where t_{ir} is the slope and s_{ir} is the y-intercept for the section of the line initiated at g_{ir-1} and terminated at g_{ir}. The piecewise linear membership function can be determined by asking the DM to specify the degree of membership in each of several values of objective functions within $z_i^{\max}$ and $z_i^{\min}$. Figure 7.5 illustrates the graph of the piecewise linear membership function.

It should be noted here that for the fuzzy equal membership functions, the DM can select left and right functions from among the same types of membership functions previously described above (excluding the hyperbolic ones).

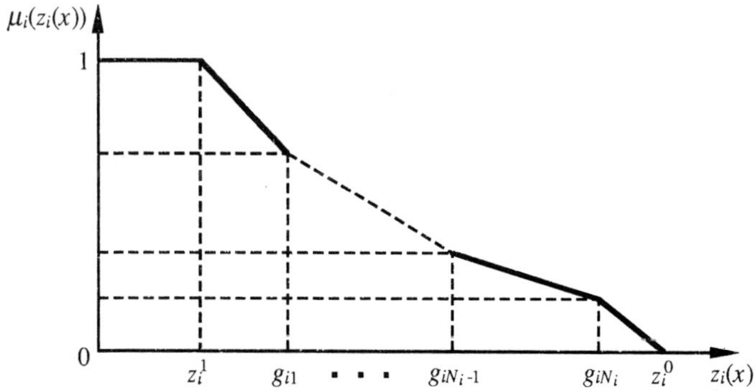

Figure 7.5 Piecewise linear membership function

7.1.2 Programs for multiobjective problems with fuzzy parameters

Interactive programming or interactive fuzzy programming processes for multiobjective linear and nonlinear programming problems with fuzzy parameters discussed in Chapters 5 and 6 include eliciting reference levels and degree α from the decision maker (DM) or eliciting a membership function for each of the objective functions together with reference membership values and degree α from the DM. Based on methods presented in Chapters 5 and 6, we have developed the corresponding interactive computer programs for deriving the (local) satisficing solution for the DM from the (local) (M-) α-Pareto optimal solution set of multiobjective linear or nonlinear programming problems with fuzzy parameters by updating the reference levels and/or the degree α.

We first give a brief explanation of the commands prepared in the main menu of our computer programs, not incorporating the fuzzy goals of the DM, for multiobjective linear and nonlinear programming problems with fuzzy parameters. Observe that each command has submenus with user-friendly prompt messages.

DATAIN: Inputs the data of the problem either in an interactive manner or by choosing the data file for the problem.

MINMAX: Displays the calculated (local) individual minimum and maximum of each of the objective functions under the given constraints for $\alpha = 0$ and $\alpha = 1$.

GO: Derives the (local) satisficing solution for the DM from among the (local) α-Pareto optimal solution set by updating the reference levels and/or the degree α.

LOOK: Displays the problem structure for review.

SUMMARY: Displays the summary of the whole interactive process.

STOP: Exits from the program.

In addition to the computer programs without the fuzzy goals of the DM, our programs incorporating the fuzzy goals of the DM include graphical representations by which the DM can visualize the shapes of the membership functions representing the fuzzy goals. Hence, it is easy for the DM to find incorrect assessments and revise them immediately before proceeding to the next stage. The main menu of our computer program, incorporating both the fuzzy parameters and fuzzy goals, has several commands and each command has submenus with user-friendly prompt messages. The functions of the commands prepared in the main menu are briefly summarized.

DATAIN: Inputs the data of the problem either in an interactive manner or by choosing the data file for the problem.

MINMAX: Displays the calculated (local) individual minimum and maximum of each of the objective functions under the given constraints for $\alpha = 0$ and $\alpha = 1$.

MF: Elicits a membership function from the DM for each of the objective functions.

GRAPH: Depicts graphically the shape of the membership function for each of the objective functions.

GO: Derives the DM's (local) satisficing solution from the (local) (M-) α-Pareto optimal solution set by updating the reference membership values and/or the degree α.

LOOK: Displays the problem structure for review.

SUMMARY: Displays the summary of the whole interactive process.

STOP: Exits from the program.

In all of our computer programs for multiobjective linear and nonlinear programming problems with fuzzy parameters, the fuzzy parameters, which reflect the experts' ambiguous understanding of the nature of parameters in the problem-formulation process, are assumed to be characterized by the fuzzy numbers whose membership functions are either linear or exponential as shown in Figure 7.6 or Figure 7.7, respectively.

Each of the membership functions for the fuzzy parameters can be determined by specifying the four points p_1, p_2, p_3, p_4 together with the types of its left and right functions (linear or exponential).

(M)-α-Pareto optimal solutions for multiobjective linear programming problems are calculated by solving the minimax problems on the basis of the simplex

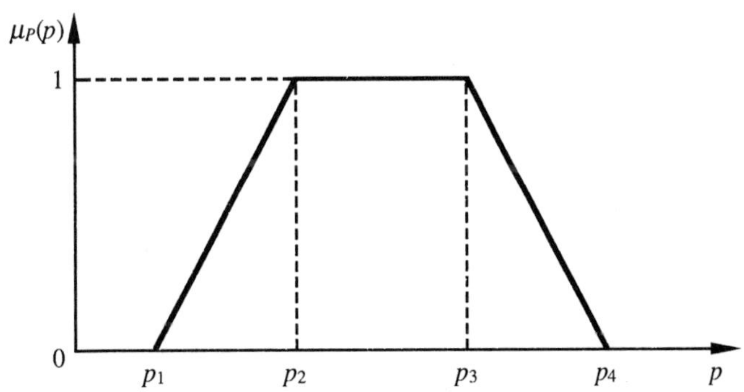

Figure 7.6 Linear membership function for fuzzy parameters

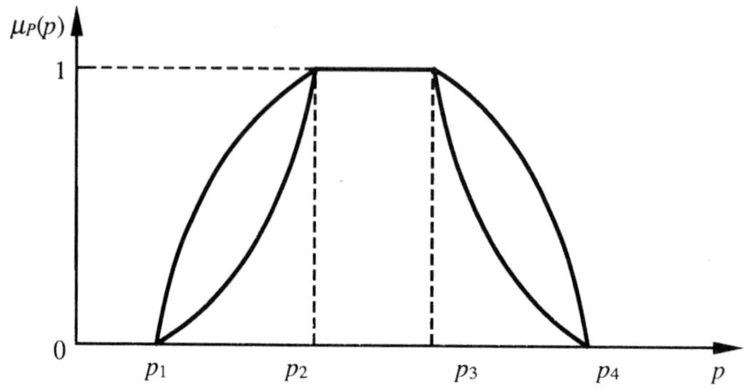

Figure 7.7 Exponential membership function for fuzzy parameters

method of linear programming as discussed in Chapter 5. For multiobjective nonlinear programing problems, (local) (M)-α-Pareto optimal solutions are obtained by solving the augmented minimax problems using one of the nonlinear programming codes discussed in the previous subsection.

As with our computer programs for multiobjective programming problems, in our computer programs for multiobjective programming problems with fuzzy parameters incorporating the fuzzy goals of the DM, the DM can select membership functions for each of the objective functions of (G) α-multiobjective programming problems in a subjective manner. Then by considering the rate of increase of membership satisfaction from among the linear, exponential, hyperbolic, hyperbolic inverse, and piecewise linear functions, the parameter values are determined by the DM. In the following, in contrast to the previous subsection, the notations for the nonlinear case are used for membership functions and objective functions. Here, except for the hyperbolic functions, it is assumed that $\mu_i(f_i(x, a_i)) = 0$ if $f_i(x) \geq f_i^0$ and $\mu_i(f_i(x, a_i)) = 1$ if $f_i(x, a_i) \leq f_i^1$, where f_i^0 is a maximum value of an unacceptable level for $f_i(x, a_i)$ and f_i^1 is a minimum value of totally desirable level for $f_i(x, a_i)$. f_i^a represents the value of $f_i(x, a_i)$ such that the degree of membership function $\mu_i(f_i(x, a_i))$ is a.

(1) Linear membership function

$$\mu_i(f_i(x, a_i)) = (f_i(x, a_i) - f_i^0) \;/\; (f_i^1 - f_i^0). \tag{7.6}$$

The linear membership function can be determined by asking the DM to specify the two points f_i^0 and f_i^1 within $f_i^{\max}$ and $f_i^{\min}$.

(2) Exponential membership function

$$\mu_i(f_i(x, a_i)) = \alpha_i \left\{1 - \exp(-\beta_i(f_i(x, a_i) - f_i^0)/(f_i^1 - f_i^0))\right\}, \tag{7.7}$$

where $\alpha_i > 1$, $\beta_i > 0$ or $\alpha_i < 0$, $\beta_i < 0$. The exponential membership function can be determined by asking the DM to specify the three points f_i^0, $f_i^{0.5}$, and f_i^1, within $f_i^{\max}$ and $f_i^{\min}$, where β_i is a shape parameter.

(3) Hyperbolic membership function

$$\mu_i(f_i(x, a_i)) = \frac{1}{2}\tanh(\alpha_i(f_i(x, a_i) - \beta_i)) + \frac{1}{2}, \tag{7.8}$$

where $\alpha_i > 0$ or $\alpha_i < 0$. The hyperbolic membership function can be determined by asking the DM to specify the two points $f_i^{0.25}$ and $f_i^{0.5}$ within $f_i^{\max}$ and $f_i^{\min}$, where α_i is a shape parameter and β_i is associated with the point of inflection.

(4) Hyperbolic inverse membership function

$$\mu_i(f_i(x, a_i)) = \tanh^{-1}(\beta_i(f_i(x, a_i) - \gamma_i)) + \frac{1}{2}, \tag{7.9}$$

where $\alpha_i > 0$ or $\beta_i < 0$. The hyperbolic inverse membership function can be determined by asking the DM to specify the three points f_i^0, $f_i^{0.25}$, and $f_i^{0.5}$ within $f_i^{\max}$ and $f_i^{\min}$, where β_i is a shape parameter and γ_i is associated with the point of inflection.

(5) Piecewise linear membership function

$$\mu_i(f_i(x, a_i)) = \sum_{j=1}^{N_i} \alpha_{ij} \mid f_i(x, a_i) - g_{ij} \mid + \beta_i f_i(x, a_i) + \gamma_i \tag{7.10}$$

where

$$\alpha_{ij} = (t_{i,j+1} - t_{ij})/2, \quad \beta_i = (t_{i,N_i+1} + t_{il})/2, \quad \gamma_i = (s_{i,N_i+1} + s_{il})/2.$$

The piecewise linear membership function can be determined by asking the DM to specify the degree of membership for several values of the objective function within f_i^0 and f_i^1.

It should be noted here that, for the fuzzy equal membership functions, the DM can select left and right functions from among the same types of membership functions described above (excluding the hyperbolic ones).

7.2 Illustrative examples with computer outputs

7.2.1 Multiobjective programming problems

We are now ready to demonstrate the interaction processes for multiobjective linear and nonlinear programming problems using our computer programs by means of illustrative examples designed to test the programs.

Example 7.1 (Linear problem)

Consider the following three-objective linear programming problem:

$$\begin{array}{lrcrcrl}
\text{minimize} & z_1(x) = & -x_1 & - & x_2 & & \\
\text{minimize} & z_2(x) = & x_1 & - & 2x_2 & & \\
\text{minimize} & z_3(x) = & x_1 & + & 2x_2 & & \\
\text{subject to} & & 3x_1 & - & 5x_2 & \leq 15 & \\
& & 3x_1 & - & 5x_2 & \leq 50 & \\
& & -x_1 & + & 3x_2 & \leq 18 & \\
& & -2x_1 & + & x_2 & \leq 5 & \\
& & x_1 & + & 4x_2 & \geq 8 & \\
& & & & x_i \geq 0, \; i = 1, 2. & &
\end{array}$$

In the following illustrations, some of the interaction processes using our computer software package for the IBM PCs are explained through the aid of some of the computer outputs.

Illustration 7.1

Using the MINMAX command, the calculated individual minimum and maximum of each of the objective functions $z_1(x)$, $z_2(x)$, and $z_3(x)$ are displayed as follows:

```
COMMAND: MINMAX                    Displays individual minimum and maximum

Individual minimum and maximum

--------+-----------------------+------------------------
        |        minimum        |         maximum
--------+-----------------------+------------------------
 z( 1)  |           -14.3333    |           -2.0000
 z( 2)  |           -11.8000    |            4.8235
 z( 3)  |             4.0000    |           19.1429
--------+-----------------------+------------------------
```

Illustration 7.2

The MF command is utilized to determine the membership functions for each of the objective functions $z_1(x)$, $z_2(x)$, and $z_3(x)$, sequentially. Here, interaction with the hypothetical DM establishes the following membership functions and corresponding assessment values:

$z_1(x)$: exponential, $(z_1^0, z_1^{0.5}, z_1^1) = (-6, -10, -11)$,

$z_2(x)$: linear, $(z_2^0, z_2^1) = (4, -11.8)$,

$z_3(x)$: hyperbolic inverse, $(z_3^0, z_3^{0.25}, z_3^{0.5}) = (19, 17, 11.5)$.

For each type of membership function, by considering the calculated individual minimum and maximum of each the objective functions, corresponding assessment values are input in a subjective manner as follows:

```
COMMAND: MF                    Elicits fuzzy goals from the DM

Input objective function number : 1

Input fuzzy goal
 (1) fuzzy max
 (2) fuzzy min
 (3) fuzzy equal
=2
Do you want membership function list ?
Input Y(yes) or N(no) : Y
List of membership function types
 (1) Linear
 (2) Exponential
 (3) Hyperbolic
 (4) Hyperbolic inverse
 (5) Piecewise linear
Input membership function type 1,2,3,4 or 5 : 2

The range of objective function 1 under the constraints
          -14.3333  <  z(1)  <              -2.0000
 Input three points such that
         m(z1)=0.0 (z1 : Unacceptable level)
         m(z2)=0.5
         m(z3)=1.0 (z3 : Totally desirable level)
  z1 = -6
  z2 = -10
  z3 = -11
 Another Mset ?  Input Y(yes) or N(no) : Y
```

Illustration 7.3

The shape of the fuzzy min membership function for $z_1(x)$ is shown graphically with the GRAPH command. Thus, the DM can check the properties of the membership functions visually.

Illustration 7.4

Using the GO command, the minimax problem is solved for the initial reference membership levels, and the DM is supplied with the corresponding M-Pareto optimal solution and the trade-off rates between the membership functions. Since

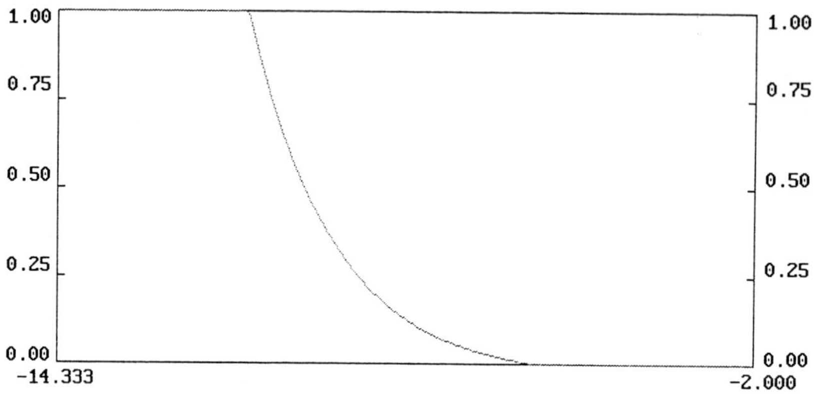


```
Graph of  Membership function  M(z(1))
          Membership function type -- Exponential

Hit any key to exit.
```

the DM is not satisfied with the current membership values, the DM updates the reference membership levels as follows:

```
COMMAND: GO                    [ ITERATION: 1 ]

Initiate an interaction with all the initial
reference membership values are 1

 Reference values rm(i) (i=1,...,3) :
 rm( 1) : 1.000000
 rm( 2) : 1.000000
 rm( 3) : 1.000000
---------------------------
    Pareto optimum
---------------------------+------------------------------
   Membership functions    |     Objective functions
---------------------------+------------------------------
m( 1) =              0.4118  |  z( 1) =            -9.728
m( 2) =              0.4118  |  z( 2) =            -2.506
m( 3) =              0.4118  |  z( 3) =            13.81
----------------------------------------------------------
                  Decision variables
----------------------------------------------------------
   x( 1) =           5.65         x( 2) =          4.078
----------------------------------------------------------
Trade-offs among membership functions
 -dm( 2)/dm(1) =              0.8556
 -dm( 3)/dm(1) =              0.1826
----------------------------------------------------------
Are you satisfied with the current membership values
of the Pareto optimal solution ?
```

```
=no

COMMAND: GO                     [ ITERATION: 2 ]

 Reference values rm(i) (i=1,...,3) :
 rm( 1) : 0.450000
 rm( 2) : 0.500000
 rm( 3) : 0.400000
```

The same procedure continues in this manner until the DM is satisfied with the current values of the membership functions. In this example, at the fifth iteration, the satisficing solution of the DM is obtained as follows:

```
Are you satisfied with the current membership values
of the Pareto optimal solution ?

=yes
                     The following values are your satisficing solution

--------------------------+------------------------------
   Membership functions   |     Objective functions
--------------------------+------------------------------
m( 1) =            0.4224 |  z( 1) =             -9.763
m( 2) =            0.4923 |  z( 2) =             -3.779
m( 3) =            0.3923 |  z( 3) =              14.28
---------------------------------------------------------
                Decision variables
---------------------------------------------------------
   x( 1) =         5.249          x( 2) =         4.514
---------------------------------------------------------
```

Example 7.2 (Nonlinear problem)

Consider the following three-objective nonlinear programming problem:

$$\begin{array}{rrl} \text{fuzzy max} & f_1(x) = & 2x_1^2 + 4(x_2 - 20)^2 + 3(x_3 - 15)^2 \\ \text{fuzzy min} & f_2(x) = & (x_1 + 10)^2 + 2(x_2 - 25)^2 + 3(x_3 + 5)^2 \\ \text{fuzzy equal} & f_3(x) = & 3(x_1 + 15)^2 + 2(x_2 + 10)^2 + (x_3 + 20)^2 \\ \text{subject to} & & (x_1 + 5)^2 + (x_2 + 8)^2 + (x_3 - 10)^2 \leq 200, \\ & & 0 \leq x_i \leq 10, \quad i = 1, 2, 3. \end{array}$$

By considering the calculated individual minimum and maximum of each of the objective functions $f_1(x)$, $f_2(x)$, and $f_3(x)$ displayed with the MINMAX command, the MF command is utilized to determine the membership functions for each of the objective functions $f_1(x)$, $f_2(x)$, and $f_3(x)$, sequentially. Here, interaction with the hypothetical DM establishes the following membership functions and corresponding assessment values:

$f_1(x)$: exponential, $(f_1^0, f_1^{0.5}, f_1^1) = (950, 1800, 2200)$,

$f_2(x)$: hyperbolic, $(f_2^{0.25}, f_2^{0.5}) = (1900, 1750)$,

$$f_3(x) \begin{cases} \text{left: exponential, } (f_3^0, f_3^{0.5}, f_3^1) = (1300, 1700, 1900). \\ \text{right: linear, } (f_3^0, f_3^1) = (2500, 1900). \end{cases}$$

For each type of membership function, corresponding assessment values are input in a subjective manner by considering the calculated individual minimum and maximum of each the objective functions, and the shape of the membership functions are checked graphically with the GRAPH command.

Then, using the GO command, the augmented minimax problem is solved for the initial reference membership levels, and the DM is supplied with the corresponding (local) (M-) Pareto optimal solution and the trade-off rates between the membership functions. Since the DM is not satisfied with the current membership values, the DM updates the reference membership levels.

The same procedure continues in this manner until the DM is satisfied with the current values of the membership functions. In this example, at the fifth iteration, the (local) satisficing solution of the DM is obtained as follows:

```
Are you satisfied with the current membership values
of the Pareto optimal solution ?

=yes

                          The following values are your satisficing solution


---------------------------+------------------------------
   Membership functions    |      Objective functions
---------------------------+------------------------------
m( 1) =             0.8002 |  f( 1) =          2063.4139
m( 2) =             0.6802 |  f( 2) =          1646.9619
m( 3) =             0.8602 |  f( 3) =          1853.0568
----------------------------------------------------------
                   Decision variables
----------------------------------------------------------
   x( 1) =          4.3348        x( 2) =         0.0225
----------------------------------------------------------
   x( 3) =          3.0358
----------------------------------------------------------
```

7.2.2 Multiobjective programming problems with fuzzy parameters

We now demonstrate the interaction processes for multiobjective linear and nonlinear programing problems with fuzzy parameters using our corresponding computer programs with several illustrative numerical examples designed to test each of the programs.

Example 7.3 (Linear problem)

Consider the following three-objective linear programming problem with fuzzy parameters:

$$
\begin{array}{lrcrcrcrl}
\text{minimize} & z_1 = & C_{11}x_1 & - & 4x_2 & + & 3x_3 & & \\
\text{minimize} & z_2 = & x_1 & + & C_{22}x_2 & - & 3x_3 & & \\
\text{minimize} & z_3 = & -x_1 & + & x_2 & + & C_{33}x_3 & & \\
\text{subject to} & & A_{11}x_1 & + & A_{12}x_2 & + & 3x_3 & \leq 12 \\
 & & x_1 & + & 2x_2 & + & A_{23}x_3 & \leq B_2 \\
 & & x_i \geq 0, & & i = 1, 2, 3 & & & &
\end{array}
$$

The membership functions for the fuzzy numbers C_{11}, C_{22}, C_{33}, A_{11}, A_{12}, A_{23}, and B_2 are explained in Table 7.1, where L and E represent, respectively, linear and exponential membership functions.

Table 7.1 Fuzzy numbers for Example 7.3

M	p_1	p_2	p_3	p_4	left	right
C_{11}	-2.00	-1.00	-1.00	0.00	E	E
C_{22}	0.00	2.00	2.00	3.00	E	E
C_{33}	-1.00	-0.50	-0.50	0.00	E	L
A_{11}	0.00	3.00	3.00	4.00	L	L
A_{12}	0.50	1.00	1.00	1.50	E	L
A_{23}	0.50	1.00	1.00	1.50	E	E
B_2	8.00	12.00	12.00	14.00	L	E

In the following illustrations, interaction processes using our computer programs for the IBM PCs are explained through the aid of some computer outputs.

Illustration 7.5

Using the MINMAX command, the calculated individual minimum and maximum of each of the objective functions z_1, z_2, and z_3 for $\alpha = 0$ and $\alpha = 1$ are displayed as follows:

```
COMMAND: MINMAX                Displays individual minimum and maximum
```

Individual minimum and maximum for α = 1

	minimum	maximum
z(1)	-24.0000	12.0000
z(2)	-12.0000	12.0000
z(3)	-4.0000	6.0000

Individual minimum and maximum for α = 0

	minimum	maximum
z(1)	-28.0000	12.0000
z(2)	-12.0000	21.0000
z(3)	-16.0000	7.0000

Illustration 7.6

Using the GO command, the minimax problem is solved for the initial reference levels and the degree α, and the DM is supplied with the corresponding α-Pareto optimal solution and the trade-off rates between the objective functions and the degree α. Since the DM is not satisfied with the current values of the objective functions, the DM updates the reference levels as follows:

```
COMMAND: GO                  [ ITERATION: 1 ]

Initiate an interaction with all the initial
reference membership values are 1

 Reference values rz(i) (i=1,...,3) :
rz( 1) :   -28.000000
rz( 2) :   -12.000000
rz( 3) :   -16.000000
--------------------------------------

The degree α of the α level sets
for the fuzzy parameters : 0.900000

--------------------------------------
```

```
     Pareto optimum
----------------------------------------
     Objective functions
----------------------------------------
z( 1) =            -11.76
z( 2) =             4.237
z( 3) =            0.2374
------------------------------------------------------------
                      Decision variables
------------------------------------------------------------
   x( 1) =            2.264    x( 2) =              3.11
   x( 3) =            1.023
------------------------------------------------------------
Trade-offs among objective functions
 -dz( 2)/dz(1) =            0.8254
 -dz( 3)/dz(1) =            0.6397
-------------------------------------------------------------
Trade-offs between α and objective functions
  -dz/dα =               5.3232
-------------------------------------------------------------




Are you satisfied with the current objective values
of the Pareto optimal solution ?

=no

COMMAND: GO                  [ ITERATION: 2 ]

 Reference values rz(i) (i=1,...,3) :
rz( 1) :     -5.000000
rz( 2) :      0.000000
rz( 3) :     -3.000000
-------------------------------------------------------------
The degree α of the α level sets
for the fuzzy parameters : 0.900000
-------------------------------------------------------------
```

In this example, at the sixth iteration, the satisficing solution of the DM is derived and all the interactive processes are summarized in Table 7.2.

Table 7.2 Interactive processes for Example 7.3

Iteration	1	2	3	4	5	6
$\bar{z}_1$	-28	-5	-3	-2	-2	-3
$\bar{z}_2$	-12	0	-1	-2.5	-2.5	-3
$\bar{z}_3$	-16	-3	-3	-3.5	-3.5	-1
α	0.9	0.9	0.9	0.9	0.8	0.8
z_1	-11.7626	-4.0907	-2.2996	-0.8760	-1.1732	-2.8582
z_2	4.2374	0.9093	-0.2996	-1.3760	-1.6734	-2.8582
z_3	0.2374	-2.0907	-2.2996	-2.3760	-2.6734	-0.8582
x_1	2.2644	2.6178	2.4556	2.2613	2.4317	1.3756
x_2	3.1103	1.2757	1.0347	0.8933	0.9556	2.0743
x_3	1.0234	1.2590	1.4777	1.6953	1.7899	2.3253
$-\partial z_2/\partial z_1$	0.8254	0.8254	0.8254	0.8254	0.7844	0.7844
$-\partial z_3/\partial z_1$	0.6397	0.6397	0.6397	0.6397	0.5681	0.5681
$-\partial z_1/\partial \alpha$	5.3232	3.4343	3.1321	2.9291	3.0433	3.5908

Example 7.4 (Linear Problem with fuzzy goals)

Consider the following three-objective linear programming problem with fuzzy parameters:

$$\begin{array}{lrcrcrcrl} \text{fuzzy min} & z_1 = & C_{11}x_1 & - & 64x_2 & + & x_3 & & \\ \text{fuzzy max} & z_2 = & -3x_1 & + & C_{22}x_2 & + & x_3 & & \\ \text{fuzzy equal} & z_3 = & 5x_1 & + & x_2 & + & C_{33}x_3 & & \\ \text{subject to} & & A_{11}x_1 & + & A_{12}x_2 & + & 3x_3 & \leq 12 \\ & & x_1 & + & 2x_2 & + & A_{23}x_3 & \leq B_2 \\ & & x_i \geq 0, & & i = 1, 2, 3 & & & \end{array}$$

The membership functions for the fuzzy numbers C_{11}, C_{22}, C_{33}, A_{11}, A_{12}, A_{23}, and B_2 are explained in Table 7.3, where L and E represent, respectively, linear and exponential membership functions.

In applying our computer program to this problem, suppose that the interaction with the hypothetical DM establishes the following membership functions

Table 7.3 Fuzzy numbers for Example 7.4

M	p_1	p_2	p_3	p_4	left	right
C_{11}	0.00	2.00	2.00	2.50	L	E
C_{22}	-1.25	-0.75	-0.75	-0.25	E	E
C_{33}	-0.25	0.00	0.00	1.00	E	E
A_{11}	0.00	3.00	3.00	4.00	L	L
A_{12}	0.50	1.00	1.00	1.50	E	L
A_{23}	0.50	1.00	1.00	1.50	E	E
B_2	8.00	12.00	12.00	14.00	L	E

and the corresponding assessment values for the three objective functions in the α-MOLP:

$$z_1: \text{ exponential } (z_{1R}^0, z_{1R}^{0.5}, z_{1R}^1) = (30, -10, -25),$$

$$z_2: \text{ hyperbolic inverse } (z_{2L}^0, z_{1R}^{0.25}, z_{2L}^{0.5}) = (-8, -7.5, -6.5),$$

$$z_3 \begin{cases} \text{left: exponential } (z_{3L}^0, z_{3L}^{0.5}, z_{3L}^1) = (12, 14, 15). \\ \text{right: linear } (z_{3R}^0, z_{3R}^1) = (18, 15). \end{cases}$$

In the following illustrations, interaction processes using our computer programs for the IBM PCs are explained through the aid of some computer outputs.

Illustration 7.7

Using the MINMAX command, the calculated individual minimum and maximum of each of the objective functions z_1, z_2, and z_3 for $\alpha = 0$ and $\alpha = 1$ are displayed as follow:

```
COMMAND: MINMAX                    Displays individual minimum and maximum


Individual minimum and maximum for α = 1

--------+-----------------------+-----------------------
        |        minimum        |        maximum
--------+-----------------------+-----------------------
 z( 1)  |           -24.0000    |              8.0000
 z( 2)  |           -12.0000    |              4.0000
 z( 3)  |            -4.0000    |             20.0000
--------+-----------------------+-----------------------


Individual minimum and maximum for α = 0

--------+-----------------------+-----------------------
        |        minimum        |        maximum
--------+-----------------------+-----------------------
 z( 1)  |           -28.0000    |             35.0000
 z( 2)  |           -42.0000    |              4.0000
 z( 3)  |            -1.0000    |             70.0000
--------+-----------------------+-----------------------
```

Illustration 7.8

The MF command is utilized to determine the membership functions for each of the objective functions z_1, z_2, and z_3 sequentially. Here the interaction with the hypothetical DM establishes that the first membership function should be exponential, the second hyperbolic inverse, and the third exponential and linear. For each type of membership function, by considering the calculated individual minimum and maximum of each of the objective functions, the corresponding assessment values are input in a subjective manner in the following way:

```
COMMAND: MF                   Elicits fuzzy goals from the DM

Input objective function number : 1

Input fuzzy goal
 (1) fuzzy max
 (2) fuzzy min
 (3) fuzzy equal
=1
Do you want membership function list ?
Input Y(yes) or N(no) : Y
List of membership function types
 (1) Linear
 (2) Exponential
 (3) Hyperbolic
 (4) Hyperbolic inverse
 (5) Piecewise linear
Input membership function type 1,2,3,4 or 5 : 2

The range of objective function 1 under the constraints
          -24.0000  <  z(1) <               8.0000
 Input three points such that
          m(z1)=0.0 (z1 : Unacceptable level)
          m(z2)=0.5
          m(z3)=1.0 (z3 : Totally desirable level)
  z1 = 30
  z2 = -10
  z3 = -25
 Another Mset ?  Input Y(yes) or N(no) :Y
```

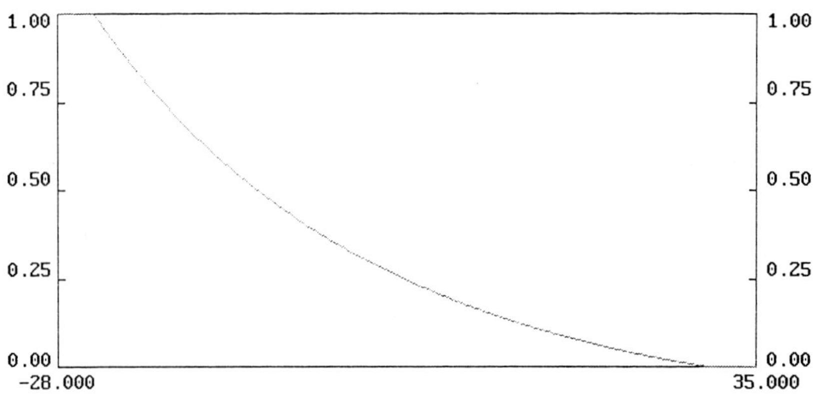


```
Graph of  Membership function  M(z(1))
          Membership function type -- Exponential

Hit any key to exit.
```

Illustration 7.9

With the GRAPH command, the shape of the membership function for z_1 is shown graphically. Thus, the DM can check the properties of the membership functions visually.

Illustration 7.10

Using the GO command, the minimax problem is solved for the initial reference membership values and the degree α, and the DM is supplied with the corresponding M-α-Pareto optimal solution and the trade-off rates between the membership functions and the degree α. Since the DM is not satisfied with the current values of the membership functions, the DM updates the reference membership values as follows:

```
COMMAND: GO                        [ ITERATION: 1 ]

Initiate an interaction with all the initial
reference membership values are 1

 Reference values rm(i) (i=1,...,3) :
 rm( 1) : 1.000000
 rz( 2) : 1.000000
 rz( 3) : 1.000000
--------------------------------------
The degree α of the α level sets
for the fuzzy parameters : 0.750000
--------------------------------------
M-α-Pareto optimum
---------------------------+------------------------------
    Membership functions   |      Objective functions
---------------------------+------------------------------
m( 1) =              0.4913  |  z( 1) =             -9.642
m( 2) =              0.4913  |  z( 2) =              -6.54
m( 3) =              0.4913  |  z( 3) =              13.98
----------------------------------------------------------
                Decision variables
----------------------------------------------------------
   x( 1) =           2.039          x( 2) =          3.556
   x( 3) =           1.525
----------------------------------------------------------
Trade-offs among membership functions
 -dm( 2)/dm(1) =              0.1589
 -dm( 3)/dm(1) =              0.3625
----------------------------------------------------------
Trade-offs between α and objective functions
  -dm/dα  =               -8.4098
----------------------------------------------------------
Are you satisfied with the current membership values
of the M-α-Pareto optimal solution ?
```

Table 7.4 Interactive processes for Example 7.4

Iteration	1	2	3	4	5
$\bar{\mu}_1$	1	0.45	0.4	0.4	0.4
$\bar{\mu}_2$	1	0.5	0.5	0.5	0.6
$\bar{\mu}_3$	1	0.55	0.55	0.55	0.65
α	0.75	0.75	0.75	0.65	0.65
μ_1	0.4913	0.4331	0.3888	0.4727	0.3773
μ_2	0.4912	0.4830	0.4888	0.5726	0.5772
μ_3	0.4912	0.5330	0.5388	0.6226	0.6272
z_1	-9.6425	-7.1239	-5.0258	-8.8649	-4.4564
z_2	-6.5401	-6.5773	-6.5513	-6.1726	-6.1526
z_3	13.9773	14.0832	14.0972	14.0972	14.3035
x_1	2.0385	2.1716	2.2677	2.1092	2.3074
x_2	3.5563	2.9893	2.5144	3.3520	2.3443
x_3	1.5249	1.5760	1.6302	1.8010	1.9210
$-\partial\mu_2/\partial\mu_1$	0.1589	0.1755	0.1901	0.0606	0.0722
$-\partial\mu_3/\partial\mu_1$	0.3625	0.4205	0.4593	0.1589	0.1894
$-\partial\mu_1/\partial\alpha$	8.4098	7.2504	6.3777	22.1895	17.3201

```
=no

COMMAND: GO                     [ ITERATION: 2 ]

 Reference values rm(i) (i=1,...,3) :
rm( 1) : 0.450000
rm( 2) : 0.500000
rm( 3) : 0.550000
-----------------------------------------------------

The degree α of the α level sets
for the fuzzy parameters : 0.750000
-----------------------------------------------------
```

In this example, at the fifth iteration, the satisficing solution of the DM is derived and all the interactive processes are summarized in Table 7.4.

Example 7.5 (Nonlinear Problem)

Consider the following three-objective nonlinear programming problem with fuzzy parameters:

$$
\begin{array}{lll}
\text{minimize} & f_1(x, A_1) = & A_{11}x^2 + (x_2 + 5)^2 + 2(x_3 - A_{12})^2 \\
\text{minimize} & f_2(x, A_2) = & (x_1 + A_{21})^2 + A_{22}(x_2 - 55)^2 + 3(x_3 + 20)^2 \\
\text{minimize} & f_3(x, A_3) = & A_{31}(x_1 - 55)^2 + A_{32}(x_2 + A_{33})^2 + (x_3 + 20)^2 \\
\text{subject to} & & B_{11}x_1^2 + B_{12}x_2^2 + B_{13}x_3^2 \leq 100 \\
& & 0 \leq x_i \leq 10, \quad i = 1, 2, 3.
\end{array}
$$

Table 7.5 Fuzzy numbers for Example 7.5

M	p_1	p_2	p_3	p_4	left	right
A_{11}	3.8	4.0	4.0	4.3	L	E
A_{12}	57.0	59.0	60.0	63.5	E	E
A_{21}	18.0	19.5	20.0	2.2.5	E	L
A_{22}	1.75	2.0	2.0	2.25	L	E
A_{31}	2.3	2.5	2.5	2.75	E	L
A_{32}	1.25	1.4	1.5	1.7	L	L
B_{11}	0.9	1.0	1.0	1.1	E	E
B_{12}	0.8	0.95	1.0	1.2	E	E
B_{13}	0.85	1.0	1.0	1.15	E	L

Table 7.6 Interactive processes for Example 7.5

Iteration	1	2	3	4	5	6
$\bar{f}_1$	5025	6800	6500	6700	6300	6500
$\bar{f}_2$	5650	6700	6300	6500	6600	6700
$\bar{f}_3$	6062.5	6800	6700	6900	6900	7000
α	0.9	0.9	0.7	0.65	0.65	0.65
f_1	6438.96	7008.58	6823.52	6787.90	6492.66	6556.71
f_2	7063.96	6908.58	6623.52	6587.90	6792.66	6756.71
f_3	7476.46	7008.58	7023.52	6987.90	7092.66	7056.71
x_1	6.3782	7.7803	7.3317	7.3351	7.0352	7.1465
x_2	6.5187	6.1078	6.8347	6.8771	6.5879	6.6174
x_3	4.4076	2.1969	2.5747	2.6287	3.8383	3.5815
$-\partial f_2/\partial f_1$	1.1521	0.9217	0.9384	0.9448	1.0935	1.0629
$-\partial f_3/\partial f_1$	1.3422	0.9205	1.0323	1.0392	1.1903	1.1506
$-\partial f_1/\partial \alpha$	797.25	835.90	719.52	705.95	694.09	696.61

The membership functions for the fuzzy numbers $A_{11}, \ldots, A_{32}$ and B_{11}, B_{12}, B_{13} in this example are explained in Table 7.5 where L and E represent, respectively, linear and exponential membership functions.

In this example, at the sixth iteration, the satisficing solution of the DM is derived and all the interactive processes are summarized in Table 7.6.

Example 7.6 (Nonlinear Problem with fuzzy goals)

Consider the following three-objective nonlinear programming problem with fuzzy parameters:

$$\begin{array}{lrl}
\text{fuzzy min} & f_1(x, A_1) = & (x_1+5)^2 + A_{11}x_2^2 + 2(x_3 - A_{12})^2 \\
\text{fuzzy min} & f_2(x, A_2) = & A_{21}(x_1 - 45)^2 + (x_2+15)^2 + 3(x_3 + A_{22})^2 \\
\text{fuzzy equal} & f_3(x, A_3) = & A_{31}(x_1+20)^2 + A_{32}(x_2 - 45)^2 + (x_3+15)^2 \\
\text{subject to} & & B_{11}x_1^2 + B_{12}x_2^2 + B_{13}x_3^2 \le 100 \\
 & & 0 \le x_i \le 10, \quad i = 1, 2, 3.
\end{array}$$

The membership functions for the fuzzy numbers $A_{11}, \ldots, A_{32}$ and B_{11}, B_{12}, B_{13} in this example are explained in Table 7.7 where L and E represent, respectively, linear and exponential membership functions.

In applying our computer program to this problem, suppose that the interaction with the hypothetical DM establishes the following membership functions and the corresponding assessment values for the three objective functions.

f_1: exponential, $(f_1^0, f_1^{0.5}, f_1^1) = (5400, 5000, 3300)$,
f_2: exponential, $(f_2^0, f_2^{0.5}, f_2^1) = (6900, 4600, 3900)$,

$$f_3 \begin{cases} \text{left: hyperbolic inverse, } (f_3^0, f_3^{0.5}, f_3^1, b_3) = (7800, 8200, 10000, 9000). \\ \text{right: exponential, } (f_3^0, f_3^{0.5}, f_3^1) = (13300, 11000, 10000). \end{cases}$$

In this example, at the fifth iteration the satisficing solution of the hypothetical DM is derived. All the interactive processes are summarized in Table 7.8.

Table 7.7 Fuzzy numbers for Example 7.6

M	p_1	p_2	p_3	p_4	left	right
A_{11}	3.8	4.0	4.0	4.3	L	E
A_{12}	48.5	50.0	50.0	52.0	E	E
A_{21}	1.85	2.0	2.0	2.2	E	L
A_{22}	18.2	20.0	20.0	22.5	L	E
A_{31}	2.9	3.0	3.0	3.15	E	L
A_{32}	4.7	5.0	5.0	5.35	L	L
B_{11}	0.9	1.0	1.0	1.1	E	E
B_{12}	0.8	1.0	1.0	1.2	E	E
B_{13}	0.85	1.0	1.0	1.15	E	L

Table 7.8 Interactive processes for Example 7.6

Iteration	1	2	3	4	5
$\bar{\mu}_1$	1	0.5	0.55	0.55	0.65
$\bar{\mu}_2$	1	0.6	0.50	0.50	0.55
$\bar{\mu}_3$	1	0.8	0.75	0.75	0.75
α	0.9	0.9	0.9	0.7	0.7
μ_1	0.5678	0.4558	0.5543	0.5979	0.6392
μ_2	0.5678	0.5558	0.5043	0.5479	0.5392
μ_3	0.5678	0.7558	0.7543	0.7979	0.7392
f_1	4917.71	5048.17	4934.94	4877.16	4816.81
f_2	4475.23	4496.28	4591.57	4510.35	4526.11
f_3	10828.78	10422.53	10425.39	10342.49	10455.02
x_1	8.4567	8.0461	7.9090	8.1024	8.1775
x_2	5.2592	6.0275	6.0594	6.0594	5.8785
x_3	1.8210	1.2689	1.8437	1.9837	2.2585
$-\partial\mu_2/\partial\mu_1$	0.6433	0.4845	0.7575	0.6824	0.7638
$-\partial\mu_3/\partial\mu_1$	1.5303	1.2998	1.6415	1.9187	2.1046
$-\partial\mu_i/\partial\alpha$	0.2143	0.2390	0.2164	0.2193	0.2074

7.3 Installation of PC versions for linear models

The following executable files are included in the distribution diskettes with data files for test problems discussed in the previous section.

(1) **IFMOLP.EXE:** interactive computer program for multiobjective linear programming problems incorporating the fuzzy goals of the DM.

(2) **IMOLPFP.EXE:** interactive computer program for multiobjective linear programming problems with fuzzy parameters.

(3) **IFMOLPFP.EXE:** interactive computer program for multiobjective linear programming problems with fuzzy parameters incorporating the fuzzy goals of the DM.

Each executable file can be installed in an IBM-PC XT/AT or a compatible computer under MS-DOS or PC-DOS 3.1 or higher with preferably 640k of memory. Although it is possible to run each executable file from floppy diskette, it is preferable to install it on a hard disk. With some knowledge of the DOS directory structure, the installation procedure is just to create a new user directory such as IFMOLP, IMOLPFP, or IFMOLPFP and to copy the corresponding files to the directory. Then, having established the path in your AUTOEXEC.BAT file, you can run each executive code file by entering the command IFMOLP, IMOLPFP, or IFMOLPFP from the DOS prompt. (Even if you are not familiar with DOS, a more experienced DOS user around you would create it for you.) After pressing any key at the invitation screen, the main menu with several commands is displayed. Using the data files EX71.DAT and EX71.MEM and following the user-friendly prompt messages, you can experience the interaction processes for Example 7.1. Similarly, data files EX73.DAT for Example 7.3 or EX74.DAT and EX74.MEM for Example 7.4 can be used to implement the interaction processes.

CHAPTER 8

SOME APPLICATIONS

We are now ready to demonstrate some application aspects of interactive fuzzy programming methods presented in the previous chapters. As examples of Japanese case studies, we present some applications of interactive fuzzy multiobjective programming methods to the operation of a packaging system in automated warehouses, pass scheduling for hot tandem mills, and regional planning.

8.1 Packaging system in automated warehouses

Consider an application of our interactive fuzzy multiobjective linear programming method presented in Section 3.5 to the optimal operation of a packaging system in automated warehouses. The decision maker (DM) (manager of warehouses) is supposed to decide the items for shipping next day by considering the total volume to be packed, the priority for packaging, and the balance of the inventory, etc. The problem formulation was originally given by Hamada (1976), and a three-objective linear programming problem formulation was developed by Nakayama, Karasawa, and Dohi (1980) with a numerical example solved by the interactive multiobjective decision-making method called the interactive relaxation method. Then, by considering the fuzzy goals of the DM, both the fuzzy multiobjective linear programming method and the interactive fuzzy multiobjective linear programming method were applied successfully to derive a satisficing solution for the DM for the formulated problem. (Sakawa 1983b, Sakawa, Yano, and Yumine 1987).

In this section, we first present the three-objective linear programming formulation as an optimal operation for a packaging system in automated warehouses according to Nakayama et al. (1980). Then an application for our interactive fuzzy multiobjective linear programming method to the formulated problem is demonstrated along with the computer outputs (Sakawa, Yano, and Yumine 1987).

8.1.1 Problem formulation

As evaluation criteria for the packaging schedule in automated warehouses, for making effective use of the facilities of the warehouse, and for satisfying the requirements of customer, the following items are taken into consideration:

(1) priority for packaging

(2) total volume to be packaged

(3) work balance of incoming and outgoing for packaging

(4) balance of inventory

(5) shortage of packaged stocks, and

(6) shortage of unpackaged stocks.

Three conflicting objective functions and several constraints can be derived on the basis of these six items.

To make a mathematical formulation, let

$$x_{jlk} = \text{volume of goods to be packaged}$$

denote the decision variable, where

j = priority of packaging for each packaging style,
l = packaging style ($l = 1, 2$ and $l = 3$: unpackaged),
k = index of rack.

The standard number of packages B_{jl} in each batch is given by the cost of stocking and arrangement in stock management. To make the number of packages as close as possible to the standard number, the first objective function to be maximized is formulated as

$$f_1 = \sum_j \sum_l \sum_k x_{jlk}/B_{jl} \tag{8.1}$$

with which the following constraint is imposed for setting the upper limit of the number to be packaged:

$$\sum_k x_{jlk} \leq B_{jl} \quad (j = 1, \ldots, J_M^R,\ l = 1, 2) \tag{8.2}$$

where

J_M^R = number of items for each packaging style.

To keep the priority for packaging, if some goods with lower priority are packaged before ones with higher priority, the following penalty is imposed and considered as the second objective function to be minimized:

$$f_2 = \sum_{j > R_l^e} \sum_l \sum_k (j - R_l^e) x_{jlk} \tag{8.3}$$

where

R_l^e = number of items to be packaged on the next day.

By considering the balance of stocks among the racks, the following constraint is imposed to make the stocks of each of the goods among the racks well-balanced:

$$y_{jlk} + x_{jlk} \leq y_{jlk'} + x_{jlk'} \quad \text{for } y_{jlk} \leq y_{jlk'} \tag{8.4}$$

$$(l = 1, 2, j - 1, \ldots, J_M^R, \ k = 1, \ldots, N_j^k - 1)$$

where

y_{jlk} = stock volume of each item,
N_j^k = number of standard stocking racks.

Observing that this constraint represents the limits of balance, the third objective function to be minimized is expressed as

$$f_3 = \sum_j \sum_l \left\{ (x_{jlk^1} - x_{jlk^2}) N_j^k / \sum_k y_{jlk} \right\} \tag{8.5}$$

where

k^1 = the rack with the maximum volume of stocks,
k^2 = the rack with the minimum volume of stocks.

Other constraints to be considered can be formulated in the following way.

The balance of operation between racks is kept under the following constraint:

$$\sum_j \sum_l x_{jlk} \leq (w_1 + w_2)(1 + \varepsilon_w)/K_M \quad (k = 1, \ldots, K_M) \tag{8.6}$$

where

w_l = facilities of packaging system,
ε_w = degree of imbalance of packaging between racks,
K_M = total number of racks.

The facilities of the packaging systems are considered as follows:

$$\alpha w_l \leq \sum_j \sum_k x_{jlk} \leq w_l \quad (l = 1, 2) \tag{8.7}$$

where

α = constant such that $0 < \alpha < 1$.

The guarantee of stocks of unpackaged goods are expressed as

$$\beta + \left(\sum_k y_{jlk} + \sum_k x_{jlk} \right) / a_{jl} \leq \left(\sum_k y_{j'3k} - \sum_l \sum_k x_{j'lk} \right) / a_{j'3} \tag{8.8}$$

$$(j = 1, \ldots, J_M^R, \ l = 1, 2)$$

where

β = days for keeping packaged goods in the warehouse,

a_{jl} = average flow of packages,
j' = priority of packaging for the other style of same goods.

According to the above considerations, it is required of the DM (the manager of the warehouse) to decide the best daily operating schedule on the basis of much data including the number of goods to be stocked and shipped.

8.1.2 Numerical example

One example, according to Nakayama et al. (1980), supposes that the DM would like to make an optimal packaging schedule of two kinds of items, A and B, in two packaging styles from two racks. The data for the constants and the decision variables of the formulated problem are given in Table 8.1.

The three-objective linear programming formulation for this numerical data can be described as follows:

$$\begin{aligned}\text{maximize} \quad z_1(x) &= 1000 f_1 \\ &= x_1 + x_2 + 5x_3 + 5x_4 + 4x_5 + 4x_6 + 2x_7 + 2x_8\end{aligned}$$

Table 8.1 Data for numerical example

			y_{jlk}(pallets)			a_{jl}(pallets)			Variables	
Goods	j	l						B_{jl}	x_{jlk}	
			$k=1$	$k=2$	$l=3$		$l=3$		$k=1$	$k=2$
	1	1	900	600		500		1000	x_1	x_2
A					6000		500			
	2	2	1700	1800		500		200	x_3	x_4
	2	1	700	900		200		250	x_5	x_6
B					3000		200			
	1	2	400	600		200		500	x_7	x_8

$J_M^R = 2, R_1^e = 1, R_2^e = 1, N_j^K = K_M = 2$

$\omega_1 = 1500$(pallets/day), $\omega_2 = 1000$(pallets/day), $\beta = 3$(days)

$$\begin{aligned}
\text{minimize} \quad z_2(x) &= f_2 \\
&= x_3 + x_4 + x_5 + x_6 \\
\text{minimize} \quad z_3(x) &= 4200 f_3 \\
&= 5.6x_1 - 5.6x_2 - 2.4x_3 + 2.4x_4 \\
&\quad -5.25x_5 + 5.25x_6 - 8.4x_7 + 8.4x_8 \\
\text{subject to} \quad & x_1 + x_2 \le 1000,\ x_3 + x_4 \le 200, \\
& x_5 + x_6 \le 250,\ x_7 + x_8 \le 500; \\
& -x_1 + x_2 \le 300,\ x_3 - x_4 \le 100, \\
& x_5 - x_6 \le 200,\ x_7 - x_8 \le 200; \\
& x_1 + x_3 + x_5 + x_7 \le 1500, \\
& x_2 + x_4 + x_6 + x_8 \le 1500; \\
& 500 \le x_1 + x_2 + x_5 + x_6 \le 1500, \\
& 300 \le x_3 + x_4 + x_7 + x_8 \le 1000; \\
& 2x_1 + 2x_2 + x_3 + x_4 \le 3000, \\
& x_1 + x_2 + 6x_3 + 6x_4 \le 1000; \\
& 3x_5 + 3x_6 + x_7 + x_8 \le 800, \\
& x_5 + x_6 + 2x_7 + 2x_8 \le 1400, \\
& x_i \ge 0,\ i = 1, 2, ..., 8.
\end{aligned}$$

In the following illustrations, some of the interaction processes using our interactive fuzzy multiobjective linear programming software package for IBM PCs are explained through the aid of some of the computer outputs.

Illustration 8.1

Using the MINMAX command, the calculated individual minimum and maximum of each of the objective functions $z_1(x)$, $z_2(x)$, and $z_3(x)$ are displayed as follows:

```
COMMAND: MINMAX                     Displays individual minimum and maximum

Individual minimum and maximum

--------+-------------------------+------------------------
        |        minimum          |        maximum
--------+-------------------------+------------------------
 z( 1)  |            1100.0000    |            2400.0000
 z( 2)  |               0.0000    |             323.5294
 z( 3)  |           -4650.0000    |           10325.0000
--------+-------------------------+------------------------
```

Illustration 8.2

The MF command is utilized to determine the membership functions for each of the objective functions $z_1(x)$, $z_2(x)$, and $z_3(x)$, sequentially. The DM establishes the following membership functions and corresponding assessment values:

$$z_1(x): \text{ piecewise linear, } (z_1^0, z_1^1) = (1400, 2400)$$
$$\text{where } (z_1^{0.2}, z_1^{0.5}) = (1700, 1950),$$

$$z_2(x): \text{ exponential, } (z_2^0, z_2^{0.5}, z_2^1) = (200, 60, 0),$$

$$z_3(x): \text{ linear, } (z_3^0, z_3^1) = (0, -4650).$$

For each type of membership function, corresponding assessment values are input in a subjective manner by considering the calculated individual minimum and maximum of each the objective functions together with the rate of increase of membership of satisfaction in the following way:

```
COMMAND: MF                    Elicits fuzzy goals from the DM

Input objective function number : 1

Input fuzzy goal
 (1) fuzzy max
 (2) fuzzy min
 (3) fuzzy equal
=1
Do you want membership function list ?
Input Y(yes) or N(no) : Y
List of membership function types
 (1) Linear
 (2) Exponential
 (3) Hyperbolic
 (4) Hyperbolic inverse
 (5) Piecewise linear
Input membership function type 1,2,3,4 or 5 : 5

The range of objective function 1 under the constraints
         1100.0000  <  z(1)  <           2400.0000
 Input two points such that
          m(z1)=0.0 (z1 : Unacceptable level)
          m(z2)=1.0 (z2 : Totally desirable level)
  z1 = 1400
  z2 = 2400
 Input the number of piecewise points : 2
 Input the objective function value in descending order
1950
```

```
1700
 Input corresponding membership value in descending order
0.5
0.2

Another Mset ?  Input Y(yes) or N(no)
```

Illustration 8.3

Using the GO command, the minimax problem is solved for the initial reference membership levels and the DM is supplied with the corresponding Pareto optimal solution and the trade-off rates between the membership functions. Since the DM is not satisfied with the current membership values, the DM updates the reference membership levels as follows:

```
COMMAND: GO                     [ ITERATION: 1 ]

Initiate an interaction with all the initial
reference membership values are 1

 Reference values rm(i) (i=1,...,3) :
 rm( 1) : 1.000000
 rm( 2) : 1.000000
 rm( 3) : 1.000000
---------------------------
    Pareto optimum
--------------------------+-----------------------------
   Membership functions    |     Objective functions
--------------------------+-----------------------------
m( 1) =          0.6986  |  z( 1) =            2129
m( 2) =          0.6986  |  z( 2) =           32.19
m( 3) =          0.6986  |  z( 3) =           -3248
-----------------------------------------------------------
                Decision variables
-----------------------------------------------------------
   x( 1) =          350      x( 2) =             650
   x( 3) =            0      x( 4) =               0
   x( 5) =        32.19      x( 6) =               0
   x( 7) =        333.3      x( 8) =           166.7
-----------------------------------------------------------
      Pareto optimality test
  eps( 1) =          0.0000
  eps( 2) =          0.0000
  eps( 3) =        280.4919
------------------------------
    Pareto optimum
```

```
---------------------------+------------------------------
   Membership functions    |      Objective functions
---------------------------+------------------------------
m( 1) =             0.6986 | z( 1) =                 2129
m( 2) =             0.6986 | z( 2) =                32.19
m( 3) =             0.7589 | z( 3) =                -3529
----------------------------------------------------------
                  Decision variables
----------------------------------------------------------
   x( 1) =            350      x( 2) =                650
   x( 3) =              0      x( 4) =                  0
   x( 5) =          32.19      x( 6) =                  0
   x( 7) =            350      x( 8) =                150
----------------------------------------------------------
Trade-offs among membership functions
 -dm( 2)/dm(1) =             0.4363
 -dm( 3)/dm(1) =             0.1935
----------------------------------------------------------
Are you satisfied with the current membership values
of the Pareto optimal solution ?

=no

COMMAND: GO                      [ ITERATION: 2 ]

 Reference values rm(i) (i=1,...,3) :
 rm( 1) : 1.000000
 rm( 2) : 1.000000
 rm( 3) : 1.000000

 rm( 1) : 0.900000
 rm( 2) : 0.700000
 rm( 3) : 0.600000
```

The same procedure continues in this manner until the DM is satisfied with the current values of the membership functions. In this example, at the fourth iteration, the satisficing solution of the DM is obtained as follows:

$$z_1(x) = \quad 2178.41, \quad \mu_1(z_1(x)) = 0.7538$$

$$z_2(x) = \qquad 44.60, \quad \mu_2(z_2(x)) = 0.6038$$

$$z_3(x) = -3594.16, \quad \mu_3(z_3(x)) = 0.7729$$

$$x = (350, 650, 0, 0, 44.6023, 0, 350, 150)$$

The optimal operation problem of a packaging system in automated warehouses, discussed thus far, was further generalized to a multiobjective linear programming problem with fuzzy parameters in which some of the parameters are considered as fuzzy numbers. The interactive fuzzy multiobjective linear programming method discussed in Chapter 5 was applied both considering the fuzzy goals and fuzzy parameters. Details can be found in Sakawa and Yano (1989d).

8.2 Pass scheduling for hot tandem mills

In industrial fields, substantial efforts have been made to improve both the quality of products and productivity. Such is the situation in steel plants, where the urgent need is to reduce cost while assuring the high quality of products. Recent advances in computer technology have made it possible to implement an optimization strategy which employs a rather complicated nonlinear mathematical model. A lot of discussion has arisen as to how we can optimize process operations or develop better process control strategies. However, it has also been pointed out that the conventional single-objective optimization method is somewhat lacking in the flexibility required to cope with various situations. Therefore, interactive aspects are needed in the optimization process whereby some optimization parameters or strategies may be adjusted according to the situation.

In 1987, in co-operation with Kobe Steel Ltd., one of the major steel companies in Japan, the author considered the pass scheduling problem for hot tandem mills from the viewpoint of multiobjective optimization. The interactive fuzzy multiobjective nonlinear programming method discussed in Section 4.5 was applied successfully to the formulated problem (Sakawa, Narazaki, Konishi, Nose, and Morita 1987; Sakawa, Narazaki, Nose, and Konishi 1987). In this section, we introduce the major results according to our papers.

8.2.1 Outline of process and mathematical formulation

A tandem mill consists of six or seven tandem stands, and the term "pass scheduling" means the scheduling of the rolling process which decides how much and how fast the thickness of a strip should be reduced at each stand. Several comments have been made on this problem from the viewpoint of addressing a single criterion such as cost. However, other existing objectives in addition to cost, such as shape, should be taken into consideration as objectives to be optimized. Hence, we regard the pass scheduling problem as a multiobjective optimization problem, where the criteria are cost, productivity, shape, thickness, and temperature. Furthermore, the interactive feature is added to the optimization process so that the DM may clarify subjective preferences and pursue their realizations. Therefore, in addition to the conventional multiobjective optimization methods, we employ the interactive fuzzy multiobjective nonlinear programming method (Sakawa and Yano 1985b, 1986f). This method is capable of capturing and expressing the DM's subjective preferences, such as "I feel quite satisfied about this objective but rather unsatisfied about the others," in the context of a mathematical programming technique.

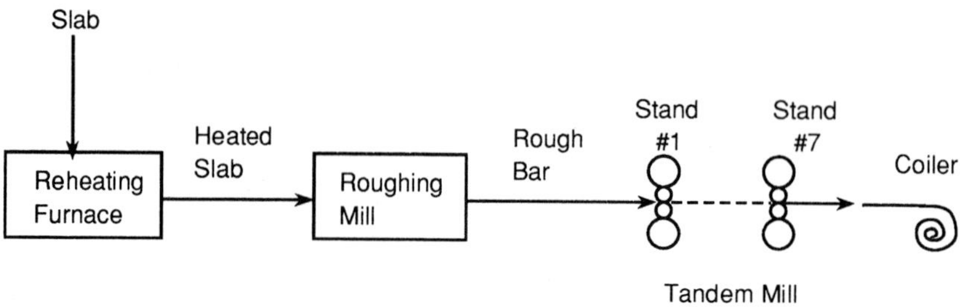

Figure 8.1 Outline of hot rolling process

Figure 8.1 shows the outline of a hot rolling process. A slab is heated to about $1200°C$ in the reheating furnace, is rolled by the roughing mill and becomes what is called a "rough bar." Then it is sent to the finishing tandem mill which consists of seven tandem stands. A rough bar is rolled and thinned, successively, by those stands. The strip is coiled after rolling and transferred to the next process.

The following items should be considered in connection with the operation of the hot tandem mill.

(1) At the exit of the seventh stand, called a "finishing stand," the thickness and temperature of a rolled strip must have the values within a permissible range of the targetted one.

(2) The rolled strip should be as flat as possible.

(3) The cost for producing one coil should be minimized.

(4) The time required for rolling should be as short as possible.

As for the flatness of a strip, the typical shape of a strip is shown in Figure 8.2, where the center of the width is slightly thicker than either end. The height of the arch shape is called the "crown" of the strip. One method to measure the flatness is to calculate dH/H, where dH denotes the difference in height between the end and the center of the width.

Figure 8.3 shows the block diagram of an optimization process. The diagram consists of two parts. One of them denotes an optimization procedure using the GRG (generalized reduced gradient) method (Lasdon et al. 1975), and the other denotes what is called a "mill setting model." A mill setting model is a mathematical model of the rolling process, which calculates the process characteristics, constraints, and objective functions by using the values of the variables that are searched and decided upon during the GRG procedure. Membership functions are introduced to evaluate the objective functions subjectively, and the optimization process is carried out in an interactive manner until the DM feels satisfied with the calculated result.

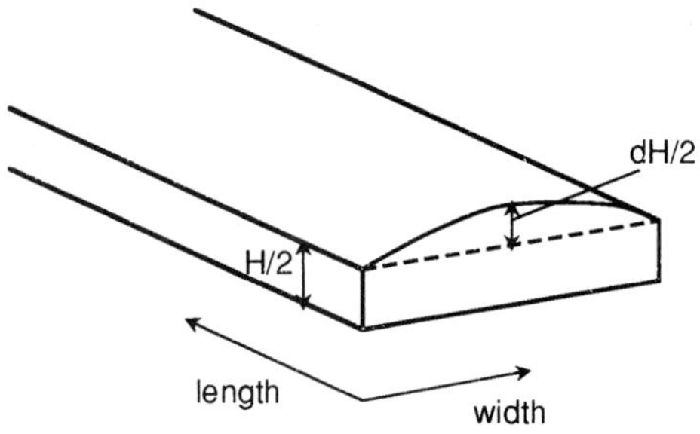

Figure 8.2 Shape of a strip

In the following, more detailed explanations of mathematical formulations are given.

First, it should be noted here that the process characteristics of the top and bottom part of a strip are calculated separately. This is necessary because, at the entrance of the first stand, the temperature of the bottom part of a strip is lower than that of the top part. We should also roll the bottom part at a higher speed than the top part to keep the finished temperature of the bottom part as hot as that of the top part.

Given the thickness and temperature of a rough bar, together with the reference thickness and temperature of a rolled strip, our pass scheduling problem is to decide the reduction rate and rolling speed at each stand by considering the seven objective functions formulated below.

Eighteen decision variables have been chosen for this problem. They are concerned with reduction rates, rolling speed at the first stand, roll bending

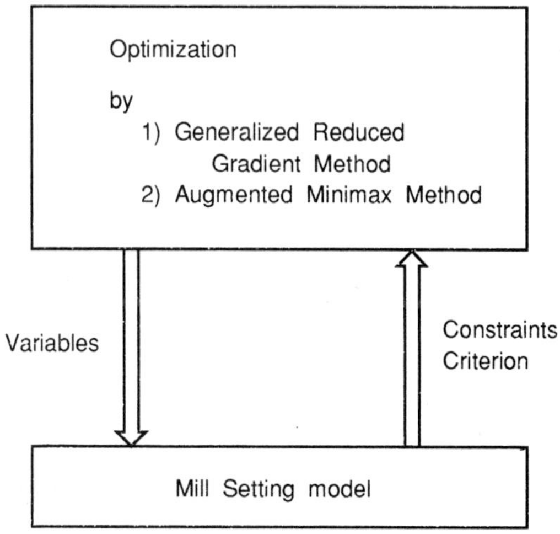

Figure 8.3 Block diagram of optimization

forces, etc., and an auxiliary variable for the augmented minimax problem. The roll bending forces are forces which adjust the deformation of the rolls and are used to control the flatness of the rolled strip.

The following seven conflicting objective functions are derived on the basis of the four items discussed above.

The objective function representing the flatness of the top part of a strip is formulated as

$$f_1 = \frac{1}{4}\sum_{i=4}^{7}\left|\frac{Cr_i^{\text{top}}}{h_i^{\text{out}}} - \frac{Cr_{i-1}^{\text{top}}}{h_{i-1}^{\text{out}}}\right| \tag{8.9}$$

where

Cr_i^{top} = crown [mm] of the top part of a strip at the exit of ith stand,
h_i^{out} = thickness at the exit of ith stand.

Similarly, the objective function for the flatness of the bottom part of a strip is given by

$$f_2 = \frac{1}{4}\sum_{i=4}^{7}\left|\frac{Cr_i^{\text{bottom}}}{h_i^{\text{out}}} - \frac{Cr_{i-1}^{\text{bottom}}}{h_{i-1}^{\text{out}}}\right| \tag{8.10}$$

where

Cr_i^{bottom} = crown [mm] of the bottom part of a strip at the exit of the ith stand.

The rolling time itself is also considered to be an objective function, namely,

$$f_3 = \text{ Rolling time [sec.].} \tag{8.11}$$

The cost function is considered to be

$$f_4 = f_3 \cdot \left(\frac{1}{7}\sum_{i=1}^{7}\frac{KW_i^{\text{top}} + KW_i^{\text{bottom}}}{2}\right) \Big/ \ (3600 \times \text{Weight}) \ [\text{kw/hr} \cdot \text{ton}] \tag{8.12}$$

where

KW_i^{top} = consumed power at the ith stand for rolling the top part of a strip [kw],
KW_i^{bottom} = that for rolling the bottom part of a strip [kw],
Weight = weight of a coil [ton].

The objective function representing the thickness of a rolled strip at the seventh stand is expressed as

$$f_5 = |h_{\text{aimed}} - h_7^{\text{out}}| \tag{8.13}$$

where

h_{aimed} = aimed value of thickness.

The temperature of a rolled strip is evaluated by the following two objective functions:

$$f_6 = |T_{\text{aimed}} - T_{\text{top}}| \tag{8.14}$$

$$f_7 = |T_{\text{aimed}} - T_{\text{bottom}}| \tag{8.15}$$

where

T_{aimed} = aimed value of temperature,
T_{top} = temperature of the top part of a strip at the exit of seventh stand,
T_{bottom} = temperature of the bottom part of a strip at the exit of seventh stand.

There are 71 constraints which are imposed to express the limits of the capabilities of equipment or to assure the quality of a rolled strip within a permissible range.

A mill setting model is a mathematical model of the process characteristics. Based on the values of the variables, the mill setting model calculates the following at each stand:

(1) the exit thickness, temperature, and flatness of the rolled strip

(2) the roll gap

(3) the rolling speed

(4) the rolling load

(5) the power required for rolling

(6) the constraints and objective functions.

The model consists of a set of nonlinear equations. The detailed mathematical expressions can be found in The Iron and Steel Institute of Japan (1984).

The interactive fuzzy multiobjective nonlinear programming method optimizes, in some sense, the degrees of satisfaction for the objectives, which are evaluated by using membership functions. The objective functions are evaluated in terms of membership values which measure the degree of satisfaction of the DM. The optimization occurs in a way that generates a pass schedule where the membership value of each objective is as close as possible to the reference membership value specified by the DM. The shape of the membership function is shown in Figure 8.4. The y-axis denotes a membership value, namely, a real number in [0,1], and the x-axis denotes physical quantities such as thickness or temperature. The membership value "1" represents complete satisfaction. There exist two specified parameters, THRMIN and THRMAX.

The membership functions are used in the constraints $g_{65} - g_{71}$. These constraints have the following form:

$$g_i = x_{18} - (r_k - \mu_k(f_k)) \tag{8.16}$$

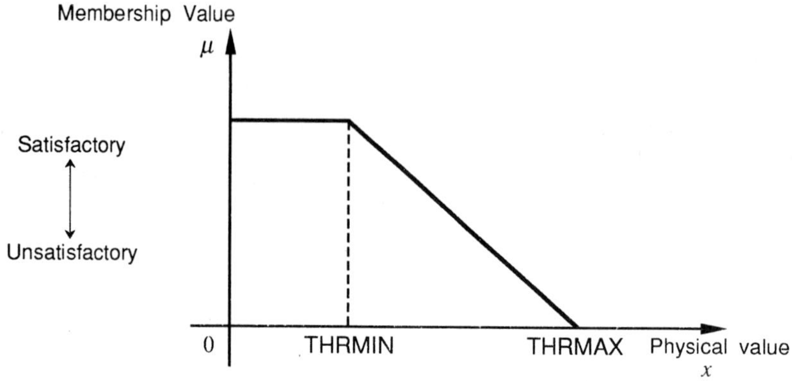

Figure 8.4 Shape of membership function

where

$k = 1, 2, ..., 7$
$i = 64 + k$
f_k = objective function
μ_k = membership function for f_k
r_k = the reference membership value
x_{18} = an auxiliary variable for the augmented minimax problem.

Figure 8.5 shows the flow of an optimization process. The DM designates the "satisfaction level," which means the membership values to be attained by the optimization process. Usually, in multiobjective optimization problems, the DM should allow some objective function to become less satisfactory to obtain a better result for a certain objective function. Here, an interactive aspect turns out to be indispensable in the optimization process.

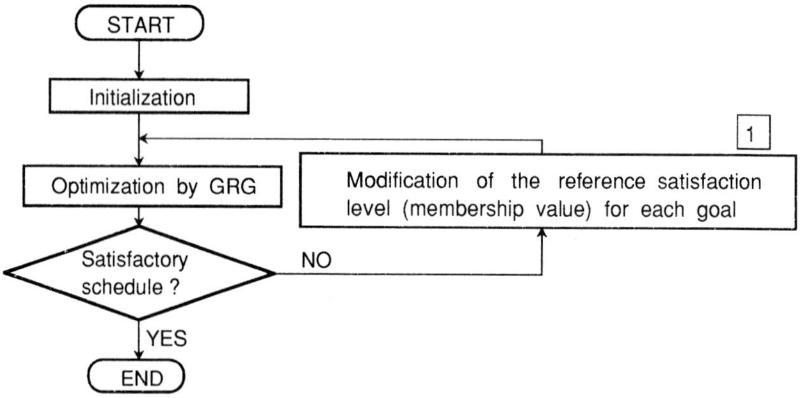

Figure 8.5 Flow of optimization process

In Block 1, we set the thickness and temperature of a rough bar and the aimed value of the thickness and temperature of a rolled strip. At the same time, the DM specifies the reference satisfaction degree r_k for each membership function. In Block 2, the GRG algorithm searches the optimum value for the 18 variables to attain the desired satisfaction levels for seven objectives. However, it is not always possible to realize the desired satisfaction levels for all the objective functions. Here it becomes necessary to have an interaction with the DM in our optimization process, which is represented by Block 3. The DM modifies the reference membership values of the objectives based on the calculated result. Generally, if a better result is expected for a certain objective, the DM must allow some other objectives to yield a lower satisfaction level. The problem is how the reference membership level should be modified for each objective function. It is well known that the effective information for that purpose is given by the trade-off ratios. The interactive optimization process is repeated until the DM can obtain a satisfactory result.

8.2.2 Calculation results

Figure 8.6 shows three cases of reduction patterns, and the corresponding satisfaction degrees are shown in Table 8.2.

Case 1 gives the best result for the flatness of the bottom part, while, from the viewpoint of rolling time, Case 3 yields the most preferable result by sacrificing the shape and temperature. Case 2 offers an unsatisfactory result for the rolling time, but it provides the best schedule for the temperature and the shape of the top part.

This kind of conflict among the objectives is one of the features of the multiobjective optimization problem. It is up to the DM to decide which objective should be improved at the expense of others.

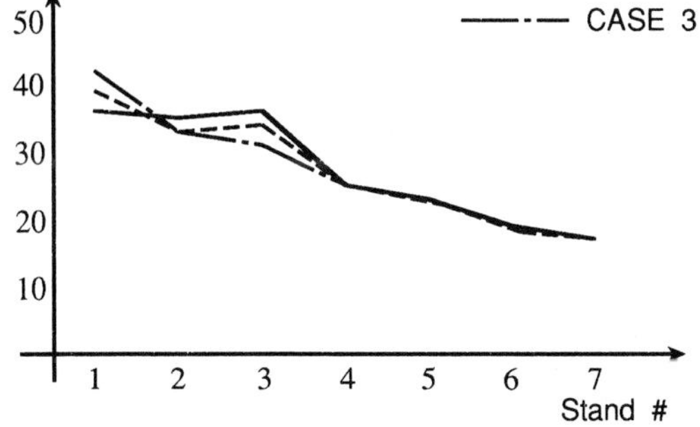

Figure 8.6 Pattern of reduction rate (35mm → 4mm)

Table 8.2 Membership values for three cases

Objective	Case 1	Case 2	Case 3
f_1 (Shape, top)	0.883	0.894	0.821
f_2 (Shape, bottom)	0.923	0.903	0.826
f_3 (Rolling time)	0.735	0.620	0.797
f_4 (Cost)	0.796	0.796	0.796
f_5 (Thickness)	0.948	0.900	0.900
f_6 (Temperature, top)	0.800	0.897	0.676
f_7 (Temperature, bottom)	0.739	0.792	0.556

Table 8.3 gives an example of interactive optimization. Based on the result in the second column, the DM sets the reference membership values as in the third column, and the obtained result is shown in the fourth column, where the rolling time and cost are improved by sacrificing other objectives.

So far we have discussed an application of the interactive fuzzy multiobjective nonlinear programming method to the multiobjective pass scheduling problem for the hot tandem mill. Our method is proven to be useful in obtaining a satisfactory pass schedule because the multiobjective characteristics of the problem are taken into consideration. Furthermore, the DM can interactively reflect subjective preferences for the objectives, such as the shape, temperature, thickness, rolling time, and required rolling power. The optimization procedure yields quite a different result according to the choice of the variables. We chose the reduction rates as variables relating to the decision of the exit thickness. Based on our experience, this choice worked more effectively than choosing the roll gaps as the alternative variables. This is because the roll gaps tend to have difficulty in moving from the initial point due to the locality of the optimization search. Therefore, it has less ability to generate various reduction patterns according to the DM's preferences.

Table 8.3 Interactive optimization

Objective function	Initial membership value	Reference membership value	Obtained membership value
f_1	0.891	0.890	0.872
f_2	0.926	0.920	0.901
f_3	0.665	0.800	0.672
f_4	0.795	0.800	0.796
f_5	0.926	0.900	0.900
f_6	0.857	0.850	0.800
f_7	0.979	0.900	0.974

8.3 Regional planning

Since the 1970s, as negative consequences of rapid growth after World War II, new social problems, especially environmental problems, have become increasingly serious issues to the people of the world. Under these circumstances, an application of the interactive fuzzy multiobjective nonlinear programming method discussed in Section 4.5 to an industrial area in Japan was considered (Sakawa and Yano 1985d) as an extension of our previous results (Sakawa and Seo 1980, 1982, 1983). In this section, we introduce the industrial pollution control problem for Osaka City in Japan in the 1970s according to our paper (Sakawa and Yano 1985d).

8.3.1 Problem formulation

The middle part of Osaka Prefecture is one of the most highly industrialized areas in Japan. Osaka City, which is the second largest industrial and commercial area in Japan, contains many small rivers which are branches of the Yodo River. The Yodo River is an important source of drinking water for Osaka's residents, but water pollution in the Yodo River basin has become increasingly serious. Air pollution is also at critical levels in the greater Osaka area. In addition, the water supply capacity is limited as, more obviously, is the available for land use since this is one of Japan's most populous areas.

The industrial pollution control problem for Osaka City is formulated as the following three-objective optimization problem (Sakawa and Seo 1980, 1982, 1983; Sakawa and Yano 1985d):

$$\left.\begin{aligned} \text{maximize} \quad & f_1 = \sum_{j=1}^{n} A_j K_j^{1-b_j} L_j^{b_j} \\ \text{minimize} \quad & f_2 = \sum_{j=1}^{n} (\omega_{1j}/k_j) K_j \\ \text{minimize} \quad & f_3 = \sum_{j=1}^{n} (\omega_{2j}/k_j) K_j \end{aligned}\right\} \tag{8.17}$$

subject to

$$\sum_{j=1}^{n} (\gamma_{ij}/k_j) K_j \leq \Gamma_i, \ i = 1, 2 \tag{8.18}$$

$$q_2 \leq \left(\sum_{j=1}^{n} K_j \right) \bigg/ \left(\sum_{j=1}^{n} L_j \right) \leq q_1 \tag{8.19}$$

$$\alpha K_{j0} \leq K_j \leq \beta K_{j0}, \ j = 1, \cdots, n \tag{8.20}$$

$$\alpha' L_{j0} \leq L_j \leq \beta' L_{j0}, \ j = 1, \cdots, n \tag{8.21}$$

where

j an industry ($j = 1, \ldots, n; \ n = 20$)

K_j capital value (book value of tangible fixed assets) in industry j

K_{j0} actual capital value in industry j

L_j number of employees in industry j

L_{j0} actual number of employees in industry j

ω_{ij} unit load of chemical oxygen demand (COD) ($i = 1$) or sulphur dioxide (SO_2) ($i = 2$) per industrial shipment in industry j

γ_{ij} resource coefficient for land ($i = 1$) or water ($i = 2$) per industrial shipment in industry j

k_j capital coefficient, namely capital value per unit of shipment in industry j

Γ_i restriction for land ($i = 1$) or water ($i = 2$)

q_i upper ($i = 1$) or lower ($i = 2$) bound for the overall capital intensity (rate of total capital value to total number of employees)

A_j, b_j parameters of the production function for each industry j, and

$\alpha, \beta, \alpha', \beta'$ parameters which represent friction (resistance) in the transfer of capital and labor.

The objective function f_1 is a Cobb-Douglas type of production function which is homogeneous of degree one, and thus, if to each factor the value of its marginal product is paid, total output is distributed between capital and labor in the production $1 - b_j$ and b_j, respectively. This value should be maximized to increase the total production. The objective function f_2 is the total amount of COD and should be minimized to decrease the water pollution. The objective function f_3 is the total amount of SO_2 and should be minimized to decrease the air pollution. Constraints (8.18) are resource constraints, each of which is a land or water resource constraint. Constraint (8.19) is the technical constraint, which shows capital intensity as a whole. This has been utilized to indicate the direction of technological changes occurring as a result of the reformation of the industrial structure. Constraints (8.20) and (8.21) are frictional constraints. Because drastic changes in the industrial structure are not desirable, frictional coefficients are imposed to provide upper and lower bounds for each decision variable. The problem is to find the compromise or satisficing allocation of production factors (capital and labor) to each industry under constraints (8.18) to (8.21).

The resource restrictions Γ_1 and Γ_2 in the constraints (8.18) were assumed to be $\Gamma_1 = 232,200$, $\Gamma_2 = 200,000$. The parameters q_1 and q_2 were supposed to be 1.4 and 0.9, respectively. The parameters for capital and labor, α, α', and β, β', were assumed to be $\alpha = \alpha' = 0.903$, $\beta = \beta' = 1.070$. The calculated values

Table 8.4 Parameters A_j, b_j, and k_j

Industry	A_j	b_j	k_j
1	10.9000	0.1145	0.1195
2	8.6200	0.1391	0.1160
3	15.3900	0.1566	0.0716
4	6.1000	0.1779	0.1599
5	9.9900	0.1723	0.0926
6	5.4600	0.1540	0.1868
7	7.2200	0.2291	0.1824
8	7.9100	0.1294	0.1400
9	6.7300	0.1479	0.1735
10	9.5200	0.1737	0.1125
11	15.2200	0.1445	0.0670
12	6.1300	0.1865	0.1926
13	6.4900	0.1216	0.1746
14	8.1800	0.0870	0.1077
15	6.8500	0.1981	0.1486
16	7.4300	0.2000	0.1659
17	9.6700	0.1588	0.1020
18	7.3600	0.1841	0.1491
19	7.0000	0.2107	0.1394
20	8.4700	0.1677	0.1228

of parameters for A_j, b_j, k_j and ω_{ij}, γ_{ij} are shown in Table 8.4 and Table 8.5, respectively.

The code numbers of the industrial classification are explained in Table 8.6. The sources for these data have been obtained mainly from the Statistical Office of Osaka Prefecture (1975), the Ministry of International Trade and Industry (1973), and the Osaka Bureau of Trade and Industry (1973).

8.3.2 Computational result

Now consider an application of the interactive fuzzy multiobjective nonlinear programming method to the formulated multiobjective nonlinear programming problem. In this subsection, we explain the interactive computational experience under our interactive fuzzy multiobjective nonlinear programming program. In this interaction, the initial values of the decision variables $x = (K_1, \ldots, K_{20}, L_1, \ldots, L_{20})$ are set at their lower bounds.

After the calculated individual minimum and maximum of each of the objective functions under the given constraints are shown to the DM, interaction is assumed to establish the following membership function and the corresponding assessment values for the three objective functions.

Table 8.5 Parameters ω_{ij} and γ_{ij}

Industry	COD	SO_2	Land	Water
1	0.07875	0.00822	0.0244	0.0407
2	0.03111	0.02235	0.0718	0.1292
3	0.03110	0.02235	0.0219	0.0072
4	0.00142	0.00076	0.1024	0.0324
5	0.00142	0.00076	0.0244	0.0121
6	0.21680	0.06751	0.0487	0.1564
7	0.07133	0.05218	0.0105	0.0154
8	0.07133	0.05218	0.0429	0.0599
9	0.03466	0.01505	0.1461	0.0212
10	0.02592	0.00413	0.0553	0.0549
11	0.02592	0.00413	0.0468	0.0542
12	0.00198	0.07963	0.1087	0.0617
13	0.00587	0.02136	0.0773	0.0562
14	0.00084	0.03055	0.0354	0.0373
15	0.00116	0.00778	0.0589	0.0293
16	0.00083	0.00340	0.0464	0.0129
17	0.00105	0.00243	0.0235	0.0133
18	0.00073	0.00116	0.0702	0.0267
19	0.00367	0.00228	0.0451	0.0324
20	0.00864	0.00228	0.0354	0.0258

$$\begin{array}{lll} f_1\text{: linear,} & (f_1^0, f_1^1) & = (4800000, 5020000) \\ f_2\text{: hyperbolic,} & (f_2^{0.25}, f_2^{0.5}) & = (147000, 145000) \\ f_3\text{: exponential,} & (f_3^0, f_3^{0.5}, f_3^1) & = (110000, 104000, 102000). \end{array}$$

Then, for the initial reference membership values, the augmented minimax problem is solved, and the DM is supplied with the corresponding Pareto optimal solution and the trade-off rates between the membership functions. In this interaction, since the DM is not satisfied with the current values of the membership functions, by considering the current values of the membership functions along with the trade-off information, the DM updates the reference membership values to (0.5, 0.6, 0.55), improving both μ_2 and μ_3 at the expense of μ_1.

The same procedure continues in this manner until the DM is satisfied with the current values of the membership functions. In this example, at the fourth iteration, the satisficing solution for the DM is derived and the whole interactive processes are shown in Table 8.7.

The satisficing allocation of capital and labor to each industry corresponding to the results obtained by interaction is summarized in Table 8.8 together with the values in 1975.

Table 8.6 Classification of industries

Code	Industries	Code	Industries
1	Foods	11	Leather products
2	Textile mill products	12	Clay and stone products
3	Apparel products	13	Iron and steel
4	Lumber and products	14	Nonferrous metals
5	Furniture	15	Fabricated metal products
6	Pulp and paper products	16	Machinery
7	Printing and publishing	17	Electrical machinery
8	Chemicals and products	18	Transportation equipment
9	Coal and petroleum products	19	Precision machinery
10	Rubber products	20	Miscellaneous

Table 8.7 Interactive processes

Iteration	1	2	3	4
$\bar{\mu}_1$	1	0.5	0.49	0.48
$\bar{\mu}_2$	1	0.6	0.6	0.62
$\bar{\mu}_3$	1	0.55	0.58	0.57
μ_1	0.5251	0.4824	0.4678	0.4568
μ_2	0.5251	0.5824	0.5778	0.5968
μ_3	0.5251	0.5324	0.5578	0.5468
f_1	4915513	4906139	4902924	4900487
f_2	144817	144394	144429	144286
f_3	103865	103826	103696	103752
$-\partial\mu_2/\partial\mu_1$	2.8539	0.9633	1.0386	0.9431
$-\partial\mu_3/\partial\mu_1$	1.1151	1.3230	1.3494	1.3559

Table 8.8 Satisficing allocation of capital and labor

	1975		Proposal	
Industry	Capital	Labor	Capital	Labor
1	31653	22527	28579	25783
2	22981	17521	20749	18740
3	10114	18088	9132	19347
4	13479	8237	14417	8810
5	8581	8275	9178	8851
6	36996	16041	33403	17157
7	75595	43494	68254	47008
8	86440	34161	78047	36539
9	2004	827	1809	885
10	5161	4195	4660	4487
11	3764	5512	3776	5896
12	15538	8472	14029	9062
13	108036	28964	103740	30980
14	28750	10147	25958	10853
15	75339	52749	80583	56420
16	81541	52358	87216	56002
17	30677	26736	32812	28597
18	32687	18597	38813	19891
19	4577	4148	4896	4437
20	26266	22701	28094	24280

The satisficing values for the objective functions can be interpreted as compromised values of the DM between the conflicting objectives which are the maximization of the production function and minimization of two environmental factors (COD and SO_2). The satisficing solution for the decision variables K_j and L_j shows the satisficing allocation of capital and labor to each industry. These results show that capital values in industry as a whole are reduced compared with the values in 1975. Capital formation in the coal and petroleum industry and in the chemicals and related products industry is especially severely reduced, and the nonferrous metals industry and the fabricated metal product industry decreased their capital formation. On the other hand, in consumer industries, as well as in machine industries such as the electrical machinery industry, capital formation is promoted. Finally, it should be pointed out here that, although the actual DM for the formulated problem would select other values for the three objectives than the values which were selected by the hypothetical DM used in this interaction, the methods of iteration and calculation are essentially the same.

CHAPTER 9

FURTHER RESEARCH DIRECTIONS

In the preceding chapters, we have discussed not only some of the most important methods for both multiobjective optimization and fuzzy multiobjective optimization, but also the latest advances in the new field of interactive multiobjective optimization under fuzziness with actual applications based on the author's continuing research work. However, many other interesting related topics, including multiobjective possibilistic programming, multiobjective programming with fuzzy decision variables and fuzzy parameters, fuzzy multiobjective combinatorial programming, fuzzy regression analysis, fuzzy multiobjective dynamic optimization, and fuzzy multiobjective hierarchical optimization, still remain to be explored. In this chapter, being limited by space, we only outline the latest major research results of the first four topics because of their close relationship to our discussions in the preceding chapters.

9.1 Multiobjective possibilistic programming

Recall the following multiobjective linear programming problem with fuzzy parameters (MOLP-FP) discussed in Chapter 5:

$$\left.\begin{array}{ll} \text{minimize} & (C_1x, C_2x, ..., C_kx)^T \\ \text{subject to} & x \in X(A,B) \triangleq \{x \in R^n \mid A_jx \leq B_j,\ j=1,...,m\,;\ x \geq 0\} \end{array}\right\} \tag{9.1}$$

where $C_i = (C_{i1}, ..., C_{in})$, $A_j = (A_{i1}, ..., A_{jn})$, and B_j are fuzzy parameters which reflect the experts' ambiguous understanding of the nature of the parameters in the problem-formulation process. These fuzzy parameters are assumed to be characterized by fuzzy numbers whose membership functions are denoted by $\mu_{C_{i1}}(c_{i1})$,..., $\mu_{C_{in}}(c_{in})$, $\mu_{A_{j1}}(a_{j1})$,...,$\mu_{A_{jn}}(a_{jn})$, and $\mu_{B_j}(b_j)$, respectively.

When we try to deal with this problem as a generalization of the usual multiobjective linear programming problems, it is quite natural that the following two questions arise:

(1) How can we define the feasibility for the m constraints

$$X(A,B) \triangleq \{x \in R^n \mid A_jx \leq B_j,\, j=1,\ldots,m;\ x \geq 0\} \tag{9.2}$$

involving fuzzy numbers?

(2) How can we define the Pareto optimality concept for the k objective functions C_1x, ..., C_kx involving fuzzy numbers?

In Chapter 5, we responded to these questions by introducing the concepts of α-multiobjective linear programming (α-MOLP) and α-Pareto optimality, and we introduced the interactive multiobjective programming methods by considering not only fuzzy parameters but also the fuzzy goals of the decision maker (DM).

However, from the viewpoint of possibility and necessity introduced by Dubois and Prade (1983, 1987), a unified approach to the MOLP-FP can be made. To define the inequality relation between two fuzzy numbers M and N as a natural extension of ranking two closed intervals, Dubois and Prade (1983, 1987) introduced the following four indices.

Definition 9.1

$$\begin{aligned}
Pos(M \geq N) &\triangleq \sup\{\min(\mu_M(u), \mu_N(v)) \mid u \geq v\} \\
Pos(M > N) &\triangleq \sup_u\{\inf_v\{\min(\mu_M(u), 1 - \mu_N(v)) \mid u \leq v\}\} \\
Nes(M \geq N) &\triangleq \inf_u\{\sup_v\{\max(1 - \mu_M(u), \mu_N(v)) \mid u \leq v\}\} \\
Nes(M > N) &\triangleq \inf\{\max(1 - \mu_M(u), 1 - \mu_N(v)) \mid u \leq v\}
\end{aligned}$$

where $\mu_M(u)$ or $\mu_N(v)$ is the membership function of fuzzy numbers M or N, and Pos or Nes is short for Possibility or Necessity.

Observe that the index $Pos(M \geq N)$ or $Pos(M > N)$ means the degree of possibility that M is greater than or equal to N or M is greater than N, and the index $Nes(M \geq N)$ or $Nes(M > N)$ means the degree of necessity that M is greater than or equal to N or M is greater than N.

The condition that the degrees of these four indices are greater than or equal to a certain level of α can be equivalently expressed as the following usual constraints by using the α-level sets of fuzzy numbers M and N (Sakawa and Yano 1990a).

Theorem 9.1

$$\begin{aligned}
Pos(M \geq N) \geq \alpha &\Leftrightarrow m^R_\alpha \geq n^L_\alpha \\
Pos(M > N) \geq \alpha &\Leftrightarrow m^R_\alpha \geq n^R_{1-\alpha} \\
Nes(M \geq N) \geq \alpha &\Leftrightarrow m^L_{1-\alpha} \geq n^R_\alpha \\
Nes(M > N) \geq \alpha &\Leftrightarrow m^L_{1-\alpha} \geq n^R_{1-\alpha}
\end{aligned}$$

where m^L_α, m^R_α and n^L_α, n^R_α are the left side and the right side extreme points of the α-level sets $[m^L_\alpha, m^R_\alpha]$ and $[n^L_\alpha, n^R_\alpha]$ of fuzzy numbers M and N, respectively, i.e., $M_\alpha = [m^L_\alpha, m^R_\alpha]$ and $N_\alpha = [n^L_\alpha, n^R_\alpha]$.

First, consider the feasibility for the constrained set $X(A, B)$ of the MOLP-FP involving fuzzy parameters. Using the four indices for ranking two fuzzy numbers, we can now define four types of feasibility for the fuzzy constrained set $X(A, B)$ as follows (Dubois 1987; Sakawa and Yano 1990a).

Definition 9.2 (α**-feasibility)**

(1) *$x \in R^n$ is said to be α-Very Weak Feasible (α-VWF) to $X(A,B)$ if and only if*

$$x \in X_{VWF}(\alpha) \triangleq \{x \geq 0 \mid Pos(A_j x \leq B_j) \geq \alpha,\ j = 1,\ldots,m\}. \quad (9.3)$$

(2) *$x \in R^n$ is said to be α-Medium Weak Feasible (α-MWF) to $X(A,B)$ if and only if*

$$x \in X_{MWF}(\alpha) \triangleq \{x \geq 0 \mid Pos(A_j x < B_j) \geq \alpha,\ j = 1,\ldots,m\}. \quad (9.4)$$

(3) *$x \in R^n$ is said to be α-Medium Strong Feasible (α-MSF) to $X(A,B)$ if and only if*

$$x \in X_{MSF}(\alpha) \triangleq \{x \geq 0 \mid Nes(A_j x \leq B_j) \geq \alpha,\ j = 1,\ldots,m\}. \quad (9.5)$$

(4) *$x \in R^n$ is said to be α-Very Strong Feasible (α-VSF) to $X(A,B)$ if and only if*

$$x \in X_{VSF}(\alpha) \triangleq \{x \geq 0 \mid Nes(A_j x < B_j) \geq \alpha,\ j = 1,\ldots,m\}. \quad (9.6)$$

From Theorem 9.1, the following theorem immediately follows for $X_{VWF}(\alpha)$, $X_{MWF}(\alpha)$, $X_{MSF}(\alpha)$, and $X_{VSF}(\alpha)$.

Theorem 9.2

$$X_{VWF}(\alpha) = \{x \geq 0 \mid a^L_{j\alpha} x \leq b^R_{j\alpha},\ j = 1,\ldots,m\}, \quad (9.7)$$

$$X_{MWF}(\alpha) = \{x \geq 0 \mid a^R_{j(1-\alpha)} x \leq b^R_{j\alpha},\ j = 1,\ldots,m\}, \quad (9.8)$$

$$X_{MSF}(\alpha) = \{x \geq 0 \mid a^L_{j\alpha} x \leq b^L_{j(1-\alpha)},\ j = 1,\ldots,m\}, \quad (9.9)$$

$$X_{VSF}(\alpha) = \{x \geq 0 \mid a^R_{j(1-\alpha)} x \leq b^L_{j(1-\alpha)},\ j = 1,\ldots,m\}, \quad (9.10)$$

where $a^L_{j\alpha} = [a^L_{j1\alpha},\ldots,a^L_{jn\alpha}]$, $a^R_{j\alpha} = [a^R_{j1\alpha},\ldots,a^R_{jn\alpha}]$ in which $a^L_{jl\alpha}$, $a^R_{jl\alpha}$ and $b^L_{j\alpha}$, $b^R_{j\alpha}$ denote the left and right extreme points of the α-level sets of fuzzy numbers A_{jl} and B_j, respectively.

It should be noted here that this theorem means that the four types of α-feasibility constrained sets can be equivalently represented as the usual linearly constrained sets.

From the definition of α-feasibility, the following inclusion properties immediatcly follow:

$$(X_{MWF}(\alpha) \cup X_{MSF}(\alpha)) \subset X_{VWF}(\alpha) \quad (9.11)$$

$$X_{VSF}(\alpha) \subset (X_{MWF}(\alpha) \cap X_{MSF}(\alpha)). \quad (9.12)$$

To define the extended Pareto optimality concept for the k objective functions $C_1x, \ldots, C_kx$ involving fuzzy numbers, consider the following multiobjective linear programming problems where only the objective functions involve fuzzy numbers:

$$\left.\begin{array}{ll} \text{minimize} & (C_1x, C_2x, ..., C_kx)^T \\ \text{subject to} & x \in X \triangleq \{x \in R^n \mid a_jx \leq b_j,\ j = 1, ..., m;\ x \geq 0\}. \end{array}\right\} \quad (9.13)$$

Using the four indices for ranking two fuzzy numbers, four types of β-Pareto optimality and four types of γ-Pareto optimality are defined as follows (Sakawa and Yano 1990a).

Definition 9.3 (β-Pareto optimal solution)

(1) $x^* \in X$ *is said to be a β-Very Weak Pareto* (β-VWP) *optimal solution to (9.13) if and only if there does not exist another $x \in X$ such that*

$$Pos(C_ix \leq C_ix^*) \geq \beta,\ i = 1, \ldots, k \quad (9.14)$$

with strict inequality holding for at least one i.

(2) $x^* \in X$ *is said to be a β-Medium Weak Pareto* (β-MWP) *optimal solution to (9.13) if and only if there does not exist another $x \in X$ such that*

$$Pos(C_ix < C_ix^*) \geq \beta,\ i = 1, \ldots, k \quad (9.15)$$

with strict inequality holding for at least one i.

(3) $x^* \in X$ *is said to be a β-Medium Strong Pareto* (β-MSP) *optimal solution to (9.13) if and only if there does not exist another $x \in X$ such that*

$$Nes(C_ix \leq C_ix^*) \geq \beta,\ i = 1, \ldots, k \quad (9.16)$$

with strict inequality holding for at least one i.

(4) $x^* \in X$ *is said to be a β-Very Strong Pareto* (β-VSP) *optimal solution to (9.13) if and only if there does not exist another $x \in X$ such that*

$$Nes(C_ix < C_ix^*) \geq \beta,\ i = 1, \ldots, k \quad (9.17)$$

with strict inequality holding for at least one i.

Definition 9.4 (γ-Pareto optimal solution)

(1) $x^* \in X$ *is said to be a γ-Very Weak Pareto* (γ-VWP) *optimal solution to (9.13) if and only if there does not exist another $x \in X$ such that*

$$Pos(C_ix \geq C_ix^*) \leq \gamma,\ i = 1, \ldots, k \quad (9.18)$$

with strict inequality holding for at least one i.

(2) $x^* \in X$ *is said to be a* γ*-Medium Weak Pareto* (γ-MWP) *optimal solution to (9.13) if and only if there does not exist another* $x \in X$ *such that*

$$Pos(C_i x > C_i x^*) \leq \gamma, \; i = 1, \ldots, k \tag{9.19}$$

with strict inequality holding for at least one i.

(3) $x^* \in X$ *is said to be a* γ*-Medium Strong Pareto* (γ-MSP) *optimal solution to (9.13) if and only if there does not exist another* $x \in X$ *such that*

$$Nes(C_i x \geq C_i x^*) \leq \gamma, \; i = 1, \ldots, k \tag{9.20}$$

with strict inequality holding for at least one i.

(4) $x^* \in X$ *is said to be a* γ*-Very Strong Pareto* (γ-VSP) *optimal solution to (9.13) if and only if there does not exist another* $x \in X$ *such that*

$$Nes(C_i x > C_i x^*) \leq \gamma, \; i = 1, \ldots, k \tag{9.21}$$

with strict inequality holding for at least one i.

It should be noted that the inequalities in the four β-Pareto definitions are quite opposite to the inequalities in the four γ-Pareto definitions.

From the definition of the four indices for ranking fuzzy numbers, it can be easily shown that the four types of γ-Pareto optimality, i.e., γ-VWP, γ-MWP, γ-MSP, and γ-VSP are equivalent to the four types of (1-β)-Pareto optimality, i.e., (1-β)-VSP, (1-β)-MWP, (1-β)-MSP, and (1-β)-VWP, respectively.

Moreover, from the definition of the four types of β-Pareto optimality and γ-Pareto optimality, the following properties hold when $0 \leq \beta_1 < \beta_2 \leq 1$ and $0 \leq \gamma_1 < \gamma_2 \leq 1$:

$$X_{VWP}(\beta_1) \subset X_{VWP}(\beta_2), \; X_{VWP}(\gamma_1) \supset X_{VWP}(\gamma_2) \tag{9.22}$$
$$X_{MWP}(\beta_1) \subset X_{MWP}(\beta_2), \; X_{MWP}(\gamma_1) \supset X_{MWP}(\gamma_2) \tag{9.23}$$
$$X_{MSP}(\beta_1) \subset X_{MSP}(\beta_2), \; X_{MSP}(\gamma_1) \supset X_{MSP}(\gamma_2) \tag{9.24}$$
$$X_{VSP}(\beta_1) \subset X_{VSP}(\beta_2), \; X_{VSP}(\gamma_1) \supset X_{VSP}(\gamma_2). \tag{9.25}$$

From these properties, it can be understood that all of the γ-Pareto optimal solution sets are reduced with the greater values of γ. Conversely, all of the β-Pareto optimal solution sets are enlarged with the greater values of β. Therefore, if we interpret the values of β or γ as the degree of Pareto optimality, γ-Pareto optimality is preferable to β-Pareto optimality. From this point of view, in the following, we consider the four types of γ-Pareto optimality for multiobjective linear programming problems with fuzzy parameters.

Similar to the case of four types of feasibility, using the properties of the γ-level sets, these four types of Pareto optimality can be represented as follows (Sakawa and Yano 1990a).

Theorem 9.3

(1) $x^* \in X$ *is* γ*-VWP to (9.13) if and only if there does not exist another* $x \in X$ *such that*

$$c_{i\gamma}^R x \leq c_{i\gamma}^L x^*, \ i = 1, \ldots, k \tag{9.26}$$

with strict inequality holding for at least one i.

(2) $x^* \in X$ *is* γ*-MWP to (9.13) if and only if there does not exist another* $x \in X$ *such that*

$$c_{i\gamma}^R x \leq c_{i(1-\gamma)}^R x^*, \ i = 1, \ldots, k \tag{9.27}$$

with strict inequality holding for at least one i.

(3) $x^* \in X$ *is* γ*-MSP to (9.13) if and only if there does not exist another* $x \in X$ *such that*

$$c_{i(1-\gamma)}^L x \leq c_{i\gamma}^L x^*, \ i = 1, \ldots, k \tag{9.28}$$

with strict inequality holding for at least one i.

(4) $x^* \in X$ *is* γ*-VSP to (9.13) if and only if there does not exist another* $x \in X$ *such that*

$$c_{i(1-\gamma)}^L x \leq c_{i(1-\gamma)}^R x^*, \ i = 1, \ldots, k \tag{9.29}$$

with strict inequality holding for at least one i.

where $c_{i\gamma}^L = [c_{i1\gamma}^L, \ldots, c_{ik\gamma}^L]$, $c_{i\gamma}^R = [c_{i1\gamma}^R, \ldots, c_{ik\gamma}^R]$ *and* $c_{il\gamma}^L$ *or* $c_{il\gamma}^R$ *denote the left and right extreme points of the* γ*-level sets of fuzzy numbers* C_{il}*, respectively.*

The following theorem, which relates four types of Pareto optimal solution sets to the usual Pareto optimal solution set, can be easily obtained (Sakawa and Yano 1990a).

Theorem 9.4

$$X_{VWP}(\gamma) \supset \bigcup_{Q \in [C_\gamma^L, C_\gamma^R]} X^P(Q) \quad (0 \leq \gamma \leq 1) \tag{9.30}$$

$$X_{MWP}(\gamma) \supset \bigcup_{Q \in [C_{1-\gamma}^R, C_\gamma^R]} X^P(Q) \quad (0 \leq \gamma \leq 0.5) \tag{9.31}$$

$$X_{MWP}(\gamma) \subset \bigcap_{Q \in [C_\gamma^R, C_{1-\gamma}^R]} X^P(Q) \quad (0.5 \leq \gamma \leq 1) \tag{9.32}$$

$$X_{MSP}(\gamma) \supset \bigcup_{Q \in [C_\gamma^L, C_{1-\gamma}^L]} X^P(Q) \quad (0 \leq \gamma \leq 0.5) \tag{9.33}$$

$$X_{MSP}(\gamma) \subset \bigcap_{Q \in [C_\gamma^L, C_{1-\gamma}^L]} X^P(Q) \quad (0.5 \leq \gamma \leq 1) \tag{9.34}$$

$$X_{VSP}(\gamma) \subset \bigcap_{Q \in [C_{1-\gamma}^L, C_{1-\gamma}^R]} X^P(Q) \quad (0 \leq \gamma \leq 1) \tag{9.35}$$

where $Q = (q_{il})$ *is a set of* $(k \times n)$ *matrices,* $X^P(Q)$ *denotes a Pareto optimal solution set to the multiobjective linear problem* $\min\{Qx \mid x \in X\}$*, and* $X_{VWP}(\gamma), \ldots, X_{VSP}(\gamma)$ *are four types of* γ*-Pareto optimal solution sets to* (9.13)*, respectively.*

Moreover, C^L_γ or C^R_γ denotes the $(k \times n)$ matrix whose (i,l) elements are the left side or right side extreme points of the γ-level sets of the fuzzy number C_{il}, and $Q \in [C^L_\gamma, C^R_\gamma]$ means $q_{il} \in [c^L_{il\gamma}, c^R_{il\gamma}]$, $i = 1, ..., k$, $l = 1, ..., n$.

It is significant to note here that, among four types of γ-Pareto optimality, $X_{MWP}(\gamma)$ for $0.5 \leq \gamma \leq 1$, $X_{MSP}(\gamma)$ for $0.5 \leq \gamma \leq 1$, and $X_{VSP}(\gamma)$ for $0 \leq \gamma \leq 1$ are included in the intersection of the usual Pareto optimal solution set $X^P(Q)$. Hence, in general, γ-MWP set $X_{MWP}(\gamma)$ $(0.5 \leq \gamma \leq 1)$, γ-MSP set $X_{MSP}(\gamma)$ $(0.5{\leq}\gamma{\leq}1)$, and γ-VSP set $X_{VSP}(\gamma)$ $(0{\leq}\gamma{\leq}1)$ may become a null set. Therefore, from this point of view, the γ-Very Weak Pareto optimal solution (γ-VWP) would be recommended as the generalized concept of Pareto optimality for the multiobjective linear programming problems where only the objective functions involve fuzzy numbers.

Now we are ready to introduce new solution concepts for multiobjective linear programming problems with fuzzy parameters by combining the extended feasibility and Pareto optimality proposed thus far.

Realizing that γ-VWP has a preferable property compared with the four types of β-Pareto optimality and the three types of γ-Pareto optimality, we focus on the following (α, γ)-Very Weak Pareto optimality concept (Sakawa and Yano 1990a).

Definition 9.5 ((α,γ)-Very Weak Pareto optimal solution)

$x^* \in X_{VWF}(\alpha)$ is said to be an (α,γ)-Very Weak Pareto optimal solution ((α,γ)-VWP) to the MOLP-FP if and only if there does not exist another $x \in X_{VWF}(\alpha)$ such that

$$Pos(C_i x \geq C_i x^*) \leq \gamma, \ i = 1, \dots, k, \tag{9.36}$$

where

$$X_{VWF}(\alpha) \triangleq \{x \geq 0 \mid Pos(A_j x \leq B_j) \geq \alpha, \ j = 1, \dots, m\}. \tag{9.37}$$

Figure 9.1 illustrates the set of (α,γ)-Pareto optimal solutions $X_{VW}(\alpha,\gamma)$ in objective space.

From Figure 9.1, $X_{VW}(\alpha,\gamma)$ consists of an infinite number of points, and $X_{VW}(\alpha,\gamma)$ includes the Pareto optimal solution set $X^P_\alpha(C^L_\gamma)$ and $X^P_\alpha(C^R_\gamma)$ corresponding to the following two types of usual multiobjective linear programming problems:

$$\left.\begin{array}{ll} \text{minimize} & C^L_\gamma x \\ \text{subject to} & x \in X_{VWF}(\alpha) \end{array}\right\} \tag{9.38}$$

$$\left.\begin{array}{ll} \text{minimize} & C^R_\gamma x \\ \text{subject to} & x \in X_{VWF}(\alpha). \end{array}\right\} \tag{9.39}$$

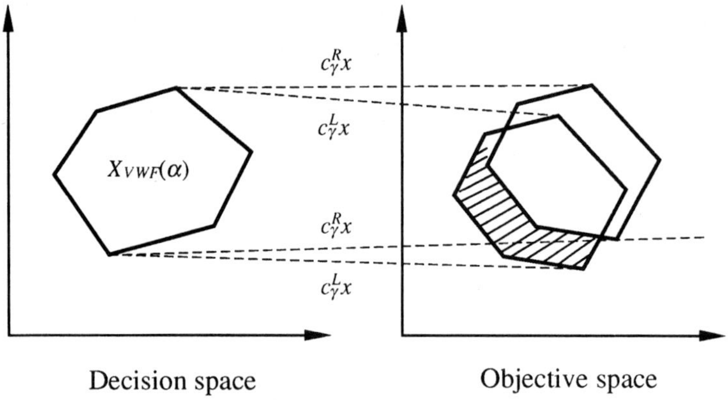

Figure 9.1 (α, γ)-Very Weak Pareto optimal solution set

It is significant to note here that $X_\alpha^P(C_\gamma^L)$ and $X_\alpha^P(C_\gamma^R)$ can be interpreted respectively as the sets of the most optimistic solutions and the most pessimistic solutions among $X_{VW}(\alpha, \gamma)$.

Now it is appropriate to consider the relationships among (α, γ)-VWP and the recent extended Pareto optimality concepts for multiobjective programming problems with fuzzy parameters.

As discussed in Chapter 5, Sakawa and Yano (1986d, 1986g, 1987c, 1988a, 1990b) recently introduced the concept of α-Pareto optimal solutions to the MOLP-FP based on the α-level sets for fuzzy numbers. They also developed interactive decision-making methods, both without and with the fuzzy goals of the decision maker, for deriving a satisficing solution efficiently from an α-Pareto optimal solution set. Since all α-Pareto optimal solutions can be obtained by solving (9.38) with $\gamma = \alpha$ from the definition of α-Pareto optimal solutions, the following relation holds:

$$X_S(\alpha) \subset X_{VW}(\alpha, \alpha) \tag{9.40}$$

where $X_S(\alpha)$ is an α-Pareto optimal solution set to the MOLP-FP.

More recently, Luhandjula (1987) and Luhandjula and Sakawa (1990) proposed the concepts of α-possibly feasible, β-possibly efficient, and the (α, β)-satisfying solution to the MOLP-FP based on the possibility theory of Zadeh (1978). Since the concept of α-possibly feasible is the same as the concept α-VWF and the definition of β-possibly efficient under the minimization problem coincides with β-VWP, the following relation holds:

$$X_L(\alpha, \beta) = X_{VW}(\alpha, \beta) \tag{9.41}$$

where $X_L(\alpha, \beta)$ is the (α, β)-satisficing solution set proposed by Luhandjula and Sakawa.

It is significant to point out here that all the results presented in this section have recently been extended not only to multiobjective linear fractional possibilistic programming but also to multiobjective nonlinear possibilistic programming. Interested readers might refer to Sakawa and Yano (1990a, 1991b) for details. However, urgent development of fuzzy interactive programming methods using the proposed extended Pareto optimality concepts together with the applications to the human-centered systems in most realistic situations under fuzziness should be required in the near future.

Finally, if we look at other recent developments of possibilistic programming, although most of them have no explicit consideration of the Pareto optimality concept, we can find various approaches to single- and/or multiobjective linear programming problems with fuzzy parameters (e.g. Orlovski 1984; Tanaka and Asai 1974a, 1974b; Tanaka, Ichihashi, and Asai 1984, 1986; Verdegay 1984; Ramik 1985, 1987; Slowinski 1986, 1990; Luhandjula 1987a, 1987b, Buckley 1988a, 1988b, 1989, 1990; Campos and Verdegay 1989; Delgado, Verdegay, and Vila 1989, 1990; Rommelfanger et al. 1989; Inuiguchi, Ichihashi, and Tanaka 1989; Inuiguchi and Ichihashi 1989, 1990). In addition to these original papers, recent comprehensive survey papers of Dubois (1987), Luhandjula (1989), and Inuiguchi, Ichihashi, and Tanaka (1990) would be very helpful for interested readers.

9.2 Multiobjective programming with fuzzy variables and parameters

As a generalized fuzzy version of the multiobjective linear programming problems with fuzzy parameters, it seems to be appropriate to consider multiobjective linear programming problems with fuzzy decision variables and fuzzy parameters. Using the four indices for ranking two fuzzy numbers proposed by Dubois and Prade (1983, 1987) via the concept of possibility and necessity, four types of fuzzy α-feasibility and four types of fuzzy γ-Pareto optimality are defined, and then a fuzzy (α, γ)-Pareto optimal solution concept is introduced (Sakawa and Yano 1990d).

Consider the following multiobjective linear programming problems with fuzzy decision variables and fuzzy parameters:

$$\left.\begin{array}{rl} \text{minimize} & (C_1 \otimes X, C_2 \otimes X, ..., C_k \otimes X)^T \\ \text{subject to} & X \in \tilde{X}(A, B, F) \\ & \triangleq \{X \in F(R^n) \mid A_j \otimes X \leq B_j,\ j = 1, ..., m\} \end{array}\right\} \quad (9.42)$$

where $F(R^n)$ is an n-dimensional vector of the fuzzy numbers whose support sets are assumed to be nonnegative, and fuzzy decision variables $X = (X_1, \ldots, X_n)^T$ and fuzzy parameters $C_i = (C_{i1}, ..., C_{in})$, $A_j = (A_{j1}, ..., A_{jn})$, B_j are assumed to be characterized by L-L fuzzy numbers whose membership functions are defined by

$$\left.\begin{array}{ll} \mu_{X_l}(x_l) = L(|x_l - x_l^m|/x_l^s), & \mu_{C_{il}}(c_{il}) = L(|c_{il} - c_{il}^m|/c_{il}^s), \\ \mu_{A_{jl}}(a_{jl}) = L(|a_{jl} - a_{jl}^m|/a_{jl}^s), & \mu_{B_j}(b_j) = L(|b_j - b_j^m|/b_j^s), \end{array}\right\} \quad (9.43)$$

where $x_l^m, c_{il}^m, a_{jl}^m, b_j^m$ and $x_l^s, c_{il}^s, a_{jl}^s, b_j^s$ are mean values and spread parameters of the membership functions of fuzzy decision variables and fuzzy parameters, and $L(\cdot)$ is the shape function.

For convenience in our subsequent discussions, we assume the shape function satisfies the following property:

$$L^{-1}(\gamma) + L^{-1}(1-\gamma) \le L^{-1}(0), \quad 0 \le \gamma \le 1. \tag{9.44}$$

As an example of the sufficient condition for (9.44), we have

$$0 \le L(x) \le 1 - x, \quad 0 \le x \le 1.$$

Then, since the support sets of fuzzy decision variables X_l, $l = 1, \ldots, n$, are assumed to be nonnegative, it follows that

$$x_l^m - L^{-1}(0)x_l^s \ge 0, \quad l = 1, \ldots, n. \tag{9.45}$$

Now consider the feasibility for the following constrained set involving fuzzy decision variables and fuzzy parameters.

$$\tilde{X}(A, B, F) = \{X \in F(R^n) \mid A_j \otimes X \le B_j, \ j = 1, \ldots, m\}. \tag{9.46}$$

Since the fuzzy decision variables are assumed to be L-L fuzzy numbers, the feasible region $\tilde{X}(A, B, F)$ can be defined on the $(2 \times n)$-dimensional real number space, where (x_l^m, x_l^s), $l = 1, \ldots, n$, are decision variables.

As discussed in Section 2.3, as a direct consequence of Nguyen's theorem, the following relation holds (recall Theorem 2.3).

Theorem 9.5

$$[A_j \otimes X]_\alpha = \left[\sum_{l=1}^{n} A_{jl} \otimes X_l\right]_\alpha = [Y_{j\alpha}^L, Y_{j\alpha}^R], \tag{9.47}$$

where

$$\begin{aligned} Y_{j\alpha}^L \triangleq \sum_{l=1}^{n} \Big\{ \min \big\{ &(a_{jl}^m - L^{-1}(\alpha)a_{jl}^s)(x_l^m - L^{-1}(\alpha)x_l^s), \\ &(a_{jl}^m - L^{-1}(\alpha)a_{jl}^s)(x_l^m + L^{-1}(\alpha)x_l^s) \big\} \Big\} \\ Y_{j\alpha}^R \triangleq \sum_{l=1}^{n} \Big\{ \max \big\{ &(a_{jl}^m + L^{-1}(\alpha)a_{jl}^s)(x_l^m + L^{-1}(\alpha)x_l^s), \\ &(a_{jl}^m + L^{-1}(\alpha)a_{jl}^s)(x_l^m - L^{-1}(\alpha)x_l^s) \big\} \Big\}. \end{aligned}$$

For simplicity in the following, assume that the right and left extreme points of the support of A_{jl} have the same sign and define the following index sets

$$\begin{aligned} L_1(j) &= \{l \in N \mid a_{jl}^m - L^{-1}(0)a_{jl}^s \ge 0\}, \\ L_2(j) &= \{l \in N \mid a_{jl}^m + L^{-1}(0)a_{jl}^s < 0\}, \end{aligned}$$

where $N = \{1, 2, \ldots, n\}$.

Using these index sets, it holds that

$$\begin{aligned}
Y_{j\alpha}^L &= \sum_{l \in L_1(j)} (a_{jl}^m - L^{-1}(\alpha)a_{jl}^s)(x_l^m - L^{-1}(\alpha)x_l^s) \\
&\quad + \sum_{l \in L_2(j)} (a_{jl}^m - L^{-1}(\alpha)a_{jl}^s)(x_l^m + L^{-1}(\alpha)x_l^s), \\
Y_{j\alpha}^R &= \sum_{l \in L_1(j)} (a_{jl}^m + L^{-1}(\alpha)a_{jl}^s)(x_l^m + L^{-1}(\alpha)x_l^s) \\
&\quad + \sum_{l \in L_2(j)} (a_{jl}^m + L^{-1}(\alpha)a_{jl}^s)(x_l^m - L^{-1}(\alpha)x_l^s).
\end{aligned}$$

Now, we can define the four types of feasibility for the fuzzy constrained set using the four indices for ranking two fuzzy numbers.

Definition 9.6 (Fuzzy α-feasibility)

(1) $X \in F(R^n)$ *is said to be fuzzy* α*-Very Weak Feasible* (α*-VWF*) *if and only if*

$$X \in \tilde{X}_{VWF}(\alpha) \triangleq \{X \in F(R^n) \mid Pos(A_j \otimes X \leq B_j) \geq \alpha,\ j = 1, \ldots, m\}. \tag{9.48}$$

(2) $X \in F(R^n)$ *is said to be fuzzy* α*-Medium Weak Feasible* (α*-MWF*) *if and only if*

$$X \in \tilde{X}_{MWF}(\alpha) \triangleq \{X \in F(R^n) \mid Pos(A_j \otimes X < B_j) \geq \alpha,\ j = 1, \ldots, m\}. \tag{9.49}$$

(3) $X \in F(R^n)$ *is said to be fuzzy* α*-Medium Strong Feasible* (α*-MSF*) *if and only if*

$$X \in \tilde{X}_{MSF}(\alpha) \triangleq \{X \in F(R^n) \mid Nes(A_j \otimes X \leq B_j) \geq \alpha,\ j = 1, \ldots, m\}. \tag{9.50}$$

(4) $X \in F(R^n)$ *is said to be fuzzy* α*-Very Strong Feasible* (α*-VSF*) *if and only if*

$$X \in \tilde{X}_{VSF}(\alpha) \triangleq \{X \in F(R^n) \mid Nes(A_j \otimes X < B_j) \geq \alpha,\ j = 1, \ldots, m\}. \tag{9.51}$$

Using the left and right extreme points $Y_{j\alpha}^L$ and $Y_{j\alpha}^R$ of the α-level sets, the above four types of fuzzy α-feasible regions can be represented as the usual sets in the $(2 \times n)$-dimensional real number space whose decision variables (x_l^m, x_l^s) are the mean values and the spread parameters of the membership functions of the fuzzy decision variables X_l, $l = 1, \ldots, n$.

Theorem 9.6

$$\begin{aligned}
\tilde{X}_{VWF}(\alpha) &= \{(x^m, x^s) \in R^{2n} \mid Y^L_{j\alpha} \leq b^R_{j\alpha},\ j = 1, \ldots, m,\ x^s \geq 0\}, \\
\tilde{X}_{MWF}(\alpha) &= \{(x^m, x^s) \in R^{2n} \mid Y^R_{j(1-\alpha)} \leq b^R_{j\alpha},\ j = 1, \ldots, m,\ x^s \geq 0\}, \\
\tilde{X}_{MSF}(\alpha) &= \{(x^m, x^s) \in R^{2n} \mid Y^L_{j\alpha} \leq b^L_{j(1-\alpha)},\ j = 1, \ldots, m,\ x^s \geq 0\}, \\
\tilde{X}_{VSF}(\alpha) &= \{(x^m, x^s) \in R^{2n} \mid Y^R_{j(1-\alpha)} \leq b^L_{j(1-\alpha)},\ j = 1, \ldots, m,\ x^s \geq 0\},
\end{aligned}$$

where $b^L_{j\alpha}$ and $b^R_{j\alpha}$ are the left and right extreme points of the α-level set of the fuzzy number B_j.

To introduce a generalized Pareto optimality concept for the k objective functions involving fuzzy decision variables and fuzzy parameters, consider the following multiobjective linear programming problems. Only the objective functions involve fuzzy numbers, and the feasible region of the mean values and the spread parameters of fuzzy decision variables are known, i.e.,

$$\left.\begin{array}{ll} \text{minimize} & (C_1 \otimes X, C_2 \otimes X, \ldots, C_k \otimes X) \\ \text{subject to} & X = (x^m, x^s) \in S \subset R^{2n}, \end{array}\right\} \tag{9.52}$$

where S is the set of pairs of the mean values and the spread parameters of the fuzzy decision variables.

Similar to the definition of the fuzzy α-feasibility, we can define the four types of fuzzy γ-Pareto optimality as follows.

Definition 9.7 (Fuzzy γ-Pareto optimality)

(1) *$X^* \in S$ is said to be fuzzy γ-Very Weak Pareto (γ-VWP) optimal if and only if there does not exist another $X \in S$ such that*

$$Pos(C_i \otimes X \geq C_i \otimes X^*) \leq \gamma,\ i = 1, \ldots, k, \tag{9.53}$$

with strict inequality holding for at least one i.

(2) *$X^* \in S$ is said to be fuzzy γ-Medium Weak Pareto (γ-MWP) optimal if and only if there does not exist another $X \in S$ such that*

$$Pos(C_i \otimes X > C_i \otimes X^*) \leq \gamma,\ i = 1, \ldots, k, \tag{9.54}$$

with strict inequality holding for at least one i.

(3) *$X^* \in S$ is said to be fuzzy γ-Medium Strong Pareto (γ-MSP) optimal if and only if there does not exist another $X \in S$ such that*

$$Nes(C_i \otimes X \geq C_i \otimes X^*) \leq \gamma,\ i = 1, \ldots, k, \tag{9.55}$$

with strict inequality holding for at least one i.

(4) *$X^* \in S$ is said to be fuzzy γ-Very Strong Pareto (γ-VSP) optimal if and only if there does not exist another $X \in S$ such that*

$$Nes(C_i \otimes X > C_i \otimes X^*) \leq \gamma, \; i = 1, \ldots, k, \tag{9.56}$$

with strict inequality holding for at least one i.

Similar to Theorem 9.6, the γ-level sets of the fuzzy objective functions $C_i \otimes X$ and $C_i \otimes X^*$ are represented as follows.

Theorem 9.7

$$\left.\begin{array}{lll} [C_i \otimes X]_\gamma & = \left[\sum\limits_{l=1}^{n} C_{il} \otimes X_l\right]_\gamma & = [F_{i\gamma}^L, F_{i\gamma}^R]; \\ [C_i \otimes X^*]_\gamma & = \left[\sum\limits_{l=1}^{n} C_{il} \otimes X_l^*\right]_\gamma & = [G_{i\gamma}^L, G_{i\gamma}^R], \end{array}\right\} \tag{9.57}$$

where

$$\begin{aligned} F_{i\gamma}^L \triangleq & \sum_{l=1}^{n} \Big\{\min\{(c_{il}^m - L^{-1}(\gamma)c_{il}^s)(x_l^m - L^{-1}(\gamma)x_l^s), \\ & \qquad (c_{il}^m - L^{-1}(\gamma)c_{il}^s)(x_l^m + L^{-1}(\gamma)x_l^s)\}\Big\}, \\ F_{i\alpha}^R \triangleq & \sum_{l=1}^{n} \Big\{\max\{(c_{il}^m + L^{-1}(\gamma)c_{il}^s)(x_l^m + L^{-1}(\gamma)x_l^s), \\ & \qquad (c_{il}^m + L^{-1}(\gamma)c_{il}^s)(x_l^m - L^{-1}(\gamma)x_l^s)\}\Big\}, \\ G_{i\gamma}^L \triangleq & \sum_{l=1}^{n} \Big\{\min\{(c_{il}^m - L^{-1}(\gamma)c_{il}^s)(x_l^{m*} - L^{-1}(\gamma)x_l^{s*}), \\ & \qquad (c_{il}^m - L^{-1}(\gamma)c_{il}^s)(x_l^{m*} + L^{-1}(\gamma)x_l^{s*})\}\Big\}, \\ G_{i\alpha}^R \triangleq & \sum_{l=1}^{n} \Big\{\max\{(c_{il}^m + L^{-1}(\gamma)c_{il}^s)(x_l^{m*} + L^{-1}(\gamma)x_l^{s*}), \\ & \qquad (c_{il}^m + L^{-1}(\gamma)c_{il}^s)(x_l^{m*} - L^{-1}(\gamma)x_l^{s*})\}\Big\}. \end{aligned}$$

Also, assume that the right and left extreme points of the support of C_{il} have the same sign and define the following index sets:

$$\begin{aligned} L_3(i) &= \{l \in N \mid c_{il}^m - L^{-1}(0)c_{il}^s \geq 0\}, \\ L_4(i) &= \{l \in N \mid c_{il}^m + L^{-1}(0)c_{il}^s < 0\}. \end{aligned}$$

Then we have

$$\begin{aligned}
F_{i\gamma}^L &= \sum_{l\in L_3(i)} (c_{il}^m - L^{-1}(\gamma)c_{il}^s)(x_l^m - L^{-1}(\gamma)x_l^s) \\
&+ \sum_{l\in L_4(i)} (c_{il}^m - L^{-1}(\gamma)c_{il}^s)(x_l^m + L^{-1}(\gamma)x_l^s), \\
F_{i\alpha}^R &= \sum_{l\in L_3(i)} (c_{il}^m + L^{-1}(\gamma)c_{il}^s)(x_l^m + L^{-1}(\gamma)x_l^s) \\
&+ \sum_{l\in L_4(i)} (c_{il}^m + L^{-1}(\gamma)c_{il}^s)(x_l^m - L^{-1}(\gamma)x_l^s), \\
G_{i\gamma}^L &= \sum_{l\in L_3(i)} (c_{il}^m - L^{-1}(\gamma)c_{il}^s)(x_l^{m*} - L^{-1}(\gamma)x_l^{s*}) \\
&+ \sum_{l\in L_4(i)} (c_{il}^m - L^{-1}(\gamma)c_{il}^s)(x_l^{m*} + L^{-1}(\gamma)x_l^{s*}), \\
G_{i\alpha}^R &= \sum_{l\in L_3(i)} (c_{il}^m + L^{-1}(\gamma)c_{il}^s)(x_l^{m*} + L^{-1}(\gamma)x_l^{s*}) \\
&+ \sum_{l\in L_4(i)} (c_{il}^m + L^{-1}(\gamma)c_{il}^s)(x_l^{m*} - L^{-1}(\gamma)x_l^{s*}).
\end{aligned}$$

Using the right and left extreme points of the γ-level sets of the objective functions involving fuzzy variables and parameters, the following theorem holds.

Theorem 9.8

(1) *$X^* \in S$ is said to be fuzzy γ-VWP if and only if there does not exist another $X \in S$ such that $F_{i\gamma}^R \leq G_{i\gamma}^L$, $i = 1, \ldots, k$, with strict inequality holding for at least one i.*

(2) *$X^* \in S$ is said to be fuzzy γ-MWP if and only if there does not exist another $X \in S$ such that $F_{i\gamma}^R \leq G_{i(1-\gamma)}^R$, $i = 1, \ldots, k$, with strict inequality holding for at least one i.*

(3) *$X^* \in S$ is said to be fuzzy γ-MSP if and only if there does not exist another $X \in S$ such that $F_{i(1-\gamma)}^L \leq G_{i\gamma}^L$, $i = 1, \ldots, k$, with strict inequality holding for at least one i.*

(4) *$X^* \in S$ is said to be fuzzy γ-VSP if and only if there does not exist another $X \in S$ such that $F_{i(1-\gamma)}^L \leq G_{i(1-\gamma)}^L$, $i = 1, \ldots, k$, with strict inequality holding for at least one i.*

Among the four types of fuzzy γ-Pareto optimality, it can be shown that $\tilde{X}_{MWP}(\gamma)$, $\tilde{X}_{MSP}(\gamma)$, and $\tilde{X}_{VSP}(\gamma)$ are included in the intersection of the usual Pareto optimal solution set. In general, however, it is not easy to obtain the intersection of the usual Pareto optimal solution set. On the other hand, since the fuzzy γ-Very Weak Pareto optimal solution (γ-VWP) set $\tilde{X}_{VWP}(\gamma)$ includes the union of the usual Pareto optimal solution set, it is easy to find out the elements of $\tilde{X}_{VWP}(\gamma)$. From this point of view, $\tilde{X}_{VWP}(\gamma)$ would be recommended as the generalized concept of Pareto optimality for our problem.

Now we can introduce new solution concepts for multiobjective linear programming problems with fuzzy decision variables and fuzzy parameters by combining fuzzy α-feasibility and fuzzy γ-Pareto optimality concepts. Realizing that fuzzy γ-VWP has a preferable property compared with the other three types of fuzzy γ-Pareto optimality, we focus on the following fuzzy (α, γ)-Very Weak Pareto optimal set $\tilde{X}_{VW}(\alpha, \gamma)$.

Definition 9.8 (Fuzzy (α, γ)-Very Weak Pareto optimal solution)

$X^* \in \tilde{X}_{VWF}(\alpha)$ is said to be a fuzzy (α, γ)-Very Weak Pareto optimal solution (fuzzy (α, γ)-VWP) if and only if there does not exist another $X \in \tilde{X}_{VWF}(\alpha)$ such that

$$Pos(C_i \otimes X \geq C_i \otimes X^*) \geq \gamma, \ i = 1, \dots, k, \tag{9.58}$$

where

$$\tilde{X}_{VWF}(\alpha) \triangleq \{X \in F(R^n) \mid Pos(A_j \otimes X \leq B_j \otimes X^*) \leq \alpha, \ j = 1, \dots, m\} \tag{9.59}$$

with strict inequality holding for at least one i.

It can be shown that part of $\tilde{X}_{VW}(\alpha, \gamma)$ can be obtained by solving the following two types of usual multiobjective linear programming problems where the decision variables consist of the mean values and the spread parameters of the fuzzy decision variables:

$$\left.\begin{array}{ll} \text{minimize} & \underline{Q}(x^m, x^s)^T \\ \text{subject to} & (x^m, x^s) \in \tilde{X}_{VWF}(\alpha) \triangleq \{Y_{j\alpha}^L \leq b_{j\alpha}^R, \ j = 1, \dots, m, \ x^s \geq 0\} \end{array}\right\} \tag{9.60}$$

where the (i, l) element of the matrix $\underline{Q}$ is

$$\begin{array}{ll} q_{il} = c_{il}^m - L^{-1}(\gamma) c_{il}^s, & l \in L_3(i) \text{ and } 1 \leq l \leq n, \\ q_{il} = (c_{il}^m - L^{-1}(\gamma) c_{il}^s)(-L^{-1}(\gamma)), & l \in L_3(i) \text{ and } n+1 \leq l \leq 2n, \\ q_{il} = c_{il}^m - L^{-1}(\gamma) c_{il}^s, & l \in L_4(i) \text{ and } 1 \leq l \leq n, \\ q_{il} = (c_{il}^m - L^{-1}(\gamma) c_{il}^s) L^{-1}(\gamma), & l \in L_4(i) \text{ and } n+1 \leq l \leq 2n; \end{array}$$

$$\left.\begin{array}{ll} \text{minimize} & \bar{Q}(x^m, x^s)^T \\ \text{subject to} & (x^m, x^s) \in \tilde{X}_{VWF}(\alpha) = \{Y_{j\alpha}^L \leq b_{j\alpha}^R, \ j = 1, \dots, m, \ x^s \geq 0\} \end{array}\right\} \tag{9.61}$$

where the (i, l) element of the matrix $\bar{Q}$ is

$$\begin{array}{ll} q_{il} = c_{il}^m + L^{-1}(\gamma) c_{il}^s, & l \in L_3(i) \text{ and } 1 \leq l \leq n, \\ q_{il} = (c_{il}^m + L^{-1}(\gamma) c_{il}^s) L^{-1}(\gamma), & l \in L_3(i) \text{ and } n+1 \leq l \leq 2n, \\ q_{il} = c_{il}^m + L^{-1}(\gamma) c_{il}^s, & l \in L_4(i) \text{ and } 1 \leq l \leq n, \\ q_{il} = (c_{il}^m + L^{-1}(\gamma) c_{il}^s)(-L^{-1}(\gamma)), & l \in L_4 \text{ and } n+1 \leq l \leq 2n. \end{array}$$

Two types of fuzzy (α, γ)-Pareto optimal solution sets obtained by solving these two problems can be interpreted as the most optimistic solution set and the most pessimistic solutions set among $\tilde{X}_{VW}(\alpha, \gamma)$. Observe that by setting $x^s = 0$ in these problems (this means the decision variables are no longer fuzzy numbers but ordinary numbers) these two problems reduce to the linear programming problem for obtaining the most optimistic and the most pessimistic (α, γ)-Pareto optimal solution sets discussed in the previous section (Sakawa and Yano 1990a).

It is appropriate to note here that all the results presented in this section have recently been extended to multiobjective linear fractional programming problems with fuzzy variables and parameters (Sakawa and Yano 1991a). A possible generalization to the nonlinear case along this line might be found in the near future. However, development of fuzzy interactive programming methods using the proposed fuzzy Pareto optimality concepts, together with the applications to the real world decision situations under fuzziness, would be urgently required.

9.3 Fuzzy multiobjective 0-1 programming through neural computation

In general, a multiobjective linear 0-1 programming problem is formulated as

$$\left.\begin{array}{ll} \text{minimize} & (c_1 x, c_2 x, \ldots, c_k x)^T \\ \text{subject to} & x \in X \triangleq \{Ax \leq b,\ x_j \in \{0,1\},\ j = 1, ..., n\} \end{array}\right\} \tag{9.62}$$

where $c_i = (c_{i1}, \ldots, c_{in}), i = 1, \ldots, k,\ x = (x_1, \ldots, x_n)^T,\ b = (b_1, \ldots, b_m)^T$, and $A = (a_{ij})$ is an $m \times n$ matrix.

For such a multiobjective 0-1 programming problem, it may be possible to soften the rigid requirement of the decision maker (DM) to strictly minimize the k objective functions under the constraints.

Similar to the fuzzy multiobjective linear programming approach of Zimmermann (1978), in the framework of the fuzzy decision of Bellman and Zadeh (1970), Zimmermann and Pollatschek (1984) incorporated the fuzzy goals of the decision maker (DM) into this problem. Their fuzzy version of this problem can be written as

$$\left.\begin{array}{l} c_1 x \precsim g_1 \\ \cdots\cdots\cdots \\ c_k x \precsim g_k \\ Ax \leq b, \\ x_j \in \{0,1\}, \quad j = 1, \ldots, n. \end{array}\right\} \tag{9.63}$$

This fuzzy multiobjective 0-1 programming problem, with the symbol $\precsim$ denoting essentially smaller than or equal to, is interpreted as the k objective functions $c_1 x, \ldots, c_k x$ should be essentially smaller than or equal to the corresponding aspiration levels $g_1, \ldots, g_k$ of the DM.

Such a fuzzy inequality concept can be treated by introducing membership functions which assume the value 1 if the conditions are satisfied and 0 if the conditions are violated beyond their limits. The simplest version of such a membership function is a linear one defined by

$$\mu_i(c_i x) = \begin{cases} 1 & ; \quad c_i x \leq g_i \\ 1 - \dfrac{c_i x_i - g_i}{d_i} & ; \quad g_i \leq c_i x \leq g_i + d_i \\ 0 & ; \quad c_i x \geq g_i + d_i \end{cases} \tag{9.64}$$

where each d_i is the subjectively chosen constant expressing a limit of the admissible violation of the ith fuzzy inequality.

Following the fuzzy decision of Bellman and Zadeh (1970) together with the linear membership functions, the problem of finding the maximum decision is

$$\underset{x \in X}{\text{maximize}} \min_{i=1,\ldots,k} \mu_i(c_i x). \tag{9.65}$$

By introducing the auxiliary variable λ, this problem is equivalently transformed to the following linear 0-1 programming problem with $0 \leq \lambda \leq 1$:

$$\left.\begin{array}{ll} \text{minimize} & \lambda \\ \text{subject to} & \lambda \leq \mu_i(c_i x),\ i = 1, \ldots, k \\ & Ax \leq b \\ & x_j \in \{0, 1\},\ j = 1, \ldots, n \\ & 0 \leq \lambda \leq 1. \end{array}\right\} \tag{9.66}$$

By slightly modifying this crisp equivalent problem, Zimmermann and Pollatschek (1984) suggested the branch-and-bound technique based algorithm for solving it together with some numerical experience. Along similar lines, Fabian and Stoica (1984) considered linear integer programming with fuzzy goals of the DM. At a similar period of time, Ignizio and Daniels (1983) proposed a solution procedure for solving fuzzy multiobjective integer programming problems via fuzzy generalized networks.

However, it should be stressed here that, although numerous solution procedures have been developed in recent years, a general solution technique for integer programming is yet to be developed. For special types of integer programming problems having their variables constrained by upper and/or lower bounds, the "branch-and-bound" technique, initially developed by Land and Doig (1960), is perhaps the most general approach for obtaining the optimal solution.

Recently, as an alternative approach, the Hopfield neural network model (Hopfield 1984, Hopfield and Tank 1985) has been successfully applied to finding reasonable solutions to combinatorial optimization problems such as traveling salesperson problems (Hopfield and Tank 1985) and Hitchcock problems (Takeda and Goodman 1986). However, in the original model, the formulated minimization problem of an energy function can never escape from the local minimum when an energy function has multiple minima. By adding a Gaussian noise in the original neural networks and using so-called annealing and sharpening schemes, it is possible to get the global minimum of the energy function by means of the modified neural network model, called the Gaussian machine (Akiyama et al. 1989).

At present, although a neural computation approach is mainly focused on 0-1 programming problems that only include equality constraints, with some devices, it is also applicable to the problems which include not only equality constraints but also inequality constraints (Sakawa 1990; Sakawa and Yano 1990e). In the remainder of this subsection, realizing that both the original and modified Hopfield neural networks are essentially an analog implementation of the penalty and barrier methods for constrained optimization, we suggest applying them to solving fuzzy multiobjective linear 0-1 programming problems via their spontaneous energy minimization processes.

Before presenting the possible neural computational methods for obtaining reasonable solutions for the fuzzy multiobjective 0-1 programming problems discussed above, it is appropriate to make a brief review of the Hopfield model and Gaussian machine in the field of neural computation.

In the original Hopfield model (Hopfield 1984, Hopfield and Tank 1985) with symmetrically interconnected networks, neurons change their states in such a manner that they minimize an energy function defined by

$$E = -\frac{1}{2}\sum_i \sum_j T_{ij} V_i V_j - \sum_i I_i V_i \tag{9.67}$$

and stop at a local minima of this quantity. On the basis of this fact, Hopfield and Tank (1985) demonstrated that a certain class of combinatorial optimization problems can be solved using their network model. However, when an energy function has multiple minima, the possibility of the convergence to one of the local minima is very high. To escape from a local minima of an energy function of a network, a Gaussian machine (Akiyama et al. 1989) is proposed by incorporating both annealing and sharpening techniques. Essentially, a Gaussian machine can be viewed as a modified version of the original Hopfield model by adding a Gaussian noise to each neuron i.

In the Hopfield model, each neuron i receives input $T_{ij}V_j$ from other neurons j and a bias input I_i associated with itself, namely, each neuron's net input U_i is defined by

$$U_i = \sum_{j=1}^{N} T_{ij} V_j + I_i, \tag{9.68}$$

where T_{ij} represents the effectiveness of a synapse and the ith neuron output V_i is governed by a nonlinear sigmoid input/output function defined as

$$V_i = \frac{1}{2}\left\{1 + \tanh\left(\frac{U_i}{\mu_o}\right)\right\}. \tag{9.69}$$

In the Gaussian machine (Akiyama et al. 1989), each neuron's net input U_i is defined by

$$U_i = \sum_{j=1}^{N} T_{ij} V_j + I_i + \varepsilon, \tag{9.70}$$

where the added noise term ε is a random variable with mean 0 and variation σ^2 obeying a Gaussian distribution. Standard deviation σ is defined as

$$\sigma = kT \tag{9.71}$$

where T is a temperature parameter of a network as defined in the Boltzmann machines (Ackley et al. 1985), and the constant k is usually assumed to be $k = \sqrt{8/\pi}$.

It should be noted here that, in the Gaussian machine, if ε tends to 0, it becomes the Hopfield and Tank model, and it behaves like a Boltzmann machine if μ_0 tends to 0.

When solving combinatorial optimization problems using a Gaussian machine, both sharpening and annealing techniques are essential; sharpening reduces the value of μ_0 and annealing decreases the value of temperature T. Using both techniques, it becomes possible to escape from a local minima of an energy function of a network.

For simulating neural networks on a digital computer, the following discrete model is adopted:

$$\text{Input of neuron } i: \quad U_{xi}(k) = \sum_y \sum_j T_{xi,yj} V_{xi}(k-1) + I_{xi} \tag{9.72}$$

$$\text{Output of neuron } i: \quad V_{xi}(k) = \frac{1}{2}\left\{1 + \tanh\left(\frac{U_{xi}(k)}{\mu_0}\right)\right\}. \tag{9.73}$$

We are now ready to present the neural computation method for solving fuzzy multiobjective 0-1 programming problems. As discussed in the previous subsection, by adopting the fuzzy decision of Bellman and Zadeh (1970) together with linear membership functions for representing the fuzzy goals of the decision maker (DM), the problem to be solved is

$$\left.\begin{array}{ll} \text{minimize} & \lambda \\ \text{subject to} & \lambda \leq \mu_i(c_i x),\ i = 1, \ldots, k \\ & Ax \leq b \\ & x_j \in \{0, 1\},\ j = 1, \ldots, n \\ & 0 \leq \lambda \leq 1. \end{array}\right\} \tag{9.74}$$

To solve this problem through neural computation, it is quite important to introduce the slack variable for each inequality constraint and then transform the inequality constraint to the equality constraint as follows:

$$\lambda + s_i = \mu_i(c_i x),\ 0 \leq s_i \leq 1,\ i = 1, \ldots, k \tag{9.75}$$

$$Ax + tb = b, \ t = (t_1, \ldots, t_m), \ 0 \leq t_i \leq 1. \tag{9.76}$$

Then, for the problem under consideration, it is appropriate to define the energy function as follows:

$$\begin{aligned} E = \ & \frac{A}{2} \sum_j x_j(1 - x_j) - \frac{B}{2} \lambda^2 \\ & + \sum_i \frac{C_i}{2} \{\lambda + s_i - \mu_i(c_i x)\}^2 + \frac{D}{2}(b - Ax - tb)^2. \end{aligned} \tag{9.77}$$

Using this energy function and linear membership functions for the fuzzy goals of the DM, together with appropriately selected values of A, B, C_i, and D, it becomes possible to obtain relatively satisfactory solutions with the Gaussian machine. However, it should be pointed out here that no sharpening scheme is required for the slack variables because they are not 0-1 variables. Further details including numerical examples can be found in Sakawa (1990) and Sakawa and Mitani (1991).

Further extensions to fuzzy multiobjective combinatorial optimization problems combining our interactive fuzzy programming methods discussed in Chapters 3, 4, 5, and 6 with the Gaussian machines are now under investigation and will be reported elsewhere. However, the energy functions of these networks have several parameters that need to be chosen properly for the network to operate satisfactorily. If the energy functions and/or the corresponding parameter values are selected incorrectly, the networks never converge to an optimal solution. Therefore, further investigations are required to determine not only the appropriate energy functions but also the proper parameter values.

9.4 Fuzzy regression analysis

When modeling the real systems such as economic systems, social systems, and environmental systems, the inherent ambiguity or fuzziness of a human's subjective judgment is influential. Therefore, exact modeling of these systems may be very difficult. Alternatively, an application of the concept of fuzzy set theory to modeling such systems seems to be much more preferable to conventional modeling approaches. From this point of view, Tanaka, Uejima, and Asai (1982) first introduced the concept of fuzzy linear regression analysis. Since then, fuzzy linear regression analysis has been developed in many directions together with some applications (e.g. Heshmaty and Kandel 1985; Tanaka 1987; Tanaka, Hayashi, and Watada 1989; Sakawa and Yano 1989c, 1992c). In contrast to the usual linear regression analysis in which deviations between the observed values and the estimated values are supposed to be caused by observation errors, in fuzzy linear regression analysis, deviations are assumed to be dependent upon the fuzziness of the system structure.

In this section, from the viewpoint of possibility and necessity introduced by Dubois and Prade (1983, 1987), we present a unified approach to the fuzzy linear regression analysis in which not only input data is a vector of nonfuzzy numbers and output data is a fuzzy number but also both input data and output data as fuzzy numbers are considered (Sakawa and Yano 1989c, 1992a, 1992b, 1992c).

9.4.1 Fuzzy linear regression analysis for fuzzy output data

First, consider a fuzzy linear regression model when fuzzy data (x_i, Y_i), $i = 1, ..., k$, are given as pairs of a vector of nonfuzzy input data $x_i = (x_{i1}, ..., x_{in})$ and the corresponding fuzzy output data Y_i. In contrast to the conventional linear regression analysis, if we assume that the deviations between the observed values and the estimated values are caused by the indefiniteness or ambiguity of the system structure, the fuzzy linear regression model, whose coefficients involve ambiguity or fuzziness, can be represented as follows:

$$Y = Ax = A_0 + A_1 x_1 + \cdots + A_n x_n \tag{9.78}$$

where $A = (A_0, A_1, \ldots, A_n)$, $x = (1, x_1, \ldots, x_n)^T$. It is assumed that Y and A_j, $j = 0, \ldots, n$ are fuzzy numbers.

When we try to deal with this problem as a generalization of the ordinary linear regression model, it is necessary to consider how to interpret the fuzzy equality relation. For that purpose, using the possibility and necessity concept introduced by Dubois and Prade (1983, 1987), a unified interpretation can be made. To define the equality relation between two fuzzy numbers M and N as a natural extension of that of two closed intervals, Dubois and Prade (1983, 1987) introduced the following three indices.

Definition 9.9

$$\begin{aligned}
Pos(M = N) &\triangleq \sup_{u \in R^1} \min \{\mu_M(u), \mu_N(u)\} \\
Nes(M \subset N) &\triangleq \inf_{u \in R^1} \max \{1 - \mu_M(u), \mu_N(u)\} \\
Nes(M \supset N) &\triangleq \inf_{u \in R^1} \max \{\mu_M(u), 1 - \mu_N(u)\}
\end{aligned}$$

where $\mu_M(u)$ or $\mu_N(u)$ is the membership function of M or N and Pos or Nes is short for Possibility or Necessity.

Observe that index $Pos(M = N)$ means the degree of possibility that M is equal to N and index $Nes(M \subset N)$ or $Nes(M \supset N)$ means the degree of necessity that N includes M or M includes N.

The condition that the degrees of these three indices are greater than or equal to a certain level of α can be equivalently expressed as the following usual constraints by using the α-level sets of fuzzy numbers M and N (Sakawa and Yano 1989c, 1992c).

Theorem 9.9

$$\begin{array}{lcl} Pos(M=N)\geq\alpha & \Leftrightarrow & m_\alpha^R\geq n_\alpha^L \text{ and } m_\alpha^L\leq n_\alpha^R \\ Nes(M\subset N)\geq\alpha & \Leftrightarrow & m_{1-\alpha}^L\geq n_\alpha^L \text{ and } m_{1-\alpha}^R\geq n_\alpha^R \\ Nes(M\supset N)\geq\alpha & \Leftrightarrow & m_\alpha^L\geq n_{1-\alpha}^L \text{ and } m_\alpha^R\geq n_{1-\alpha}^R \end{array}$$

where m_α^L or n_α^L and m_α^R or n_α^R are left and right extreme points of the α-level set of fuzzy numbers M or N, respectively.

Using these three indices, it is possible to introduce the corresponding fuzzy linear regression models. To do so in the following, for simplicity, assume that the membership functions of fuzzy output data Y_i, $i=1,\ldots,k$, and the fuzzy parameters A_j, $j=0,\ldots,n$, are given as L-L fuzzy numbers defined by

$$\left.\begin{array}{lcl} \mu_{Y_i}(y_i) & = & L(\mid \bar{y}_i - y_i \mid / e_i),\ e_i>0 \\ \mu_{A_j}(a_j) & = & L(\mid \bar{a}_j - a_j \mid / c_j),\ c_j>0 \end{array}\right\} \tag{9.79}$$

where $\bar{y}_i$ and $\bar{a}_j$ are mean values, e_i and c_j are spread parameters, and $L(\cdot)$ is a shape function assumed to be strictly monotone decreasing and continuous. Such L-L fuzzy numbers Y_i and A_j are often symbolically represented by $Y_i=(\bar{y}_i,e_i)_L$, $i=1,\ldots,k$, $A_j=(\bar{a}_j,c_j)_L$, $j=0,\ldots,n$.

For notational convenience, denote the fuzzy parameter vector $A=(A_0,\ldots,A_n)$ by $(\bar{a},c)_L$ where $\bar{a}=(\bar{a}_0,\ldots,\bar{a}_n)$, $c=(c_0,\ldots,c_n)$. Then, by the operations of fuzzy numbers, the following relation holds:

$$Ax_i=(\bar{a}x_i,c|x_i|)_L \quad \text{or} \quad \sum_{j=0}^{n} A_j x_{ij} = \left(\sum_{j=0}^{n}\bar{a}_j x_{ij}, \sum_{j=0}^{n} c_j|x_{ij}|\right)_L . \tag{9.80}$$

Now, using Theorem 9.9 together with this relation and the properties of L-L fuzzy numbers, the following theorem holds for some fixed degree α.

Theorem 9.10

(1) *$Pos(Y_i=Ax_i)\geq\alpha$ if and only if*

$$\left.\begin{array}{lcl} -\bar{y}_i+\sum\limits_{j=0}^{n}\bar{a}_j x_{ij} & \leq & L^{-1}(\alpha)e_i+L^{-1}(\alpha)\sum\limits_{j=0}^{n}c_j|x_{ij}| \\ \bar{y}_i-\sum\limits_{j=0}^{n}\bar{a}_j x_{ij} & \leq & L^{-1}(\alpha)e_i+L^{-1}(\alpha)\sum\limits_{j=0}^{n}c_j|x_{ij}| \end{array}\right\}. \tag{9.81}$$

(2) *$Nes(Y_i\subset A_i)\geq\alpha$ if and only if*

$$\left.\begin{array}{lcl} -\bar{y}_i+\sum\limits_{j=0}^{n}\bar{a}_j x_{ij} & \leq & -L^{-1}(1-\alpha)e_i+L^{-1}(\alpha)\sum\limits_{j=0}^{n}c_j|x_{ij}| \\ \bar{y}_i-\sum\limits_{j=0}^{n}\bar{a}_j x_{ij} & \leq & -L^{-1}(1-\alpha)e_i+L^{-1}(\alpha)\sum\limits_{j=0}^{n}c_j|x_{ij}| \end{array}\right\}. \tag{9.82}$$

(3) $Nes(Y_i \supset Ax_i) \geq \alpha$ *if and only if*

$$\left.\begin{array}{rcl} -\bar{y}_i + \sum_{j=0}^{n} \bar{a}_j x_{ij} & \leq & L^{-1}(\alpha)e_i - L^{-1}(1-\alpha) \sum_{j=0}^{n} c_j |x_{ij}| \\ \bar{y}_i - \sum_{j=0}^{n} \bar{a}_j x_{ij} & \leq & L^{-1}(\alpha)e_i - L^{-1}(1-\alpha) \sum_{j=0}^{n} c_j |x_{ij}| \end{array}\right\}. \quad (9.83)$$

Considering the correspondence between the inequalities in the least square problems for obtaining the usual linear regression models and the inequalities of this theorem, it seems to be quite natural to adopt the following objective functions $z(c)$ for the three indices, respectively, for determining the fuzzy linear regression model in the sense that the fuzziness of the model should be optimized:

$$z(c) = \sum_{i=1}^{k} \sum_{j=0}^{n} c_j |x_{ij}|. \quad (9.84)$$

Then, for a certain degree α, the following three types of problems $\mathrm{P}_1(\alpha)$, $\mathrm{P}_2(\alpha)$, and $\mathrm{P}_3(\alpha)$ for obtaining the fuzzy linear regression models can be formulated corresponding to the three indices:

$$\begin{array}{rl} \text{minimize} & z(c) \triangleq \sum_{i=1}^{k} \sum_{j=0}^{n} c_j |x_{ij}| \\ \text{subject to} & -\bar{y}_i + \sum_{j=0}^{n} \bar{a}_j x_{ij} \leq L^{-1}(\alpha)e_i + L^{-1}(\alpha) \sum_{j=0}^{n} c_j |x_{ij}|,\ i = 1, ..., k \\ & \bar{y}_i - \sum_{j=0}^{n} \bar{a}_j x_{ij} \leq L^{-1}(\alpha)e_i + L^{-1}(\alpha) \sum_{j=0}^{n} c_j |x_{ij}|,\ i = 1, ..., k \\ & c_j \geq 0,\ j = 0, \ldots, n \end{array}$$

$$\begin{array}{rl} \text{minimize} & z(c) \triangleq \sum_{i=1}^{k} \sum_{j=0}^{n} c_j |x_{ij}| \\ \text{subject to} & -\bar{y}_i + \sum_{j=0}^{n} \bar{a}_j x_{ij} \leq -L^{-1}(1-\alpha)e_i + L^{-1}(\alpha) \sum_{j=0}^{n} c_j |x_{ij}|,\ i = 1, ..., k \\ & \bar{y}_i - \sum_{j=0}^{n} \bar{a}_j x_{ij} \leq -L^{-1}(1-\alpha)e_i + L^{-1}(\alpha) \sum_{j=0}^{n} c_j |x_{ij}|,\ i = 1, ..., k \\ & c_j \geq 0,\ j = 0, \ldots, n \end{array}$$

$$\begin{array}{rl} \text{maximize} & z(c) \triangleq \sum_{i=1}^{k} \sum_{j=0}^{n} c_j |x_{ij}| \\ \text{subject to} & -\bar{y}_i + \sum_{j=0}^{n} \bar{a}_j x_{ij} \leq L^{-1}(\alpha)e_i - L^{-1}(1-\alpha) \sum_{j=0}^{n} c_j |x_{ij}|,\ i = 1, ..., k \\ & \bar{y}_i - \sum_{j=0}^{n} \bar{a}_j x_{ij} \leq L^{-1}(\alpha)e_i - L^{-1}(1-\alpha) \sum_{j=0}^{n} c_j |x_{ij}|,\ i = 1, ..., k \\ & c_j \geq 0,\ j = 0, \ldots, n. \end{array}$$

It should be noted here that, for some fixed value of α, $\mathrm{P}_1(\alpha)$, $\mathrm{P}_2(\alpha)$, or $\mathrm{P}_3(\alpha)$ can be easily solved because it is a linear programming problem with decision variables $(\bar{a}, c_j)$, $j = 0, \ldots, n$.

So far, three types of problems have been formulated for deriving fuzzy linear regression models based on the three indices under the assumption that input data are not fuzzy numbers but output data Y_i, $i = 1, \ldots, k$, are fuzzy numbers. However, almost all the existing data thus far are not fuzzy numbers but usual numbers. Therefore, the problem for deriving the fuzzy linear regression model, where input and output data are not fuzzy numbers, would be necessary.

Observing that the nonfuzzy number y_i can be obtained as a special case of the fuzzy number $Y_i = (\bar{y}_i, e_i)_L$ by setting $y_i = \bar{y}_i$, $e_i = 0$, $i = 1, \ldots, k$, both $\mathrm{P}_1(\alpha)$ and $\mathrm{P}_2(\alpha)$ are reduced to the following linear programming problem $\mathrm{P}_4(\alpha)$:

$$\begin{array}{ll} \text{minimize} & z(c) \triangleq \sum_{i=1}^{k} \sum_{j=0}^{n} c_j |x_{ij}| \\ \text{subject to} & -y_i + \sum_{j=0}^{n} \bar{a}_j x_{ij} \leq L^{-1}(\alpha) \sum_{j=0}^{n} c_j |x_{ij}|, \ i = 1, \ldots, k \\ & y_i - \sum_{j=0}^{n} \bar{a}_j x_{ij} \leq L^{-1}(\alpha) \sum_{j=0}^{n} c_j |x_{ij}|, \ i = 1, \ldots, k \\ & c_j \geq 0, \ j = 0, \ldots, n. \end{array}$$

Further details along this line with some applications can be found in Sakawa and Yano (1989c, 1992c).

However, when solving $\mathrm{P}_1(\alpha), \mathrm{P}_2(\alpha)$, $\mathrm{P}_3(\alpha)$, or $\mathrm{P}_4(\alpha)$ by the linear programming method for obtaining fuzzy linear regression models, it is necessary for the decision maker to specify the fuzzy threshold α for the indices in advance. It should be noted that, in general, the greater value of the fuzzy threshold α gives the greater value of the fuzziness $z(c)$ to $\mathrm{P}_1(\alpha), \mathrm{P}_2(\alpha)$, or $\mathrm{P}_4(\alpha)$ and the smaller value of the fuzziness $z(c)$ to $\mathrm{P}_3(\alpha)$. From this point of view, by considering the conflict between maximizing the degree α of the fuzzy threshold, and minimizing (or maximizing) the fuzziness $z(c)$ of the fuzzy linear regression model in the problems $\mathrm{P}_1(\alpha)$ and $\mathrm{P}_2(\alpha)$ (or $\mathrm{P}_3(\alpha)$), multiobjective programming problems for obtaining the fuzzy linear regression models have been formulated by the authors. Details of the problems and the interactive algorithm for deriving the satisficing solution for the decision maker from the Pareto optimal solution set can be found in Sakawa and Yano (1992c).

9.4.2 Fuzzy linear regression analysis for fuzzy input-output data

In this subsection, as a natural extension of the fuzzy linear regression models introduced in the previous subsection, we further present the fuzzy linear regression models where both input data and output data are represented as fuzzy numbers. To be more specific, consider a fuzzy linear regression model when fuzzy input-out data (X_i, Y_i), $i = 1, ..., k$, are given as pairs of vectors of fuzzy input data $X_i = (X_{i1}, ..., X_{in})$ and the corresponding fuzzy output data Y_i.

To deal with such fuzzy input-output data, consider the fuzzy linear regression model represented by:

$$Y = A \otimes X = A_0 + A_1 \otimes X_1 + \cdots + A_n \otimes X_n \tag{9.85}$$

where $A = (A_0, A_1, \ldots, A_n)$ and $X = (1, X_1, \ldots, X_n)^T$. Here, it is assumed that Y, A_0, A_j, and X_j, $j = 1, ..., n$, are fuzzy numbers and "$\otimes$" is the product operator based on the extension principle of Zadeh (1965).

For simplicity in the following discussions, assume that the fuzzy input data X_{ij}, $i = 1, ..., k$, $j = 1, ..., n$, are given as L-L fuzzy numbers defined by

$$\mu_{X_{ij}}(x_{ij}) = L\left(\frac{|\,\bar{x}_{ij} - x_{ij}\,|}{d_{ij}}\right), \quad d_{ij} > 0 \tag{9.86}$$

and the fuzzy output data Y_i, $i = 1, ..., k$, and fuzzy parameters A_j, $j = 1, ..., n$, are also given as L-L fuzzy numbers defined by (9.79), respectively. Furthermore, assume that the α-level set of the fuzzy number X_{ij} is nonnegative for any $\alpha \in [0, 1]$.

Observe that L-L fuzzy numbers X_{ij}, Y_i, and A_j are often symbolically represented by $X_{ij} = (\bar{x}_{ij}, d_{ij})_L$, $Y_i = (\bar{y}_i, e_i)_L$, and $A_j = (\bar{a}_j, c_j)_L$.

It should be emphasized here that, although the explicit form of the membership function of $A \otimes X_i$ cannot directly be obtained, the α-level set of $A \otimes X_i$ can be obtained as a direct result of Nguyen's theorem as discussed in Section 2.3. Hence, the following theorem holds.

Theorem 9.11

$$(A \otimes X_i)_\alpha = [Y_{i\alpha}^L, Y_{i\alpha}^R] \tag{9.87}$$

where

$$\begin{aligned} Y_{i\alpha}^L &\triangleq \sum_{j=0}^{n} \Big\{\min\{(\bar{a}_j - L^{-1}(\alpha)c_j)(\bar{x}_{ij} - L^{-1}(\alpha)d_{ij}), \\ &\qquad (\bar{a}_j - L^{-1}(\alpha)c_j)(\bar{x}_{ij} + L^{-1}(\alpha)d_{ij})\}\Big\} \\ Y_{i\alpha}^R &\triangleq \sum_{j=0}^{n} \Big\{\max\{(\bar{a}_j + L^{-1}(\alpha)c_j)(\bar{x}_{ij} + L^{-1}(\alpha)d_{ij}), \\ &\qquad (\bar{a}_j + L^{-1}(\alpha)c_j)(\bar{x}_{ij} - L^{-1}(\alpha)d_{ij})\}\Big\}. \end{aligned}$$

To deal with the minimization and maximization operators in $Y_{i\alpha}^L$ and $Y_{i\alpha}^R$, $i = 1, \ldots, k$, assume that the following relations hold for any fixed degree α:

$$\begin{aligned} &\bar{a}_j - L^{-1}(\alpha)c_j \geq 0, && j \in J_1 \\ &\bar{a}_j - L^{-1}(\alpha)c_j \leq 0, \ \bar{a}_j + L^{-1}(\alpha)c_j \geq 0, && j \in J_2 \\ &\bar{a}_j + L^{-1}(\alpha)c_j \leq 0, && j \in J_3 \end{aligned}$$

where $J = \{0, \ldots, n\} = J_1 \cup J_2 \cup J_3$, $J_i \cap J_j = \emptyset$, $i\,(\neq j) = 1, 2, 3$.

Such assumptions may not be so restrictive since, in real world situations, the signs of the parameters of the linear regression models are often known in advance. On the basis of such assumptions, define the following set which depends on the index sets J_1, J_2, and J_3:

$$\begin{aligned} L(J_1, J_2, J_3) = \ &\{(\bar{a}, c) \in \mathrm{R}^{2(n+1)} \mid \bar{a}_j - L^{-1}(\alpha)c_j \geq 0, \ c_j \geq 0, j \in J_1, \\ &\bar{a}_j - L^{-1}(\alpha)c_j \leq 0, \ \bar{a}_j + L^{-1}(\alpha)c_j \geq 0, \ c_j \geq 0, j \in J_2, \\ &\bar{a}_j + L^{-1}(\alpha)c_j \leq 0, \ c_j \geq 0, j \in J_3\}. \end{aligned}$$

Then $Y_{i\alpha}^L$ and $Y_{i\alpha}^R$, $i = 1, \ldots, k$, can be expressed as the following forms, respectively:

$$\left.\begin{aligned} Y_{i\alpha}^L = & \sum_{j \in J_1} (\bar{a}_j - L^{-1}(\alpha)c_j)(\bar{x}_{ij} - L^{-1}(\alpha)d_{ij}) \\ & + \sum_{j \in J_2 \cup J_3} (\bar{a}_j - L^{-1}(\alpha)c_j)(\bar{x}_{ij} + L^{-1}(\alpha)d_{ij}) \\ Y_{i\alpha}^R = & \sum_{j \in J_1 \cup J_2} (\bar{a}_j + L^{-1}(\alpha)c_j)(\bar{x}_{ij} + L^{-1}(\alpha)d_{ij}) \\ & + \sum_{j \in J_3} (\bar{a}_j + L^{-1}(\alpha)c_j)(\bar{x}_{ij} - L^{-1}(\alpha)d_{ij}). \end{aligned}\right\} \tag{9.88}$$

By adopting the three indices for equality relations, we can formulate the following three types of problems $\text{FP}_1(\alpha)$, $\text{FP}_2(\alpha)$, and $\text{FP}_3(\alpha)$ for obtaining the fuzzy linear regression models where both input data and output data are fuzzy numbers (Sakawa and Yano 1992a):

$$\begin{aligned} &\text{minimize} && z(\bar{a}, c) \triangleq \sum_{i=1}^{k} \left(Y_{i0}^R - Y_{i0}^L\right) \\ &\text{subject to} && Pos(Y_i = A \otimes X_i) \geq \alpha, \ i = 1, \ldots, k, \\ & && (\bar{a}, c) \in L(J_1, J_2, J_3) \\ &\text{minimize} && z(\bar{a}, c) \triangleq \sum_{i=1}^{k} \left(Y_{i0}^R - Y_{i0}^L\right) \\ &\text{subject to} && Nes(Y_i \subset A \otimes X_i) \geq \alpha, \ i = 1, \ldots, k, \\ & && (\bar{a}, c) \in L(J_1, J_2, J_3) \\ &\text{maximize} && z(\bar{a}, c) \triangleq \sum_{i=1}^{k} \left(Y_{i0}^R - Y_{i0}^L\right) \\ &\text{subject to} && Nes(Y_i \supset A \otimes X_i) \geq \alpha, \ i = 1, \ldots, k, \\ & && (\bar{a}, c) \in L(J_1, J_2, J_3) \end{aligned}$$

where the value of $\alpha \, (0 \leq \alpha \leq 1)$ represents the degree of conformity between the fuzzy output data Y_i and the fuzzy linear regression model $A \otimes X_i$, and the objective function $z(\bar{a}, c)$ means the degree of fuzziness of the fuzzy linear regression model (9.85).

From Theorem 9.9 and the definition of L-L fuzzy numbers Y_i, X_{ij}, and A_j, the inequalities of the problems $\text{FP}_1(\alpha), \text{FP}_2(\alpha)$, and $\text{FP}_3(\alpha)$ can be equivalently transformed into the usual inequalities.

Theorem 9.12

(1) *$Pos(Y_i = A \otimes X_i) \geq \alpha$ if and only if*

$$\bar{y}_i + L^{-1}(\alpha)e_i \geq Y_{i\alpha}^L, \quad \bar{y}_i - L^{-1}(\alpha)e_i \leq Y_{i\alpha}^R. \tag{9.89}$$

(2) *$Nes(Y_i \subset A \otimes X_i) \geq \alpha$ if and only if*

$$\bar{y}_i + L^{-1}(1-\alpha)e_i \leq Y_{i\alpha}^R, \quad \bar{y}_i - L^{-1}(1-\alpha)e_i \geq Y_{i\alpha}^L. \tag{9.90}$$

(3) $Nes(Y_i \supset A \otimes X_i) \geq \alpha$ *if and only if*

$$\bar{y}_i + L^{-1}(\alpha)e_i \geq Y^R_{i,1-\alpha}, \quad \bar{y}_i - L^{-1}(\alpha)e_i \leq Y^L_{i,1-\alpha}. \tag{9.91}$$

It should be noted here that, for some fixed degree α, the inequalities of this theorem are linear with respect to $(\bar{a}, c)$ since $Y^L_{i\alpha}$ and $Y^R_{i\alpha}$ are the linear functions with respect to $(\bar{a}, c)$ shown in (9.88). From this fact, the problems $\mathrm{FP}_1(\alpha)$, $\mathrm{FP}_2(\alpha)$, and $\mathrm{FP}_3(\alpha)$ can be reduced to linear programming problems with respect to $(\bar{a}, c)$. Therefore, it is easy to solve these problems by using the linear programming method. Further details and a numerical example can be found in Sakawa and Yano (1992a).

However, when solving one of these three linear programming problems the value of the degree α must be selected within the interval $[0, 1]$. In general, the higher degree α of the fuzzy threshold is more preferable because α can be interpreted as the degree of the equivalence between the fuzzy output data Y_i and the estimated value $A \otimes X_i$ of the fuzzy linear regression model. From this point of view, by considering the conflict between maximizing the degree α of the fuzzy threshold and minimizing (or maximizing) the fuzziness $z(\bar{a}, c)$ of the fuzzy linear regression model in one of these three problems, the corresponding multiobjective programming problems for obtaining the fuzzy linear regression models, in which both input data and output data are fuzzy numbers, have been formulated by the authors. Details of the problems and the interactive algorithm for deriving the satisficing solution for the decision maker from the Pareto optimal solution set can be found in Sakawa and Yano (1992b). A recently edited volume on "Fuzzy Regression Analysis" which covers all of the major theoretical and practical advances in the wide range of fuzzy regression analysis (Kacprzyk and Fedrizzi 1992) is very useful for interested readers.

APPENDIX

HYPERPLANE METHODS AND TRADE-OFFS

In this Appendix, as a generalization of the well-known existing scalarizing methods for multiobjective programming problems discussed in Chapter 3 and 4, a new scalarizing method, called the hyperplane method, proposed by the authors (Yano and Sakawa 1989, Sakawa and Yano 1990c, 1992d), is introduced. It places special emphasis not only on characterizing Pareto optimal solutions but also on obtaining trade-off information.

A.1 Hyperplane Problems

Consider multiobjective nonlinear programming (MONLP) problems of the following form:

$$\left.\begin{array}{ll} \text{minimize} & f(x) \triangleq (f_1(x), f_2(x), ..., f_k(x))^T \\ \text{subject to} & x \in X \triangleq \{x \in R^n \mid g_j(x) \leq 0,\ j = 1, ..., m\} \end{array}\right\} \qquad (A.1)$$

where x is an n-dimensional vector of decision variables, $f_1(x), ..., f_k(x)$ are k distinct objective functions of the decision vector x, $g_1(x), ..., g_m(x)$ are m inequality constraints, X is the feasible set of constrained decisions, and the functions $f_i(x)$, $i = 1, ..., k$, $g_j(x)$, $j = 1, ..., m$, are assumed to be twice continuously differentiable.

Associated with the MONLP, as a generalization of the scalarizing methods discussed in Sections 3.2 and 4.2, we introduce the following generalized scalar optimization problem called the hyperplane problem denoted by $HP(Q, t_{-k})$.

$$\left.\begin{array}{ll} \underset{x \in X}{\text{minimize}} & z \\ \text{subject to} & QF(f(x)) \leq D(z, t_{-k}), \end{array}\right\} \qquad (A.2)$$

where $z \in R^1$ and Q is the $k \times k$ matrix whose ith row vector is denoted by q_i,

$$Q = \begin{bmatrix} q_1 \\ \vdots \\ q_k \end{bmatrix} = \begin{bmatrix} q_{11} \cdots q_{1k} \\ \vdots \ddots \vdots \\ q_{k1} \cdots q_{kk} \end{bmatrix}. \qquad (A.3)$$

All of the row vectors q_i, $i = 1, ..., k$, are nonnegative and at least one element is positive, i.e.,

$$q_i = (q_{i1}, ..., q_{ik}) \geq 0,\ i = 1, ..., k. \qquad (A.4)$$

$F(\cdot)$ is the k-dimensional vector function defined by

$$F(f(x)) = (F_1(f_1(x)), ..., F_k(f_k(x)))^T, \tag{A.5}$$

where $F_i(\cdot)$, $i = 1, ..., k$, is a strictly monotone increasing and continuously differentiable function on the range $\{f_i(x) \mid x \in X\}$,

$$\frac{\partial F_i(f_i(x))}{\partial f_i(x)} > 0, \ \forall \ x \in X, \ i = 1, ..., k. \tag{A.6}$$

$D(\cdot)$ is the k-dimensional vector function defined by

$$D(z, t_{-k}) \ = \ (D_1(z, t_{-k}), ..., D_k(z, t_{-k}))^T, \tag{A.7}$$

and $D_i(\cdot), i = 1, ..., k$, is continuously differentiable on (z, t_{-k}).

We now define the subsets S of space (z, t_{-k}) by

$$S \ = \ \{ (z, t_{-k}) \in R^k \mid \det A(z, t_{-k}) \neq 0 \}, \tag{A.8}$$

where $A(z, t_{-k})$ is the Jacobian matrix of $D(\cdot)$ defined by

$$A(z, t_{-k}) = \begin{bmatrix} \partial D_1/\partial z & \partial D_1/\partial t_1 & \cdots & \partial D_1/\partial t_{k-1} \\ \partial D_2/\partial z & \partial D_2/\partial t_1 & \cdots & \partial D_2/\partial t_{k-1} \\ & \cdots\cdots & \cdots\cdots & \\ \partial D_k/\partial z & \partial D_k/\partial t_1 & \cdots & \partial D_k/\partial t_{k-1} \end{bmatrix}. \tag{A.9}$$

Then it is reasonable to assume that $D(z, t_{-k})$ satisfies the following two properties.

Property A.1

There exists $(z, t_{-k}) \in S$ satisfying

$$QF(f(x)) = D(z, t_{-k}), \ \forall \ x \in X. \tag{A.10}$$

Property A.2

$$\frac{\partial D(z, t_{-k})}{\partial z} \geq 0, \ \forall \ (z, t_{-k}) \in S. \tag{A.11}$$

If we define the set

$$G = \{D(z, t_{-k}) \in R^k \mid (z, t_{-k}) \in S\}, \tag{A.12}$$

the bijection between S and G holds from the implicit function theorem since the Jacobian matrix $A(z, t_{-k})$ is nonsingular on S. Moreover, from Property A.1, it holds that

$$G \supset \{QF(f(x)) \mid x \in X\}. \tag{A.13}$$

In the following, we define the cut T of the space $t_{-k} \in R^{k-1}$ by

$$T = \{t_{-k} \in R^{k-1} \mid (z, t_{-k}) \in S\}. \tag{A.14}$$

Now it is appropriate to give the intuitive insights for the hyperplane problem $HP(Q, t_{-k})$. In the k-dimensional space of $F(\cdot)$, shown in Figure A.1, recall that hyperplane H_i with normal vectors q_i, $i = 1, \ldots, k$, can be represented by

$$H_i = \{F(f(x)) \in R^k \mid q_i F(f(x)) = c_i\}, \ i = 1, ..., k, \tag{A.15}$$

where c_i, $i = 1, ..., k$, are constants.

Then, from Property A.2, it follows that the hyperplane problem $HP(Q, t_{-k})$ may be regarded as the minimization problem which minimizes all of the constants c_i through the vector function $D(z, t_{-k})$.

Now consider the case where the matrix Q in the $HP(Q, t_{-k})$ is set as the identity matrix,

$$Q = I = \begin{bmatrix} 1 & & & 0 \\ & 1 & & \\ & & \ddots & \\ 0 & & & 1 \end{bmatrix}. \tag{A.16}$$

Then the $HP(Q, t_{-k})$ becomes the following hyperplane problem denoted by $HP(I, t_{-k})$:

$$\left.\begin{array}{ll} \underset{x \in X}{\text{minimize}} & z \\ \text{subject to} & F(f(x)) \leq D(z, t_{-k}). \end{array}\right\} \tag{A.17}$$

The relationships between an optimal solution of the $HP(I, t_{-k})$ and a Pareto optimal solution of the MONLP can be characterized by the following theorems.

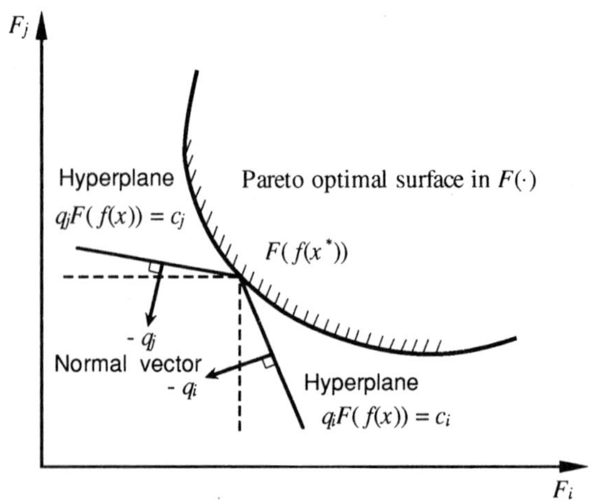

Figure A.1 Hyperplane problem

Theorem A.1

(1) *If there exists a* $t_{-k} \in T$ *such that* x^* *is a unique optimal solution to the* $HP(I, t_{-k})$*, then* x^* *is a Pareto optimal solution of the MONLP.*

(2) *If* x^* *is a Pareto optimal solution of the MONLP, then there exists a* $t_{-k} \in T$ *such that* x^* *is an optimal solution to the* $HP(I, t_{-k})$*.*

Proof

(1) Assume that x^* is a unique optimal solution to the $HP(I, t_{-k})$ and z^* is a corresponding minimum value, then it holds that $F(f(x^*)) \leq D(z^*, t_{-k})$. If x^* is not a Pareto optimal solution to the MONLP, then there exists $x \in X$ such that $f(x) \leq f(x^*)$, or equivalently $F(f(x)) \leq F(f(x^*))$. Therefore, it holds that $F(f(x)) \leq F(f(x^*)) \leq D(z^*, t_{-k})$. This contradicts the fact that x^* is a unique optimal solution to the $HP(Q, t_{-k})$.

(2) Assume that x^* is a Pareto optimal solution to the MONLP, then there exists $(z^*, t^*_{-k}) \in S$ such that $F(f(x^*)) = D(z^*, t^*_{-k})$ from Property A.1 and $x^* \in X$. If (x^*, z^*) is not an optimal solution to the $HP(I, t^*_{-k})$, then there exist z and $x \in X$ such that $z < z^*$ and $F(f(x)) \leq D(z, t^*_{-k})$. Therefore, it holds that $F(f(x)) \leq D(z, t^*_{-k}) \leq D(z^*, t^*_{-k}) = F(f(x^*))$. This means that $f(x) \leq f(x^*)$, which contradicts the assumption that x^* is a Pareto optimal solution to the MONLP.

This completes the proof of the theorem. □

Now, to investigate the relationships between the hyperplane problem $HP(I, t_{-k})$ and the minimax problems discussed in Chapters 3 and 4, assume that

$$D(z, t_{-k}) = D_{MP}(z, t_{-k}) = (z/t_1, ..., z/t_{-k})^T, \ t_i = t_j, \ i \neq j. \tag{A.18}$$

By using the substitution

$$f_i(x) = -\mu_i(f_i(x)), \ i = 1, ..., k, \tag{A.19}$$

$$F_i(\mu_i(f_i(x))) = \bar{\mu}_i - \mu_i(f_i(x)), \ i = 1, ..., k, \tag{A.20}$$

it is evident that the $HP(I, t_{-k})$ reduces to the minimax problems discussed in Chapters 3 and 4.

It is now appropriate to show that $D_{MP}(z, t_{-k})$ satisfies Properties A.1 and A.2. Let

$$S_{MP} = \left\{ (z, t_{-k}) \ \middle| \ z > 0, \ \sum_{i=1}^{k} t_i = 1, \ t_i > 0, \ i = 1, ..., k \right\}.$$

Then the Jacobian matrix $A_{MP}(\cdot)$ for $D_{MP}(\cdot)$ becomes

$$A_{MP}(z, t_{-k}) = \begin{bmatrix} t_1^{-1} & -zt_1^{-2} & & & \\ t_2^{-1} & & zt_2^{-2} & & \\ \vdots & & & \ddots & -zt_{k-1}^{-2} \\ t_k^{-1} & zt_k^{-2} & zt_k^{-2} & \cdots & zt_k^{-2} \end{bmatrix}.$$

By applying the fundamental transformations for the matrix $A_{MP}(\cdot)$ repeatedly, we obtain

$$\det A_{MP}(z, t_{-k}) = z^{k-1} \prod_{i=1}^{k} t_i^{-2}.$$

Clearly, it holds that

$$\det A_{MP}(z, t_{-k}) \neq 0, \ \forall (z, t_{-k}) \in S_{MP}.$$

On the other hand, $\{D_{MP}(z, t_{-k}) \mid (z, t_{-k}) \in S_{MP}\}$ is equivalent to the region $\{D = (D_1, ..., D_k) \in R^k \mid D_i > 0 \text{ or } D_i < 0, i = 1, ..., k\}$. This implies that $D_{MP}(\cdot)$ satisfies Property A.1 if $F(f(x)) \subset R_k^+$ or $F(f(x)) \subset R_k^-$. Moreover, it holds that

$$\frac{\partial D_{MP}}{\partial z} = (t_1^{-1}, ..., t_k^{-1}) > 0, \ \forall (z, t_{-k}) \in S_{MP}.$$

Therefore, $D_{MP}(\cdot)$ also satisfies Property A.2.

It should be emphasized here that, for generating Pareto optimal solutions using Theorem A.1, uniqueness of the solution must be verified. To circumvent the necessity to test the uniqueness of the solution to the $HP(I, t_{-k})$, we introduce the following hyperplane problems denoted by $HP(I(\rho), t_{-k})$:

$$\begin{aligned} &\underset{x \in X}{\text{minimize}} \quad z \\ &\text{subject to} \quad I(\rho)F(f(x)) \leq D(z, t_{-k}) \end{aligned} \tag{A.21}$$

where the matrix $I(\rho)$ is defined as

$$Q = I(\rho) = \begin{bmatrix} 1+\rho & & & \rho \\ & 1+\rho & & \\ \vdots & & \ddots & \vdots \\ \rho & & & 1+\rho \end{bmatrix}, \tag{A.22}$$

where ρ is a sufficiently small positive number.

Then the $HP(I(\rho), t_{-k})$ becomes

$$\left.\begin{aligned} &\underset{x \in X}{\text{minimize}} \quad z \\ &\text{subjectto} \quad F_i(f_i(x)) + \rho \sum_{j=1}^{k} F_j(f_j(x)) \leq D_i(z, t_{-k}), \ i = 1, ..., k. \end{aligned}\right\} \tag{A.23}$$

The following theorem shows that the optimal solution of the $HP(I(\rho), t_{-k})$ is Pareto optimal even if it is not unique.

Theorem A.2

(1) *If x^* is an optimal solution to the $HP(I(\rho), t_{-k})$ for a positive scalar ρ, then x^* is a Pareto optimal solution to the MONLP.*

(2) *If x^* is a Pareto optimal solution to the MONLP, then there exists a sufficiently small positive scalar ρ and $t_{-k} \in T$ such that x^* is an optimal solution to the $HP(I(\rho), t_{-k})$.*

Proof

(1) Let x^* be an optimal solution to the $HP(I(\rho), t_{-k})$ and z^* be a corresponding optimal value. Then it holds that $I(\rho)F(f(x^*)) \leq D(z^*, t_{-k})$. Assume that x^* is not a Pareto optimal solution, then there exists $x \in X$ such that $f(x) \leq f(x^*)$, or equivalently, $F(f(x)) \leq F(f(x^*))$. Since all of the matrix $I(\rho)$ are positive, it holds that

$$I(\rho)F(f(x)) < I(\rho)F(f(x^*)) \leq D(z^*, t_{-k}).$$

This inequality means that there exist z such that $z < z^*$ and $I(\rho)F(f(x)) \leq D(z, t_{-k})$ because at least one of $D_i(\cdot)$, $i = 1, ..., k$, is continuously differentiable on z from Property A.2. This is a contradiction. Hence, x^* is a Pareto optimal solution to the MONLP.

(2) Assume that (x^*,z^*) is not an optimal solution to the $HP(I(\rho), t_{-k})$ for any positive scalar ρ and t_{-k} such that $I(\rho)F(f(x^*)) = D(z^*, t_{-k})$. Then there exist $x \in X$ and $z < z^*$ such that $I(\rho)F(f(x)) \leq D(z, t_{-k})$. Since it holds that $D(z^*, t_{-k}) \geq D(z, t_{-k})$ from Property A.2 and $z^* > z$, the following relation holds:

$$I(\rho)F(f(x)) - F(f(x^*)) < 0.$$

If either any $\{F_i(f_i(x)) - F_i(f_i(x^*))\}$ is positive or all $\{F_i(f_i(x)) - F_i(f_i(x^*))\}$, $i = 1, ..., k$, are zero, this inequality would be violated for sufficiently small positive ρ. Hence,

$$F_i(f_i(x)) - F_i(f_i(x^*)) \leq 0, \ i = 1, ..., k$$

must hold. By the strict monotonicity of $F_i(\cdot)$, we have $f_i(x) - f_i(x^*) \leq 0$, which contradicts the fact that x^* is a Pareto optimal solution.

This completes the proof of the theorem. □

Now, by substituting the $D_{MP}(\cdot)$ for $D(\cdot)$ in the $HP(I(\rho), t_{-k})$, we can immediately obtain the following augmented minimax problem:

$$\left.\begin{array}{ll} \underset{x \in X}{\text{minimize}} & z \\ \text{subjectto} & I(\rho)F(f(x)) \leq D_{MP}(z, t_{-k}). \end{array}\right\} \quad (A.24)$$

Then, by using the same substitution shown in (A.19) and (A.20), the $HP(I(\rho), t_{-k})$ reduces to the augmented minimax problem discussed in Chapters 4 and 6.

A.2 Trade-offs

To develop a meaningful formula which relates the trade-off rates within the Pareto optimal solution set to Lagrange multipliers of the $HP(Q, t_{-k})$ where $Q = I$ or $Q = I(\rho)$, define the following Lagrange function $L(x, z, \lambda, t_{-k})$ for the $HP(Q, t_{-k})$:

$$L(x, z, \lambda, t_{-k}) = z + \sum_{i=1}^{k} \lambda_i \left\{ \sum_{j=1}^{k} q_{ij} F_j(f_j(x)) - D_i(z, t_{-k}) \right\}. \tag{A.25}$$

In the following, let us assume that (x^*, z^*) is a unique local optimal solution of the $HP(Q, t_{-k})$ and satisfies the following three assumptions.

Assumption A.1
The second-order sufficiency conditions are satisfied at $(x^*,\ z^*)$.
Assumption A.2
$(x^*,\ z^*)$ is a regular point of the constraints of the $HP(Q, t_{-k})$, i.e., the gradients of the active constraints are lineally independent.
Assumption A.3
There are no degenerate constraints at $(x^*,\ z^*)$, i.e., all active constraints have strictly positive corresponding Lagrange multipliers.

Then the following sensitivity theorem (for details, see, e.g., Luenberger 1973 or Fiacco 1983) based on the implicit function theorem holds.

Theorem A.3
Let $(x^*,\ z^*)$ be a unique local optimal solution of the $HP(Q, t_{-k})$ satisfying Assumptions A.1, A.2, and A.3. Let λ^* denote the Lagrange multipliers corresponding to the constraints of the $HP(Q, t_{-k})$. Then there exists a continuously differentiable vector valued function $x(t_{-k}), z(t_{-k})$, and $\lambda(t_{-k})$ defined on some neighborhood $N(t^*_{-k}) \cap T$ so that $x(t^*_{-k}) = x^*$, $z(t^*_{-k}) = z^*$, $\lambda(t^*_{-k}) = \lambda^*$, where $x(t_{-k}), z(t_{-k})$ is a unique local optimal solution of the $HP(Q, t_{-k})$ for any $t_{-k} \in N(t^*_{-k}) \cap T$ satisfying Assumptions A.1, A.2, and A.3, and $\lambda(t_{-k})$ is the Lagrange multiplier corresponding to the constraints of $HP(Q, t_{-k})$.
Moreover, the following relations hold on some neighborhood $N(t^*_{-k}) \cap T$ of t^*_{-k}:

$$\frac{\partial z(t_{-k})}{\partial t_i} = -\sum_{j=1}^{k} \lambda_j(t_{-k}) \frac{\partial D_j(z(t_{-k}), t_{-k})}{\partial t_i}, \quad i = 1, ..., k-1. \tag{A.26}$$

From Theorem A.3, if all the k constraints of $HP(Q, t_{-k})$ are active, then the following theorem holds.

Theorem A.4
Let all the assumptions in Theorem A.3 be satisfied. Also assume that all the k constraints of the $HP(Q, t_{-k})$ are active. Then, for the optimal value

function $z(t_{-k})$ and the corresponding Lagrange multipliers $\lambda(t_{-k})$ for any $t_{-k} \in N(t^*_{-k}) \cap T$, it holds that

$$\sum_{i=1}^{k} \lambda_i(t_{-k}) d\{D_i(z(t_{-k}), t_{-k})\} = 0, \tag{A.27}$$

where $d\{D_i(\cdot)\}$ denotes a total differential with respect to t_{-k}, i.e.,

$$d\{D_i(z(t_{-k}), t_{-k})\} = \sum_{j=1}^{k-1} \left\{ \frac{\partial D_i}{\partial z} \frac{\partial z}{\partial t_j} + \frac{\partial D_i}{\partial t_j} \right\} dt_j,\ i = 1, ..., k. \tag{A.28}$$

Proof

Observe that $\lambda(t_{-k}) > 0$ for $t_{-k} \in N(t^*_{-k}) \cap T$ due to the active condition and Assumption A.3. From Assumption A.1, it holds that

$$\frac{\partial L}{\partial z} = 1 - \sum_{i=1}^{k} \lambda_i(t_{-k}) \frac{\partial D_i(z(t_{-k}), t_{-k})}{\partial z} = 0. \tag{A.29}$$

Using the Jacobian matrix $A(z, t_{-k})$ of $D(z, t_{-k})$, the total differential form of $D(z, t_{-k})$ becomes

$$\begin{bmatrix} d\{D_1(z, t_{-k})\} \\ \cdots\cdots \\ d\{D_k(z, t_{-k})\} \end{bmatrix} = A \begin{bmatrix} dz \\ dt_1 \\ \cdots \\ dt_{k-1} \end{bmatrix}. \tag{A.30}$$

Multiplying both side by $\lambda(t_{-k})$ yields

$$\begin{aligned} &\sum_{i=1}^{k} \lambda_i(t_{-k}) d\{D_i(z(t_{-k}), t_{-k})\} \\ &= \left[\sum_{i=1}^{k} \lambda_i(t_{-k}) \frac{\partial D_i(z, t_{-k})}{\partial z}, \sum_{i=1}^{k} \lambda_i(t_{-k}) \frac{\partial D_i(z, t_{-k})}{\partial t_1}, ..., \sum_{i=1}^{k} \lambda_i(t_{-k}) \frac{\partial D_i(z, t_{-k})}{\partial t_{k-1}} \right] \\ &\times [\, dz, dt_1, ..., dt_{k-1} \,]^T. \end{aligned}$$

In view of (A.26) and (A.29), for the optimal value function $z(t_{-k})$ for any $t^*_{-k} \in N(t^*_{-k}) \cap T$, it follows that

$$\sum_{i=1}^{k} \lambda_i(t_{-k}) d\{D_i(z(t_{-k}), t_{-k})\} = dz - \sum_{j=1}^{k-1} \frac{\partial z(t_{-k})}{\partial t_j} dt_j. \tag{A.31}$$

This completes the proof of the theorem. □

From Theorem A.4, the following result immediately follows.

Theorem A.5

Let all the assumptions in Theorem A.4 be satisfied. Then, for the optimal solution $x(t_{-k})$ and the corresponding Lagrange multipliers $\lambda(t_{-k})$ of the $HP(Q, t_{-k})$ for any $t_{-k} \in N(t^*_{-k}) \cap T$, it holds that

$$\sum_{j=1}^{\kappa} \left\{ \sum_{i=1}^{k} (\lambda_i(t_{-k}) q_{ij}) \frac{\partial F_j}{\partial f_j} \right\} df_j(x(t_{-k})) = 0. \tag{A.32}$$

Proof

By the active condition and Theorem A.3, it follows that

$$\sum_{j=1}^{k} q_{ij} F_j(f_j(x(t_{-k}))) = D_i(z(t_{-k}), t_{-k}). \tag{A.33}$$

This implies

$$\sum_{i=1}^{k} \lambda_i(t_{-k}) d\{D_i(z(t_{-k}), t_{-k})\} = \sum_{i=1}^{k} \lambda_i(t_{-k}) \left\{ \sum_{j=1}^{k} q_{ij} \frac{\partial F_j}{\partial F_j} df_j(x(t_{-k})) \right\}.$$

This completes the proof of the theorem. □

It is now appropriate to consider the practical implications of Theorem A.5. Let

$$\bar{f}_j(t_{-k}) = f_j(x(t_{-k})), \ j = 1, ..., k. \tag{A.34}$$

Since the continuous differentiability of $\bar{f}_j(\cdot)$ follows from those of $f_j(\cdot)$ and $x(\cdot)$, (A.32) becomes

$$\sum_{j=1}^{k} \left\{ \sum_{i=1}^{k} (\lambda_i(t_{-k}) q_{ij}) \frac{\partial F_j}{\partial \bar{f}_j} d\bar{f}_j(t_{-k}) \right\} = 0. \tag{A.35}$$

Let

$$\bar{P} = \{\bar{f}(t_{-k}) \mid t_{-k} \in N(t^*_{-k})\}.$$

Then $\bar{P}$ represents a surface in the k-dimensional space and the slope at its tangent plane can be obtained by (A.35). The trade-off rates between $\bar{f}_i, i = 1, ..., k-1$, and $\bar{f}_k$ on the surface $\bar{P}$ are represented by

$$-\left.\frac{\partial \bar{f}_k(t_{-k})}{\partial \bar{f}_i(t_{-k})}\right|_{t_{-k} \in N(t^*_{-k})} = \frac{\sum_{j=1}^{k} (\lambda_j(t_{-k}) q_{ji}) \partial F_i(\bar{f}_i(t_{-k}))/\partial \bar{f}_i}{\sum_{j=1}^{k} (\lambda_j(t_{-k}) q_{jk}) \partial F_k(\bar{f}_k(t_{-k}))/\partial \bar{f}_k}, \ i = 1, ..., k-1. \tag{A.36}$$

In what follows, we show that the trade-off rates between f_i, $i = 1, ..., k-1$, and f_k on the Pareto surface

$$P = \{f(x) \mid x \in X^P(Q)\}$$

coincide with the trade-off rates between $\bar{f}_i, i = 1, ..., k-1$, and $\bar{f}_k$, where $X^P(Q)$ is the Pareto optimal solution set of the $HP(Q, t_{-k})$ for any $t_{-k} \in T$. We shall prove the following result.

$$\left.\frac{\partial f_k(x)}{\partial f_i(x)}\right|_{\substack{x = x^* \\ f \in P}} = \left.\frac{\partial \bar{f}_k(t_{-k})}{\partial \bar{f}_i(t_{-k})}\right|_{\substack{t_{-k} = t^*_{-k} \\ \bar{f} \in \bar{P}}} i = 1, ..., k-1. \tag{A.37}$$

For that purpose, define the continuously differentiable function $\bar{D}_i(t_{-k})$ by

$$\bar{D}_i(t_{-k}) = D_i(z(t_{-k}), t_{-k}), \ i = 1, ..., k, \tag{A.38}$$

where $z(t_{-k})$ is the optimal value function. Let

$$\bar{D}_{-k}(t_{-k}) = (\bar{D}_1(t_{-k}), ..., \bar{D}_{k-1}(t_{-k}))^T, \quad \bar{D}(t_{-k}) = (\bar{D}_1(t_{-k}), ..., \bar{D}_k(t_{-k}))^T, \tag{A.39}$$

and define the $(k-1) \times (k-1)$ Jacobian matrix $B_{-k}(t_{-k})$ of $\bar{D}_{-k}(t_{-k})$ by

$$B_{-k}(t_{-k}) = \begin{bmatrix} \frac{\partial \bar{D}_1}{\partial t_1} & \cdots & \frac{\partial \bar{D}_1}{\partial t_{k-1}} \\ \vdots & \ddots & \vdots \\ \frac{\partial \bar{D}_{k-1}}{\partial t_1} & \cdots & \frac{\partial \bar{D}_{k-1}}{\partial t_{k-1}} \end{bmatrix}. \tag{A.40}$$

Then the relationships between the $k \times k$ Jacobian matrix $A(z, t_{-k})$ and the $(k-1) \times (k-1)$ Jacobian matrix $B_{-k}(t_{-k})$ can be characterized by the following theorem.

Theorem A.6

Let all the assumptions in Theorem A.4 be satisfied. Then, for any $t_{-k} \in N(t^*_{-k}) \cap T$, it holds that

$$\det B_{-k}(t_{-k}) = (-1)^{k-1}\lambda_k \det A(z(t_{-k}), t_{-k}), \tag{A.41}$$

where $z(t_{-k})$ is the optimal value function and λ_k is the kth Lagrange multiplier of the kth constraint of the $HP(Q, t_{-k})$.

Proof

From the definition of $B_{-k}(t_{-k})$, it follows that

$$\begin{aligned}
\det B_{-k}(t_{-k}) &= \det \begin{bmatrix} \frac{\partial D_1}{\partial z}\frac{\partial z}{\partial t_1} + \frac{\partial D_1}{\partial t_1} & \cdots & \frac{\partial D_1}{\partial z}\frac{\partial z}{\partial t_{k-1}} + \frac{\partial D_1}{\partial t_{k-1}} \\ \vdots & \ddots & \vdots \\ \frac{\partial D_{k-1}}{\partial z}\frac{\partial z}{\partial t_1} + \frac{\partial D_{k-1}}{\partial t_1} & \cdots & \frac{\partial D_{k-1}}{\partial z}\frac{\partial z}{\partial t_{k-1}} + \frac{\partial D_{k-1}}{\partial t_{k-1}} \end{bmatrix} \\
&= \left\{\sum_{i=1}^{k-1} \frac{\partial D_i}{\partial z}\right\}\{-\lambda_1|A_{k1}| - (-1)^{k-2}\lambda_k|A_{11}|\} \\
&\quad + \sum_{i=1}^{k-1} \det \begin{bmatrix} \frac{\partial D_i}{\partial t_1} & \cdots & \frac{\partial D_i}{\partial t_{k-1}} \\ \cdots & \ddots & \cdots \\ \frac{\partial D_{k-1}}{\partial z}\frac{\partial z}{\partial t_1} + \frac{\partial D_{k-1}}{\partial t_1} & \cdots & \frac{\partial D_{k-1}}{\partial z}\frac{\partial z}{\partial t_{k-1}} + \frac{\partial D_{k-1}}{\partial t_{k-1}} \end{bmatrix}
\end{aligned}$$

where $|A_{ij}|$ denotes the determinant of the $(k-1) \times (k-1)$ matrix without the ith row and the jth column of $A(z(t_{-k}), t_{-k})$.

By similar operations, it follows that

$$\det B_{-k}(t_{-k}) = \left\{\sum_{i=1}^{k-1} \frac{\partial D_i}{\partial z}\right\}\{-\lambda_1|A_{k1}| - (-1)^{k-2}\lambda_k|A_{11}|\}$$

$$+\left\{\sum_{i=2}^{k-1}\frac{\partial D_i}{\partial z}\right\}\{\lambda_1|A_{k1}| - \lambda_2|A_{k1}| - (-1)^{k-3}\lambda_k|A_{21}| - (-1)^{k-3}\lambda_k|A_{11}|\}$$

$$+\cdots+\left\{\sum_{i=k-1}^{k-1}\frac{\partial D_i}{\partial z}\right\}\{\lambda_{k-2}|A_{k1}| - \lambda_{k-1}|A_{k1}|$$

$$-(-1)^0\lambda_k|A_{k-1,1}| - (-1)^0\lambda_k|A_{k-2,1}|\} + |A_{k1}|$$

$$= \frac{\partial D_1}{\partial z}\{-\lambda_1|A_{k1}| - (-1)^{(k-1)-1}\lambda_k|A_{11}|\}$$

$$+\cdots+\frac{\partial D_{k-1}}{\partial z}\{-\lambda_{k-1}|A_{k1}| - (-1)^{(k-1)-(k-1)}\lambda_k|A_{k-1,1}|\} + |A_{k1}|$$

$$= \left(1 - \sum_{i=1}^{k-1}\lambda_i\frac{\partial D_i}{\partial z}\right)|A_{k1}| - \lambda_k\sum_{i=1}^{k-1}(-1)^{(k-1)-i}\frac{\partial D_i}{\partial z}|A_{i1}|.$$

In view of (A.29), it holds that

$$\det B_{-k}(t_{-k}) = -\lambda_k\frac{\partial D_k}{\partial z}|A_{k1}| - \lambda_k\sum_{i=1}^{k-1}(-1)^{(k-1)-i}\frac{\partial D_i}{\partial z}|A_{i1}|$$

$$= -\lambda_k\sum_{i=1}^{k}(-1)^{(k-1)-i}\frac{\partial D_j}{\partial z}|A_{i1}|.$$

By expanding the det $A(z(t_{-k}), t_{-k})$ with respect to the first column, we obtain

$$\det A(z(t_{-k}), t_{-k}) = \sum_{i=1}^{k}(-1)^{i-1}\frac{\partial D_i}{\partial z}|A_{i1}|.$$

Therefore, (A.41) has been established. □

The following theorem shows the relationships between Pareto optimal solutions in the decision space and the parameter $t_{-k} \in T$ in the $HP(Q, t_{-k})$.

Theorem A.7

Let all the assumptions in Theorem A.5 be satisfied. Then there exist a neighborhood $N(x^*)$ of x^* and a continuously differentiable vector valued function $x(t_{-k})$ defined on some neighborhood $N(t^*_{-k})$ of $t^*_{-k} \in T$ such that

$$X^P(Q) \cap N(x^*) \subset x(N(t^*_{-k})) \subset X^P(Q). \tag{A.42}$$

Proof

From Theorem A.3, there exists a neighborhood $N(t^*_{-k})$ of t^*_{-k} such that the $HP(Q, t_{-k})$ has a unique local solution $x(t_{-k})$ for any $t_{-k} \in N(t^*_{-k})$. Hence, it follows from Theorems A.1 and A.2 that

$$x(N(t_{-k})) \subset X^P(Q).$$

To prove the first part of (A.42), we shall show that there exists a neighborhood $N(x^*)$ of x^* so that

$$\exists\ t_{-k} \in N(t^*_{-k}) \text{ such that } \hat{x} = x(t_{-k}) \text{ for } \forall \hat{x} \in X^P(Q) \cap N(x^*).$$

Let

$$N(t^*_{-k}) = \{t_{-k} \in T \mid \|t_{-k} - t^*_{-k}\|_{k-1} < \delta_t\} \tag{A.43}$$

where $\delta_t > 0$ and $\|\cdot\|_{k-1}$ is an ℓ_p-norm in R^{k-1}. By the continuity of $Q_{-k}F(f_i(\cdot))$ in x, for any $\delta_d > 0$, there exists δ_x >0 such that

$$\|x - x^*\|_n < \delta_x \to \|Q_{-k}F(f(x)) - Q_{-k}F(f(x^*))\|_{k-1} < \delta_d \tag{A.44}$$

where Q_{-k} is $a(k-1) \times k$ matrix with the kth row vector of the matrix Q eliminated. For such δ_x, define the neighborhood $N(x^*)$ of x^* by

$$N(x^*) = \{x \mid \|x - x^*\|_n < \delta_x\}. \tag{A.45}$$

Now, from Property A.1, for any $\hat{x} \in N(x*) \cap X^P(Q)$, there exists $(\hat{z}, \hat{t}_{-k})$ such that

$$QF(f(\hat{x})) = D(\hat{z}, \hat{t}_{-k}). \tag{A.46}$$

From the assumption that all the k constraints of $HP(Q, t_{-k})$ are active, it holds that

$$D(\hat{z}, \hat{t}_{-k}) = D(z(\hat{t}_{-k}), \hat{t}_{-k}). \tag{A.47}$$

Then, since the Jacobian matrix $A(z, t_{-k})$ of $D(z, t_{-k})$ is nonzero from the definition of the set S, it follows that $\hat{z} = z(\hat{t}_{-k})$. Consequently, by the definition of $\bar{D}_{-k}(\cdot)$, it holds that

$$\|D_{-k}(\hat{z}, \hat{t}_{-k}) - D_{-k}(z^*, t^*_{-k})\|_{k-1} = \|\bar{D}_{-k}(\hat{t}_{-k}) - \bar{D}_{-k}(t^*_{-k})\|_{k-1} < \delta_d. \tag{A.48}$$

Now, from Theorem A.6, observe that det $B_{-k}(t_{-k}) \neq 0$ for any $t_{-k} \in T$. From this fact and the implicit function theorem, for any $\delta_t > 0$, there exists δ_d such that

$$\|\bar{D}_{-k}(\hat{t}_{-k}) - \bar{D}_{-k}(t^*_{-k})\|_{k-1} < \delta_d \to \|\hat{t}_{-k} - t^*_{-k}\|_{k-1} < \delta_t. \tag{A.49}$$

In view of (A.44) and (A.49), it follows that $\hat{t}_{-k} \in N(t^*_{-k})$. Moreover, by the view of $N(t^*_{-k})$, the $HP(Q, \hat{t}_{-k})$ has a unique local solution $x(\hat{t}_{-k})$. Also since $\hat{x} \in X^P$, $\hat{x}$ is a local solution to the $HP(Q, \hat{t}_{-k})$ from Theorem A.1 and A.2. Therefore, we conclude that $\hat{x} = x(\hat{t}_{-k})$ as required. □

From Theorem A.7 and (A.36), we can immediately obtain the following theorem.

Theorem A.8

Let all the assumptions in Theorem A.4 be satisfied. Then the trade-off rates between the objective functions $f_i(x^*), i = 1, ..., k-1$, and $f_k(x^*)$ at the optimal *solution to the* $HP(Q, t^*_{-k})$ *on the Pareto surface* P *in the objective space can be represented by*

$$-\left.\frac{\partial f_k(x)}{\partial f_i(x)}\right|_{\substack{x=x^* \\ f\in P}} = \frac{\sum_{j=1}^{k} \lambda_j^* q_{ji} \partial F_i(f_i(x^*))/\partial f_i}{\sum_{j=1}^{k} \lambda_j^* q_{jk} \partial F_k(f_k(x^*))/\partial f_k}, \quad i = 1, ..., k-1. \tag{A.50}$$

It should be noted here that the vector function $D(\cdot)$ in the $HP(Q, t_{-k})$ does not explicitly appear in (A.50), that is, the trade-off rates on the Pareto surface can be determined by the values of Lagrange multipliers and the elements of Q and $F(\cdot)$. This means that the vector function $D(\cdot)$ influences the values of trade-off rates implicitly through the Lagrange multipliers.

We are now ready to demonstrate that the trade-off rates formula in Chapters 2, 3, and 4 can be derived from the trade-off rates formula (A.50).

In the $HP(Q, t_{-k})$, the minimax problem is obtained by using the substitution shown in (A.19), (A.20) and

$$q_{ii} = 1, \quad i = 1, ..., k, \quad q_{ij} = 0, \quad i \neq j$$

Thus, the trade-off rates formula for the minimax problem is represented by

$$-\frac{\partial \mu_k(f_k(x))}{\partial \mu_i(f_i(x))} = \frac{\lambda_i}{\lambda_k}, \quad i = 1, ..., k-1. \tag{A.51}$$

Similarly, by using the substitution,

$$q_{ii} = 1 + \rho, \quad i = 1, ..., k, \quad q_{ij} = \rho, \quad i \neq j$$

in the $HP(Q, t_{-k})$, the augmented minimax problem is obtained. Therefore, the trade-off rates formula for the augmented minimax problem is represented by

$$-\frac{\partial \mu_k(f_k(x))}{\partial \mu_i(f_i(x))} = \frac{\lambda_i + \rho}{\lambda_k + \rho}, \quad i = 1, ..., k-1. \tag{A.52}$$

Observe that this trade-off rates formula (A.52) coincides with the trade-off rates formula in Chapters 4 and 6.

So far, we have discussed the trade-off rates formula for the $HP(Q, t_{-k})$ under Assumptions A.1, A.2, and A.3. However, for the linear $HP(Q, t_{-k})$ in which all the objective functions and the constraints are linear, it should be emphasized that the second-order sufficiency conditions are not satisfied because the Hessian matrix always becomes a zero matrix.

For the linear $HP(Q, t_{-k})$, only with Assumption A.3, the following observations can be made concerning the trade-off rates formula (A.50). If Assumption A.3 is satisfied for the optimal solution to the linear $HP(Q, t_{-k})$, then from the theory of the simplex method of linear programming, it is known that there exists the neighborhood $N(t_{-k})$ of t_{-k} such that

$$z(t_{-k}) = \sum_{i=1}^{k} \pi_i D_i(z(t_{-k}), t_{-k}), \tag{A.53}$$

where $z(t_{-k})$ is the optimal value function to the linear $HP(Q, t_{-k})$ and π_i, $i = 1, ..., k$, are the simplex multipliers corresponding to the constraints of $HP(Q, t_{-k})$.

Since Theorem A.3 directly follows from (A.53), if all the constraints of the linear $HP(Q, t_{-k})$ are active, then both Theorems A.4 and A.5 are satisfied for the linear $HP(Q, t_{-k})$. Therefore, the following theorem holds.

Theorem A.9

Let x^* be a nondegenerate optimal solution to the linear $HP(Q, t^*_{-k})$ and let π^*_i, $i = 1, ..., k$, denote the simplex multipliers corresponding to the constraints of the linear $HP(Q, t^*_{-k})$. Also assume that all the k constraints of the linear $HP(Q, t^*_{-k})$ are active. Then, for the linear $HP(Q, t^*_{-k})$, the trade-off rates between the objective functions $f_i(x^*)$, $i = 1, ..., k-1$, and $f_k(x^*)$ at the optimal solution to the $HP(Q, t^*_{-k})$ on the Pareto surface P in the objective space can represented by

$$-\left.\frac{\partial f_k(x)}{\partial f_i(x)}\right|_{\substack{x=x^* \\ f\in P}} = \frac{\sum_{j=1}^{k} \pi^*_j q_{ji} \partial F_i(f_i(x^*))/\partial f_i}{\sum_{j=1}^{k} \pi^*_j q_{jk} \partial F_k(f_k(x^*))/\partial f_k}, \quad i = 1, ..., k-1. \tag{A.54}$$

Now, it is possible to investigate the relationships between the hyperplane problem $HP(I, t_{-k})$ and the linear constraint problems in Chapters 3 and 5.

By replacing $D(z, t_{-k})$ with

$$D_{CP}(z, t_{-k}) = (t_1, ..., t_{k-1}, z)^T, \tag{A.55}$$

the $HP(I, t_{-k})$ reduces to the constraint problem $CP(t_{-k})$ itself. To show that $D_{CP}(z, t_{-k})$ satisfies Properties 1 and 2, let $S_{CP} = R^k$. Then the Jacobian matrix $A_{CP}(\cdot)$ for $D_{CP}(\cdot)$ becomes

$$A_{CP}(z, t_{-k}) = \begin{bmatrix} 0 & & & & \\ \cdot & & & & \\ \cdot & & I & & \\ \cdot & & & & \\ 0 & & & & \\ 1 & 0 & \cdot & \cdot & \cdot & 0 \end{bmatrix} \tag{A.56}$$

and

$$\det A_{CP}(z, t_{-k}) = (-1)^{k-1} \neq 0, \ \forall (z, t_{-k}) \in S_{CP}.$$

This means that $D_{CP}(\cdot)$ satisfies Property 1. Furthermore, it holds that

$$\frac{\partial D_{CP}}{\partial z} = (0, ..., 0, 1)^T \geq 0.$$

This implies that $D_{CP}(\cdot)$ satisfies Property 2.

From the previous discussions, the constraint problem can be regarded as one of the special cases of the $HP(I, t_{-k})$.

Therefore, from Theorem A.9, the trade-off rates formula for the linear constrained problem is represented as follows:

$$-\frac{\partial \mu_k(f_k(x))}{\partial \mu_i(f_i(x))} = \frac{\pi_i}{\pi_k}, \quad i = 1, ..., k-1. \tag{A.57}$$

However, since it always holds that $\pi_k = 1$ in the constraint problem because of $\partial L / \partial z = 0$, (A.57) is represented as

$$-\frac{\partial \mu_k(f_k(x))}{\partial \mu_i(f_i(x))} = \pi_i, \quad i = 1, ..., k-1. \tag{A.58}$$

This formula coincides with the trade-off rates formula in Chapters 3 and 5. The trade-off rates formula in Chapter 3 is much like this discussion and, thus, is omitted.

REFERENCES

Ackley, D. H., G. E. Hinton and T. J. Sejonowski (1985). A learning algorithm for Boltzmann machines, Cognitive Science, 9, pp. 147-169.

Akiyama, Y., A. Yamashita, M. Kajiura and H. Aiso (1989). Combinatorial optimization with Gaussian machines, Proceedings of IEEE/INNS International Joint Conference on Neural Networks, 1, pp. 533-540.

Avriel, M. (1976). Nonlinear Programming: Analysis and Method, Prentice-Hall, Englewood.

Baptistella, L. F. B. and A. Ollero, (1980). Fuzzy methodologies for interactive multicriteria optimization, IEEE Transactions on Systems, Man, and Cybernetics, SMC-10, pp. 355-365.

Bazarra, M. S. and C. M. Shetty (1979). Nonlinear Programming: Theory and Algorithms, John Wiley & Sons, New York.

Bellman, R. E and L. A. Zadeh (1970). Decision making in a fuzzy environment. Management Science, 17, pp. 141-164.

Benayoun, R., J. de Montgofier, J. Tergny and O. Larichev (1971). Linear programming with multiple objective functions, Step method (STEM), Mathematical Programming, 1, pp. 366-375.

Bowman, V. J. (1976). On the relationship of the Tchebycheff norm and the efficient frontier of multi-criteria objectives, Multiple Criteria Decision Making (H. Thiriez and S. Zionts, eds.), Springer-Verlag, Berlin, pp. 76-86.

Brooke, A., D. Kendrick and A. Meeraus (1988). GAMS: A User's Guide, The Scientific Press, California.

Buckley, J. J. (1983). Fuzzy programming and the Pareto optimal set, Fuzzy Sets and Systems, 10, pp. 56-64.

Buckley, J. J. (1988a). Possibility and necessity in optimization, Fuzzy Sets and Systems, 25, pp. 1-13.

Buckley, J. J. (1988b). Possibilistic linear programming with triangular fuzzy numbers, Fuzzy Sets and Systems, 26, pp. 135-138.

Buckley, J. J. (1989). Solving possibilistic linear programming problems, Fuzzy Sets and Systems, 31, pp. 329-341.

Buckley, J. J. (1990). Multiobjective possibilistic linear programming, Fuzzy Sets and Systems, 35, pp. 23-28.

Campos, L. and J. L. Verdegay (1989). Linear programming problems and ranking of fuzzy numbers, Fuzzy Sets and Systems, 32, pp. 1-11.

Carlsson, C. (1982). Tackling an MCDM-problems with the help of some results from fuzzy set theory, European Journal of Operational Research, 10, pp. 270-281.

Carlsson, C. and P. Korhonen (1985). A parametric approach to fuzzy linear programming, Fuzzy Sets and Systems, 20, pp. 17-30.

Chankong, V. and Y. Y. Haimes (1978). The interactive surrogate worth trade-off (ISWT) methods for multiobjective decision-making, Multicriteria Problem Solving (S. Zionts, ed.), Springer-Verlag, Berlin, pp. 42-67.

Chankong, V. and Y. Y. Haimes (1982). On the characterization of noninferior solutions of the vector optimization problem, Automatica, 18, pp.697-707.

Chankong, V. and Y. Y. Haimes (1983a). Multiobjective Decision Making: Theory and Methodology, North-Holland, Amsterdam.

Chankong, V. and Y. Y. Haimes (1983b). Optimization based methods for multiobjective decision-making: an overview, Large Scale Systems, 5, pp. 1-33.

Charnes, A. and W. W. Cooper (1961). Management Models and Industrial Applications of Linear Programming, Vols. I and II, Wiley.

Charnes, A. and W. W. Cooper (1977). Goal programming and multiple objective optimizations, European Journal of Operational Research, 1, pp. 39-54.

Chanas, S. (1983). Parametric programming in fuzzy linear programming, Fuzzy Sets and Systems, 11, pp. 243-251.

Choo, E. U. and D. R. Atkins (1980). An interactive algorithm for multicriteria programming. Computer & Operations Research, 7, pp. 81-87.

Choo, E. U. and D. R. Atkins (1983). Proper efficiency in nonconvex multicriteria programming, Mathematics of Operations Research, 8, pp. 467-470.

Chvatal, V. (1983). Linear Programming, W.H. Freeman and Company.

Cochrane, J. L. and M. Zeleny (eds.) (1973). Multiple Criteria Decision Making, University of South Carolina Press, Columbia.

Cohon, J. L. (1978). Multiobjective Programming and Planning, Academic Press, New York.

Czyzak, P. and R. Slowinski (1989). Multiobjective diet optimization problem under fuzziness, The Interface between Artificial Intelligence and Operations Research in Fuzzy Environment, (eds. J. L. Verdegay and M. Delgado), Verlag TUV Rheinland, Koln, pp. 85-103.

Czyzak, P. (1990). Application of the FLIP method to farm structure optimization under uncertainty, Stochastic versus Fuzzy Approaches to Multiobjective Mathematical Programming Problems under Uncertainty, (eds. R. Slowinski and J. Teghem), Kluwer Academic Publishers, pp. 263-278.

Czyzak, P. and R. Slowinski (1991). A visual interactive method for MOLP problems with fuzzy coefficients, Preprints of IFSA'91 Brussels, Computer, Management & Systems Science, pp. 56-59.

Dantzig, G. B. (1963). Linear Programming and Extensions, Princeton University Press, Princeton.

Delgado, M., J. L. Verdegay, and M. A. Vila (1989). A general model for fuzzy linear programming, Fuzzy Sets and Systems, 29, pp. 21-29.

Delgado, M., J. L. Verdegay, and M. A. Vila, (1990). A possibilistic approach for multiobjective programming problems. Efficiency of Solutions, Stochastic versus Fuzzy Approaches to Multiobjective Mathematical Programming Problems under Uncertainty, (eds. R. Slowinski and J. Teghem), Kluwer Academic Publishers, pp. 229-248.

Dubois, D. (1987). Linear programing with fuzzy data, Analysis of Fuzzy Information, Volume 3, (ed. J. C. Bezdek), CRC Press, Inc., Florida, pp. 241-263.

Dubois, D. and H. Prade (1978). Operations on fuzzy numbers, International Journal of Systems Science, 9, pp. 613-626.

Dubois, D. and H. Prade (1980a). Systems of linear fuzzy constraints, Fuzzy Sets and Systems, 3, pp. 37-48.

Dubois, D. and H. Prade (1980b). Fuzzy Sets and Systems: Theory and Applications, Academic Press, New York.

Dubois, D. and H. Prade (1983). Ranking fuzzy numbers in the setting of possibility theory, Information Sciences, 30, pp. 183-224.

Dubois, D. and H. Prade (1987). Fuzzy numbers: An overview, Analysis of Fuzzy Information, Volume 1, (ed. J. C. Bezdek), CRC Press, Inc, Florida, pp. 3-39.

Dyer, J. S. (1972). Interactive goal programming, Management Science, 19, pp. 62-70.

Dyer, J. S. (1973). An empirical investigation of a man-machine interactive approach to the solution of the multiple criteria problem, Multiple Criteria Decision Making (eds. J. L. Cochrane and M. Zeleny), University of South Carolina Press, South Carolina, pp. 202-216.

Evans, J. P. and R. E. Steuer (1973). A revised simplex method for linear multiple objective programs, Mathematical Programming, 5, pp. 54-72.

Fabian, C. and M. Stoica (1984). Fuzzy integer programming, Fuzzy Sets and Decision Analysis, TIMS Studies in the Management Sciences Vol. 20, (eds. H.-J. Zimmermann, L. A. Zadeh, and B. R. Gaines), North-Holland, Amsterdam, pp. 123-131.

Feng, Y. (1983). A method using fuzzy mathematics to solve vectormaximum problems, Fuzzy Sets and Systems, 9, pp. 129-136.

Fiacco, A. V. (1983). Introduction to Sensitivity and Stability Analysis in Nonlinear Programming, Academic Press, New York.

Fichefet, J. (1976). GPSTEM: an interactive multiobjective optimization method, Progress in Operations Research, North-Holland, Amsterdam, Vol. 1, pp. 317-332.

Flachs, J. and M. A. Pollatschek (1978). Further results on fuzzy mathematical programming, Information and Control, 38, pp. 241-257.

Gaines, B. R. (1976). Foundations of fuzzy reasoning, International Journal of Man-Machine Studies, 8, pp. 623-668.

Gass, S. I. (1958). Linear Programming, McGraw-Hill, New York, 5th Edition (1985).

Geoffrion A. M. (1968). Proper efficiency and the theory of vector maximization. Journal of Mathematical Analysis and Applications, 22, pp. 618-630.

Geoffrion A. M., J. S. Dyer, and A. Feinberg (1972). An interactive approach for multicriterion optimization, with an application to the operation of an academic department, Management Science, 19, pp. 357-368.

Gill, P. E., W. Murray, and M. H. Wright (1981). Practical Optimization, Academic Press, London.

Goicoechea, A., D. R. Hansen, and L. Duckstein (1982). Multiobjective Decision Analysis with Engineering and Business Applications, John Wiley & Sons, New York.

Grauer M. (1983a). A dynamic interactive decision analysis and support system (DIDASS), user's guide, WP-83-60, International Institute for Applied Systems Analysis, Laxenburg, Austria.

Grauer, M. (1983b). Reference point optimization - the nonlinear case, Essays and Surveys on Multiple Criteria Decision Making (ed. P. Hansen), Springer-Verlag, Berlin, pp. 126-135.

Grauer M., A. Lewandowski and A. P. Wierzbicki (eds.) (1982). Multiobjective and Stochastic Optimization, Proceedings of an IIASA Task Force Meeting, IIASA

Collaborative Proceedings Series, CP-82-S12, International Institute for Applied Systems Analysis, Laxenburg, Austria.

Grauer, M. and A. P. Wierzbicki (eds.) (1984). Interactive Decision Analysis, Proceedings, Laxenburg, Austria, 1983, Springer-Verlag, Berlin.

Griffith, R. E and R. A. Stewart (1961). A nonlinear programming technique for the optimization of continuous processing systems. Management Science, 7, pp. 379-392.

Hadley, G. (1962). Linear Programming, Addison-Wesley, Massachusetts.

Haimes, Y. Y. (1977). Hierarchical Analyses of Water Resources Systems, McGraw-Hill, New York.

Haimes, Y. Y. and V. Chankong (1979). Kuhn-Tucker multipliers as trade-offs in multiobjective decision-making analysis, Automatica, 15, pp. 59-72.

Haimes, Y. Y. and W. A. Hall (1974). Multiobjectives in water resources systems analysis: the surrogate worth trade-off method, Water Resources Research, 10, pp. 614-624.

Haimes, Y. Y., W. A. Hall, and H. T. Freedman, (1975). Multiobjective Optimization in Water Resources Systems: The Surrogate Worth Trade-off Method, Elsevier, Amsterdam.

Haimes, Y. Y., L. Lasdon, and D. Wismer (1971). On a bicriteria formulation of the problems of integrated system identification and system optimization, IEEE Transactions on Systems, Man, and Cybernetics, SMC-1, pp. 296-297.

Haimes, Y. Y., K. Tarvainen, T. Shima, and J. Thadathil (1990). Hierarchical Multiobjective Analysis of Large-Scale Systems, Hemisphere, New York.

Hamada, T. (1976). Studies on Optimal Operations in Automated Warehouses, Osaka University, Ph.D. Thesis.

Hannan, E. L. (1981). Linear programming with multiple fuzzy goals, Fuzzy Sets and Systems, 6, pp. 235-248.

Heshmaty, B. and A. Kandel (1985). Fuzzy linear regression and its application to forecasting in uncertain environment, Fuzzy Sets and Systems, 15, pp. 159-191.

Hopfield, J. (1984). Neurons with graded response have collective computational properties like those of two-state neurons, Biophysics, Proceedings of the National Academy of Sciences USA, 81, pp. 3088-3092.

Hopfield, J. and D. Tank (1985). Neural computation of decisions in optimization problems, Biological Cybernetics, 52, pp. 141-152.

Hwang, C. L. and A. S. M. Masud (1979). Multiple Objective Decision Making: Methods and Applications, Springer-Verlag, Berlin.

Ignizio, J. P. (1976). Goal Programming and Extensions, Lexington Books, D. C. Heath and Company, Lexington.

Ignizio, J. P. (1982). Linear Programming in Single and Multiple Objective Systems, Prentice-Hall, Englewood Cliffs, New Jersey.

Ignizio, J. P. (1983). Generalized goal programming: an overview, Computer & Operations Research, 10, pp. 277-289.

Ignizio, J. P. and S. C. Daniels (1983). Fuzzy multicriteria integer programming via fuzzy generalized networks, Fuzzy Sets and Systems, 10, pp. 261-270.

Ijiri, Y. (1965). Management Goals and Accounting for Control, North-Holland, Amsterdam.

Inuiguchi, M. and H. Ichihashi (1990). Relative modalities and their use in possibilistic linear programming, Fuzzy Sets and Systems, 35, pp. 303-323.

Inuiguchi, M., H. Ichihashi, and H. Tanaka (1989). Possibilistic linear programming with measurable multiattribute value functions, ORSA Journal on Computing, 1, pp. 146-158.

Inuiguchi, M., H. Ichihashi and H. Tanaka (1990). Fuzzy programming: a survey of recent development, Stochastic versus Fuzzy Approaches to Multiobjective Mathematical Programming Problems under Uncertainty, (eds. R. Slowinski and J. Teghem), Kluwer Academic Publishers, pp. 45-68.

Ishido, K., H. Nakayama, K. Furukawa, K. Inoue, and K. Tanikawa (1986). Multiobjective management of erection for cablestayed bridge using satisficing trade-off method, Proceedings of the VII-th International Conference on Multiple Criteria Decision Making - Towards Interactive & Intelligent Decision Support Systems, Kyoto, Japan.

Jain, R. (1976). Tolerance analysis using fuzzy sets, International Journal of Systems Science, 7, pp. 1393-1401.

Jain, R (1977). A procedure for multiple-aspect decision-making using fuzzy sets, International Journal of Systems Science, 8, pp. 1-7.

Kacprzyk, J. and M. Fedrizzi (eds.) (1992). Fuzzy Regression Analysis, Omnitech Press, Warsaw and Physica-Verlag, Heidelberg.

Kacprzyk, J. and S. A. Orlovski (eds.) (1987). Optimization Models Using Fuzzy Sets and Possibility Theory, D. Reidel Publishing Company, Dordrecht.

Kacprzyk, J. and S. A. Orlovski (1987). Fuzzy optimization and mathematical programming: a brief introduction and survey, Optimization Models Using Fuzzy Sets and Possibility Theory, (eds. J. Kacprzyk and S. A. Orlovski), D. Reidel Publishing Company, Dordrecht, pp. 50-72.

Kaufmann, A. and M. M. Gupta (1985). Introduction to Fuzzy Arithmetic: Theory and Applications, Van Nostrand Reinhold, New York.

Kickert, W. J. M. (1978). Fuzzy Theories on Decision-making, Martinus Nijhoff, Leiden.

Kirkpatrick, S., C. D. Gelatt and M. P. Vecchi (1983). Optimization by simulated annealing, Science, 220, pp. 671-680.

Korhonen, P. and J. Laakso (1986). A visual interactive method for solving multiple criteria problem, European Journal of Operational Research, 24, pp. 277-287.

Kuhn, H. W. and A. W. Tucker (1951). Nonlinear programming, Proceedings of the Second Berkeley Symposium on Mathematical Statistics and Probability, (ed. J. Neyman), University of California Press, pp. 481-492.

Land, A. and A. Doig (1960). An automatic method of solving discrete programming problems, Econometrica, 28, pp. 497-520.

Lasdon, L. S. (1970). Optimization Theory for Large Systems, Macmillan, New York.

Lasdon, L. S., R. L. Fox, and M. W. Ratner (1974). Nonlinear optimization using the generalized reduced gradient method, Revue Française d'Automatique, Informatique et Research Opérationnelle, 3, pp. 73-103.

Lasdon, L. S., A. D. Waren, M. W. Ratner, and A. Jain (1975). GRG System Documentation, Technical Memorandum VIS-75-01, Case Western Reserve University.

Lasdon, L. S., A. D. Waren and M. W. Ratner (1980). GRG2 User's Guide, Technical Memorandum, University of Texas.

Leberling, H. (1981). On finding compromise solution in multicriteria problems using the fuzzy min-operator, Fuzzy Sets and Systems, 6, pp. 105-228.

Lee, S. M. (1972). Goal Programming for Decision Analysis, Auerbach, Philadelphia.

Lewandowski, A. and A. P. Wierzbicki (eds.) (1989). Aspiration Based Decision Support Systems: Theory, Software and Applications, Springer-Verlag, Berlin.

Liebman, J., L. S. Lasdon, L. Schrage, and A. D. Waren (1980). Modeling and Optimization with GINO, The Scientific Press, California.

Lightner, H. R. (1979). Multiple Criterion Optimization and Statistical Design for Electronic Circuits, Carnegie-Mellon University, Ph.D. Thesis.

Luenberger, D. G. (1973). Linear and Nonlinear Programming, Second Edition (1984), Addison-Wesley, California.

Luhandjula, M. K. (1982). Compensatory operators in fuzzy linear programming with multiple objectives, Fuzzy Sets and Systems, 8, pp. 245-252.

Luhandjula, M. K. (1984). Fuzzy approaches for multiple objective linear fractional optimization, Fuzzy Sets and Systems, 13, pp. 11-23.

Luhandjula, M. K. (1987a). Multiple objective programming problems with possibilistic coefficients, Fuzzy Sets and Systems, 21, pp. 135-145.

Luhandjula, M. K. (1987b). Linear programming with a possibilistic objective function, European Journal of Operational Research, 31, pp. 110-117.

Luhandjula, M. K. (1989). Fuzzy optimization : an appraisal, Fuzzy Sets and Systems, 30, pp. 257-282.

Luhandjula, M. K. and M. Sakawa (1990). Multiple objective linear programming problems in the presence of fuzzy coefficients, Stochastic versus Fuzzy Approaches to Multiobjective Mathematical Programming Problems under Uncertainty, (eds. R. Slowinski and J. Teghem), Kluwer Academic Publishers, pp. 301-320.

Mangasarian, O. L. (1969). Nonlinear Programming, McGraw-Hill, New York.

March, J. G. and H. A. Simon (1958). Organizations, Wiley, New York.

Masud, A. S. and C. L. Hwang (1981). Interactive sequential goal programming, Journal of the Operational Research Society, 32, pp. 391-400.

Ministry of International Trade and Industry, Tokyo, Japan (1973). 1973 Report on Industrial Land and Water Research and Statistics Department, Minister's Secretariat (in Japanese).

Monachi, D. E., C. C. Kisiel, and L. Duckstein (1973). Interactive multiobjective programming in water resources: a case study, Water Resources Research, 9, pp. 837-850.

Murtagh, B. A. and M. A. Saunders (1987). MINOS 5.1 User's Guide, Report SOL 83-20R, revised, Stanford University.

Musselman, K. and J. Talavage (1980). A trade-off cut approach to multiple objective optimization, Operations Research, 28, pp. 837-850.

Nakayama, H. and K. Furukawa (1985). Satisficing trade-off method with an application to multiobjective structural design, Large Scale Systems, 8, pp. 47-57.

Nakayama, H. and Y. Karasawa and S. Dohi (1980). Subjective programming applied to optimal operation in automated warehouses, International Journal of Systems Science, 11, pp. 513-525.

Negoita, C. V. and D. A. Ralescu (1975). Applications of Fuzzy Sets to Systems Analysis, Birkhauser, Basel.

Negoita, C. V. and D. A. Ralescu (1977). On fuzzy optimization, Kybernetes, 6, pp. 193-195.

Negoita, C. V. and M. Sularia (1976). On fuzzy mathematical programming and tolerances in planning, Econ. Comput. Econ. Cybern. Stud. Res., 1, pp. 3-14.

Nguyen, H. T. (1978). A note on the extension principle for fuzzy sets, Journal of Mathematical Analysis and Applications, 64, pp. 338-353.

Nijkamp, P. (1979). Multidimensional Spatial Data and Decision Analysis, John Wiley & Sons, Chichester.

Oppenheimer, K. R. (1978). A proxy approach to multiattribute decision making, Management Science, 24, pp. 675-689.

Orlovski, S. A. (1977). On programming with fuzzy constraint sets, Kybernetes, 6, pp. 197-201.

Orlovski, S. A. (1983). Problems of decision-making with fuzzy Information, WP-83-28, International Institute for Applied Systems Analysis, Laxenburg, Austria.

Orlovski, S. A. (1984). Multiobjective programming problems with fuzzy parameters, Control and Cybernetics, 13, pp. 175-183.

Osaka Bureau of Trade and Industry, Osaka, Japan (1973). Input-Output Table for Industrial Pollution Analysis in Kinki Area in 1973 (in Japanese).

Pierre, D. A. and M. J. Lowe (1975). Mathematical Programming via Augmented Lagrangians, Addison-Wesley Publishing Company, London.

Ramik, J. (1987). A unified approach to fuzzy optimization, Preprints of Second IFSA Congress, Tokyo, pp. 128-130.

Ramik, J. and J. Rimanek (1985). Inequality relation between fuzzy numbers and its use in fuzzy optimization, Fuzzy Sets and Systems, 16, pp. 123-138.

Rietveld, P. (1980). Multiple Objective Decision Methods and Regional Planning, North-Holland, Amsterdam.

Rommelfanger, H. (1989). Interactive decision making in fuzzy linear optimization problems, European Journal of Operational Research, 41, pp. 210-217.

Rommelfanger, H. (1990). FULPAL - An interactive method for solving (multiobjective) fuzzy linear programming problems, Stochastic versus Fuzzy Approaches to Multiobjective Mathematical Programming Problems under Uncertainty, (eds. R. Slowinski and J. Teghem), Kluwer Academic Publishers, pp. 279-299.

Rommelfanger, H., R. Hanuscheck, and J. Wolf (1989). Linear programming with fuzzy objective functions, Fuzzy Sets and Systems, 29, pp. 31-48.

Sakawa, M. (1980). Interactive multiobjective decision making by the sequential proxy optimization technique: SPOT, WP-80-66, International Institute for Applied Systems Analysis, Laxenburg, Austria.

Sakawa, M. (1981). An interactive computer program for multiobjective decision making by the sequential proxy optimization technique, International Journal of Man-Machine Studies, 14, pp. 193-213.

Sakawa, M. (1982a). Interactive multiobjective decision making by the sequential proxy optimization technique: SPOT, European Journal of Operational Research, 9, pp. 386-396.

Sakawa, M. (1982b). Interactive multiobjective optimization by the sequential proxy optimization technique (SPOT), IEEE Transactions on Reliability, R-31, pp. 461-464.

Sakawa, M. (1983a). Interactive computer programs for fuzzy linear programming with multiple objectives, International Journal of Man-Machine Studies, 18, pp. 489-503.

Sakawa, M. (1983b). Interactive fuzzy decision making for multiobjective linear programming and its application, Proceedings of IFAC Symposium on Fuzzy Information, Knowledge Representation and Decision Analysis, (ed. E. Sanchez), pp. 295-300.

Sakawa, M. (1984a). Interactive multiobjective decision-making by the fuzzy sequential proxy optimization technique: FSPOT, Fuzzy Sets and Decision Analysis, TIMS Studies in the Management Sciences Vol. 20, (eds. H.-J. Zimmermann, L. A. Zadeh, and B. R. Gaines), North-Holland, pp. 241-260.

Sakawa, M. (1984b). Interactive fuzzy decision making for multiobjective nonlinear programming problems, Interactive Decision Analysis, (eds. M. Grauer and A. P. Wierzbicki), Springer-Verlag, pp. 105-112.

Sakawa, M. (1984c). Interactive fuzzy goal programming for multiobjective nonlinear programming problems and its applications to water quality management, Control and Cybernetics, 13, pp. 217-228.

Sakawa, M. (1986). Interactive multiobjective decision-making using fuzzy satisficing intervals and its application to an urban water resources system, Large Scale Systems, 10, pp. 203-213.

Sakawa, M. (1990). Fuzzy combinatorial optimization by neural networks, Proceedings of International Computer Symposium, Taipei, Taiwan, pp. 751-756.

Sakawa, M. and T. Mitani (1991). Fuzzy 0-1 programming through neural computation, Preprints of IFSA'91 Brussels, Computer, Management & Systems Science, pp. 217-220.

Sakawa, M. and N. Mori (1983a). Interactive multiobjective decisionmaking for nonconvex problems based on the penalty scalarizing functions, European Journal of Operational Research, 17, pp. 320-330.

Sakawa, M. and N. Mori (1983b). Interactive multiobjective decision-making for nonconvex problems based on the weighted Tchebycheff norm, Large Scale Systems, 5, pp. 69-82.

Sakawa, M., H. Narazaki, M. Konishi, K. Nose, and T. Morita (1987). A fuzzy satisficing approach to multiobjective pass scheduling for hot tandem mills, Toward Interactive and Intelligent Decision Support Systems, Vol. 1, Proceedings, Kyoto, Japan (eds. Y. Sawaragi, K. Inoue, and H. Nakayama), Springer-Verlag, pp. 363-373.

Sakawa, M., H. Narazaki, K. Nose, and M. Konishi (1987). An interactive fuzzy satisficing method for multiobjective pass design in hot rolling, Preprints of Second IFSA Congress, Vol. 2, Tokyo, Japan, pp. 785-788.

Sakawa, M. and F. Seo, (1980). Interactive multiobjective decisionmaking for large-scale systems and its application to environmental systems, IEEE Transactions on Systems, Man, and Cybernetics, SMC-10, pp. 796-806.

Sakawa, M. and F. Seo, (1982). Interactive multiobjective decision making in environmental systems using sequential proxy optimization techniques (SPOT), Automatica, 18, pp. 155-165.

Sakawa, M. and F. Seo (1983). Interactive multiobjective decision-making in environmental systems using the fuzzy sequential proxy optimization technique, Large Scale Systems, 4, pp. 223-243.

Sakawa, M. and H. Yano (1984a). An interactive fuzzy satisficing method using penalty scalarizing problems, Proceedings of International Computer Symposium (ICS'84), Tamkang University, Taiwan, pp. 1122-1129.

Sakawa, M. and H. Yano (1984b). An interactive goal attainment method for multiobjective nonconvex problems, Cybernetics and Systems Research 2, (ed. R. Trappl) North-Holland.

Sakawa, M. and H. Yano (1985a). Interactive multiobjective reliability design of a standby system by the fuzzy sequential proxy optimization technique (FSPOT), International Journal of Systems Science, 16, pp. 177-195.

Sakawa, M. and H. Yano (1985b). Interactive fuzzy decision-making for multi-objective nonlinear programming using reference membership intervals, International Journal of Man-Machine Studies, 23, pp. 407-421.

Sakawa, M. and H. Yano (1985c). Interactive decision making for multiobjective linear fractional programming problems with fuzzy parameters, Cybernetics and Systems: An International Journal, 16, pp. 377-394.

Sakawa, M. and H. Yano (1985d). An interactive fuzzy satisficing method using augmented minimax problems and its application to environmental systems, IEEE Transactions on Systems, Man, and Cybernetics, SMC-15, pp. 720-729.

Sakawa, M. and H. Yano (1986a). An interactive fuzzy decisionmaking method using constraint problems, IEEE Transactions on Systems, Man, and Cybernetics, SMC-16, pp. 179-182.

Sakawa, M. and H. Yano (1986b). An interactive fuzzy satisficing method using constraint problems and its application to regional planning, Kybernetes, 15, pp. 121-129.

Sakawa, M. and H. Yano (1986c). An interactive fuzzy satisficing method for multiobjective nonlinear programming problems with fuzzy parameters, CP-86-15, International Institute for Applied Systems Analysis, Laxenburg, Austria.

Sakawa, M. and H. Yano (1986d). An interactive fuzzy satisficing method for multiobjective linear programming problems with fuzzy parameters, Proceedings of the IFAC/IFORS Symposium, Large Scale Systems: Theory and Applications, Zurich, Switzerland.

Sakawa, M. and H. Yano (1986e). An interactive method for multiobjective nonlinear programming problems with fuzzy parameters, Cybernetics and Systems '86, (ed. R. Trappl), D. Reidel Publishing Company, Dordrecht, pp. 607-614.

Sakawa, M. and H. Yano (1986f). Interactive fuzzy decision making for multiobjective nonlinear programming using augmented minimax problems, Fuzzy Sets and Systems, 20, pp. 31-43.

Sakawa, M. and H. Yano (1986g). Interactive decision making for multiobjective linear programming problems with fuzzy parameters, Large-Scale Modeling and Interactive Decision Analysis, (eds. G. Fandel, M. Grauer, A. Kurzhanski, and A. P.Wierzbicki), Springer-Verlag, pp. 88-96.

Sakawa, M. and H. Yano (1986h). Personal computer-aided interactive decision making for multiobjective linear fractional programming problems with fuzzy parameters, Proceedings of International Computer Symposium (ICS'86). National Cheng-Kung University, Taiwan, pp. 1868-1876.

Sakawa, M. and H. Yano (1987a). Interactive fuzzy decisionmaking for multiobjective nonlinear programming problems with fuzzy parameters, Proceedings of 10th World Congress on Automatic Control, Vol. 7, Munich, FRG, pp. 408-413.

Sakawa, M. and H. Yano (1987b). An interactive fuzzy satisficing method for multiobjective linear fractional programming problems with fuzzy parameters, Towards Interactive and Intelligent Decision Support Systems, Vol. 2, Proceedings, Kyoto, Japan (eds. Y. Sawaragi, K. Inoue, and H. Nakayama), Springer-Verlag, pp. 338-347.

Sakawa, M. and H. Yano (1987c). Interactive fuzzy decision making for generalized multiobjective linear programming problems with fuzzy parameters, Preprints of Second IFSA Congress, Vol. 1, Tokyo, Japan, pp. 191-194.

Sakawa, M. and H. Yano (1987d). An interactive satisficing method for multiobjective nonlinear programming problems with fuzzy parameters, Optimization Models Using Fuzzy Sets and Possibility Theory, (eds. J. Kacprzyk and S. A. Orlovski), D. Reidel Publishing Company, Dordrecht, pp. 258-271.

Sakawa, M. and H. Yano (1988a). An interactive algorithm for multiobjective linear programming problems with fuzzy coefficients, Cybernetics and Systems'88 (ed. R. Trappl), D. Reidel Publishing Company, Dordrecht, pp. 693-700.

Sakawa, M. and H. Yano (l988b). An interactive fuzzy satisficing method for multiobjective linear fractional programming problems, Fuzzy Sets and Systems, 28, pp. 129-144.

Sakawa, M. and H. Yano (1989a). Interactive decision making for multiobjective nonlinear programming problems with fuzzy parameters, Fuzzy Sets and Systems, 29, pp. 325-326.

Sakawa, M. and H. Yano (1989b). An interactive fuzzy satisficing method for multiobjective nonlinear programming problems with fuzzy parameters, Fuzzy Sets and Systems, 30, pp. 221-238.

Sakawa, M. and H. Yano (1989c). Fuzzy linear regression and its application to the sales forecasting, International Journal on Policy and Information, 13, pp. 111-125.

Sakawa, M. and H. Yano (1989d). Personal computer-aided interactive decision making for multiobjective linear programming problems with fuzzy parameters and its applications, Improving Decision Making in Organisations (eds. A. G. Lockett and G. Islei), Springer-Verlag, Berlin, pp. 462-471.

Sakawa, M. and H. Yano (1990a). Feasibility and Pareto optimality for multiobjective linear and linear fractional programming problems with fuzzy parameters, The Interface between Artificial Intelligence and Operations Research in Fuzzy Environment, (eds. J. L. Verdegay and M. Delgado), Verlag TUV Rheinland, pp. 213-232.

Sakawa, M. and H. Yano (1990b). An interactive fuzzy satisficing method for generalized multiobjective linear programming problems with fuzzy parameters, Fuzzy Sets and Systems, 35, pp. 125-142.

Sakawa, M. and H. Yano (1990c). Trade-off rates in the hyperplane method for multiobjective optimization problems, European Journal of Operational Research, 44, pp. 105-118.

Sakawa, M. and H. Yano (1990d). Feasibility and Pareto optimality for multiobjective linear programming problems with fuzzy decision variables and fuzzy parameters, Cybernetics and Systems '90, (ed. R. Trappl), Kluwer Academic Publishers, pp. 155-162.

Sakawa, M. and H. Yano (1990e). Combinatorial optimization by neural networks, Proceedings of the International Conference on Fuzzy Logic & Neural Networks, Iizuka, Japan, pp. 783-786.

Sakawa, M. and H. Yano (1990f). Interactive decision making for multiobjective linear fractional programming problems with fuzzy parameters based on solution concept incorporating fuzzy goals, Proceedings of the International Conference on Fuzzy Logic & Neural Networks, Iizuka, Japan, pp. 821-824.

Sakawa, M. and H. Yano (1990g). Interactive decision-making for multiobjective programming problems with fuzzy parameters, Stochastic versus Fuzzy Approaches to Multiobjective Mathematical Programming Problems under Uncertainty, (eds. R. Slowinski and J. Teghem), Kluwer Academic Publishers, pp. 191-227

Sakawa, M. and H. Yano (1991a). Pareto optimality for multiobjective linear fractional programming problems with fuzzy decision variables and fuzzy parameters, Preprints of IFSA'91 Brussels, Computer, Management & Systems Science, pp. 229-232.

Sakawa, M. and H. Yano (1991b). Feasibility and Pareto optimality for multiobjective nonlinear programming problems with fuzzy parameters, Fuzzy Sets and Systems, 43, pp. 1-15.

Sakawa, M. and H. Yano (1992a). Fuzzy linear regression analysis for fuzzy input-output data, Information Sciences, 63, pp. 191-206.

Sakawa, M. and H. Yano (1992b). Multiobjective fuzzy linear regression analysis for fuzzy input-output data, Fuzzy Sets and Systems, 47, pp. 173-181.

Sakawa, M. and H. Yano (1992c). Fuzzy linear regression and its applications, Fuzzy Regression Analysis, (eds. J. Kacprzyk and M. Fedrizzi), Omnitech Press, Warsaw and Physica-Verlag, Heidelberg, pp. 61-80.

Sakawa, M. and H. Yano (1992d). Generalized hyperplane methods for charactering Λ-extreme points and trade-off rates for multiobjective optimization problems, European Journal of Operational Research, 57, pp. 368-380.

Sakawa, M., H. Yano and J. Takahashi (1992). Pareto optimality for multiobjective linear fractional programming problems with fuzzy parameters, Information Sciences, 63, pp. 33-53.

Sakawa, M., H. Yano, and T. Yumine (1987). An interactive fuzzy satisficing method for multiobjective linear programming problems and its application, IEEE Transactions on Systems, Man, and Cybernetics, SMC-17, pp. 654-661.

Sakawa, M. and T. Yumine (1983). Interactive fuzzy decision-making for multiobjective linear fractional programming problems, Large Scale Systems, 5, pp. 105-114.

Sakawa, M., T. Yumine, and H. Yano (1984). An interactive fuzzy satisficing method for multiobjective nonlinear programming problems, CP-84-18, International Institute for Applied Systems Analysis, Laxenburg Austria.

Sakawa, M., T. Yumine, and H. Yano (1987). An interactive fuzzy satisficing method for multiobjective nonlinear programming problems, Analysis of Fuzzy Information, Volume III, (ed. J. C. Bezdek), CRC Press. Inc., Florida, pp. 273-283.

Seo, F. and M. Sakawa, (1988). Multiple Criteria Decision Analysis in Regional Planning: Concepts, Methods and Applications, D. Reidel Publishing Company, Dordrecht.

Slowinski, R. (1986). A multicriteria fuzzy linear programming method for water supply system development planning, Fuzzy Sets and Systems, 19, pp. 217-237.

Slowinski, R. (1987). An interactive method for multiobjective linear programming with fuzzy parameters and its application to water supply planning, Optimization Models Using Fuzzy Sets and Possibility Theory, (eds. J. Kacprzyk and S. A. Orlovski), D. Reidel Publishing Company, Dordrecht, pp. 396-414.

Slowinski, R. (1990). "FLIP": an interactive method for multiobjective linear programming with fuzzy coefficients, Stochastic versus Fuzzy Approaches to Multiobjective Mathematical Programming Problems under Uncertainty, (eds. R. Slowinski and J. Teghem), Kluwer Academic Publishers, pp. 249-262.

Slowinski, R. and J. Teghem (1988). Fuzzy versus stochastic approaches to multicriteria linear programming under uncertainty, Naval Research Logistics, 35, pp. 673-695.

Slowinski, R. and J. Teghem (1990). Comparison study of "STRANGE" and "FLIP," Stochastic versus Fuzzy Approaches to Multiobjective Mathematical Programming Problems under Uncertainty, (eds. R. Slowinski and J. Teghem), Kluwer Academic Publishers, pp. 365-396.

Slowinski, R. and J. Teghem (eds.) (1990). Stochastic versus Fuzzy Approaches to Multiobjective Mathematical Programming Problems under Uncertainty, Kluwer Academic Publishers, Dordrecht.

Sommer, G. and M. A. Pollastschek (1978). A fuzzy programming approach to an air pollution regulation problem, Progress in Cybernetics and Systems Research, (eds. R. Trappl, G.J. Klir and L. Ricciardi), pp. 303-323.

Stadler, W. (1979). A survey of multicriteria optimization or the vector maximum problem, Part 1: 1776-1960, Journal of Optimization Theory and Applications, 29, pp. 1-52.

Stadler, W. (1983a). Stability of the natural shapes of sinusoidally loaded uniform shallow arches, Quarterly Journal of Mechanics and Applied Mathematics, 36, pp. 365-386.

Stadler, W. (1983b). Instability of optimal equilibrium in the minimum mass design of uniform shallow arches, Journal of Optimization Theory and Applications, 41, pp. 299-316.

Stadler, W. (ed.) (1988). Multicriteria Optimization in Engineering and in the Sciences, Plenum Press, New York.

Statistical Office of Osaka Prefecture, Osaka, Japan (1975). Industrial Statistics Survey Result Table of Industrial Statistics in 1975 (in Japanese).

Steuer, R. E. (1977). An interactive multiple objective linear programming procedure, Multiple Criteria Decision Making, (eds. M. K. Starr and M. Zeleny), North-Holland, pp. 225-239.

Steuer, R. E. (1986). Multiple Criterion Optimization: Theory, Computation, and Application, John Wiley & Sons, New York.

Steuer, R. E. and E. U. Choo (1983). An interactive weighted Tchebycheff procedure for multiple objective programming, Mathematical Programming, 26, pp. 326-344.

Takeda, M. and J. W. Goodman (1986). Neural networks for computation: number representations and programming complexity, Applied Optics, 25, pp. 3033-3046.

Tanaka, H. (1987). Fuzzy data analysis by possibilistic linear models, Fuzzy Sets and Systems, 24, pp. 363-375.

Tanaka, H. (1988). Possibilistic linear systems and their application to the linear regression models, Fuzzy Sets and Systems, 27, pp. 257-289.

Tanaka, H. and K. Asai (1984a). Fuzzy solution in fuzzy linear programming problems, IEEE Transactions on Systems, Man, and Cybernetics, SMC-14, pp. 325-328.

Tanaka, H. and K. Asai (1984b). Fuzzy linear programming problems with fuzzy numbers, Fuzzy Sets and Systems, 13, pp. 1-10.

Tanaka, H., M. Ichihashi, and K. Asai (1984). A formulation of fuzzy linear programming problem based on comparison of fuzzy numbers, Control and Cybernetics, 13, pp. 185-194.

Tanaka, H., I. Hayashi, and J. Watada (1989). Possibilistic linear regression for fuzzy data, European Journal of Operational Research, 39, pp. 1-8.

Tanaka, H., T. Okuda, and K. Asai (1974). On fuzzy mathematical programming, Journal of Cybernetics, 3,4, pp. 37-46.

Tanaka, H., S. Uejima, and K. Asai (1982). Linear regression analysis with fuzzy model, IEEE Transactions on Systems, Man, and Cybernetics, SMC-12, pp. 903-907.

The Iron and Steel Institute of Japan (1984). Theory and Practice of Flat Rolling (in Japanese).

Verdegay, J. L. (1984). A dual approach to solve the fuzzy linear programming problem, Fuzzy Sets and Systems, 14, pp. 131-141.

Verdegay, J. L. and M. Delgado (eds.) (1990). The Interface between Artificial Intelligence and Operations Research in Fuzzy Environment, Verlag TUV Rheinland, Koln, pp.213-232.

Weistroffer, H. R. (1982). Multiple criteria decision making with interactive overachievement programming, Operations Research Letters, 1, pp. 241-245.

Weistroffer, H. R. (1983). An interactive goal-programming method for nonlinear multiple-criteria decision-making problems, Computer & Operations Research, 10, pp. 311-320.

Weistroffer, H. R. (1984). A combined over- and under-achievement programming approach to multiple objectives decision making, Large Scale Systems, 7, pp. 47-58.

Wiedey, G. and H.-J. Zimmermann (1978). Media selection and fuzzy linear programming, Journal of Operational Research Society, 29, pp. 1071-1084.

Wierzbicki, A. P. (1977). Basic properties of scalarizing functionals for multiobjective optimization, Forshung und Statistik, Ser. Optimization, 8, pp. 55-60.

Wierzbicki, A. P. (1979a). The use of reference objectives in multiobjective optimization - Theoretical implications and practical experiences, WP-79-66, International Institute for Applied Systems Analysis, Laxenburg, Austria.

Wierzbicki, A. P. (1979b). A methodological guide to multiobjective optimization, WP-79-122, International Institute for Applied Systems Analysis, Laxenburg, Austria.

Wierzbicki, A. P. (1980). A mathematical basis for satisficing decision making, WP-80-90, International Institute for Applied Systems Analysis, Laxenburg, Austria.

Wierzbicki, A. P. (1982). A mathematical basis for satisficing decision making, Mathematical Modeling, 3, pp. 391-405.

Yager, R. R., S. Ovchinnikov, R. M. Tong, and H. T. Nguyen (eds.) (1987). Fuzzy Set and Applications, Selected Papers by Zadeh, Wiley, New York.

Yano, H. and M. Sakawa (1987). Trade-off rates in the weighted Tchebycheff norm method, Large Scale Systems, 13, pp. 167-177.

Yano, H. and M. Sakawa (1989a). A unified approach for characterizing Pareto optimal solutions of multiobjective optimization problems: the hyperplane method, European Journal of Operational Research, 39, pp. 61-70.

Yano, H. and M. Sakawa (1989b). Interactive fuzzy decision making for generalized multiobjective linear fractional programming problems with fuzzy parameters, Fuzzy Sets and Systems, 32, pp. 245-261.

Yu, P. L. (1973). A class of solutions for group decision problems, Management Science, 19, pp. 936-946.

Yu, P. L. and G. Leitmann (1974). Compromise solutions, domination structures, and Salukvadze's solution, Journal of Optimization Theory and Applications, 13, pp. 362-378.

Zadeh, L. A. (1963). Optimality and non-scalar valued performance criteria, IEEE Transactions on Automatic Control, AC-8, pp. 59-60.

Zadeh, L. A. (1965). Fuzzy sets, Information and Control, 8, pp. 338-353.

Zadeh, L. A. (1975). The concept of a linguistic variable and its application to approximate reasoning, Information Science, Part 1, 8, pp. 199-249; Part 2, 8, pp. 301-359; Part 3, 9, pp. 43-80.

Zadeh, L. A. (1978). Fuzzy sets as a basis for a theory of possibility, Fuzzy Sets and Systems, 1, pp. 3-28.

Zangwill, W. I. (1969). Nonlinear Programming: A Unified Approach, Prentice-Hall, Englewood Cliffs.

Zeleny, M. (1973). Compromise Programming, Multiple Criteria Decision Making, (eds. J. L. Chochrane and M. Zeleny), University of South Carolina Press, pp. 262-301.

Zeleny, M. (1976). The theory of displaced ideal, Multiple Criteria Decision Making, Kyoto, 1975, (ed. M. Zeleny), Springer-Verlag, pp. 153-206.

Zeleny, M. (1982). Multiple Criteria Decision Making, McGraw-Hill, New York.

Zimmermann, H.-J. (1976). Description and optimization of fuzzy systems, International Journal of General Systems, 2, pp. 209-215.

Zimmermann, H.-J. (1978). Fuzzy programming and linear programming with several objective functions, Fuzzy Sets and Systems, 1, pp. 45-55.

Zimmermann, H.-J. (1983). Fuzzy mathematical programming, Computer & Operations Research, 10, pp. 291-298.

Zimmermann, H.-J. (1985). Fuzzy Set Theory and Its Application, Second Edition (1991), Kluwer Academic Publishers, Dordrecht.

Zimmermann, H.-J. (1987). Fuzzy Set, Decision Making and Expert Systems, Kluwer-Nijhoff Publishing, Dordrecht.

Zimmermann, H.-J., B. R. Gaines, and L. A. Zadeh (eds.) (1984). Fuzzy Sets and Decision Analysis, TIMS Studies in the Management Sciences, 12, North-Holland, Amsterdam.

Zimmermann, H.-J. and M. A. Pollatschek (1984). Fuzzy 0-1 linear programming, Fuzzy Sets and Decision Analysis, TIMS Studies in the Management Sciences Vol. 20, (eds. H.-J. Zimmermann, L. A. Zadeh, and B. R. Gaines), North-Holland, Amsterdam, pp. 133-145.

Zionts, S. and J. Wallenius (1976). An interactive programming method for solving the multiple criteria problem, Management Science, 22, pp. 652-663.

Zionts, S. and J. Wallenius (1983). An interactive multiple objective linear programming method for a class of underlying nonlinear utility functions, Management Science, 29, pp. 519-529.

Books on Fuzzy Sets and Multiobjective Optimization

Bezdek, J. C. (1981). Pattern Recognition with Fuzzy Objective Function Algorithms, Plenum Press, New York.

Carlsson, C. (1980). Fuzzy Set Theory for Management Decisions, Interdisciplinary Systems Research Series, Vol. 82, Verlag TUV, Rheinland, Koln.

Chankong, V., and Y. Y. Haimes (1983). Multiobjective Decision Making: Theory and Methodology, North-Holland, Amsterdam.

Charnes, A., and W. W. Cooper (1961). Management Models and Industrial Applications of Linear Programming, Vols. I and II, Wiley, New York.

Cohon, J. L. (1978). Multiobjective Programming and Planning, Academic Press, New York.

Dubois, D. and H. Prade (1980). Fuzzy Sets and Systems, Theory and Applications, Academic Press, New York.

Dubois, D. and H. Prade (1988). Possibility Theory: an Approach to Computerized Processing of Uncertainty, Plenum Press, London.

Goicoechea, A., D. R. Hansen, and L. Duckstein (1982). Multiobjective Decision Analysis with Engineering and Business Applications, John Wiley & Sons, New York.

Guddat, J., F. G. Vasquez, K. Tammer, and K. Wendler (1985). Multiobjective and Stochastic Optimization Based on Parametric Optimization, Akademic-Verlag, Berlin.

Haimes, Y. Y. (1977). Hierarchical Analyses of Water Resources Systems: Modeling and Optimization of Large-scale Systems, McGraw-Hill, New York.

Haimes, Y. Y., W. A. Hall, and H. T. Freedman (1975). Multiobjective Optimization in Water Resources Systems: The Surrogate Worth Trade-off Method, Elsevier, Amsterdam.

Haimes, Y. Y., K. Tarvainen, T. Shima, and J. Thadathil (1990). Hierarchical Multiobjective Analysis of Large-Scale Systems, Hemisphere, New York.

Hwang, C. L. and A. S. Md. Masud (1979). Multiple Objective Decision-Making –Methods and Applications, Springer-Verlag, Berlin.

Ignizio, J. P. (1976). Goal Programming and Extensions, Lexington Books, D. C. Heath and Company, Lexington.

Ignizio, J. P. (1982). Linear Programming in Single and Multiple Objective Systems, Prentice-Hall, Englewood Cliffs, New Jersey.

Ijiri, Y. (1965). Management Goals and Accounting for Control, North-Holland, Amsterdam.

Kacprzyk, J. (1983). Multistage Decision Making under Fuzziness, Interdisciplinary Systems Research Series, Vol. 79, Verlag TUV, Rheinland, Koln.

Kandel, A. (1982). Fuzzy Techniques in Pattern Recognition, John Wiley & Sons, New York.

Kandel, A. (1985). Fuzzy Mathematical Techniques with Applications, Addison Wesley, California.

Kaufmann, A. (1975). Introduction to the Theory of Fuzzy Subsets, Vol. 1, Academic Press, New York.

Kaufmann, A. and M. M. Gupta (1985). An Introduction to Fuzzy Arithmetics, Van Nostrand Reinhold Company, New York.

Kickert, W. J. U. (1978). Fuzzy Theories on Decision-making, Martinus, Nijhoff, Leiden.

Klir, G. J. and T. A. Folger (1988). Fuzzy Sets, Uncertainty and Information, Prentice-Hall, Englewood Cliffs.

Lee, S. M. (1972). Goal Programming for Decision Analysis, Auerbach, Philadelphia.

Negoita, C. V. (1982). Fuzzy Systems, Abacus Press, Tunbridge Wells.

Negoita, C. V. and D. A. Ralescu (1975). Applications of Fuzzy Sets to Systems Analysis, Birkhauser, Basel.

Nijkamp, P. (1977). Theory and Application of Environmental Economics, North-Holland, Amsterdam.

Nijkamp, P. (1979). Multidimensional spatial Data and Decision Analysis, Wiley, New York.

Nijkamp, P. (1980). Environmental Policy Analysis, Operational Methods and Models, Wiley, New York.

Novak, V. (1989). Fuzzy Sets and their Applications, Adam Hilger, Bristol.

Pedrycz, W. (1989). Fuzzy Control and Fuzzy Systems, John Wiley & Sons, New York.

Rietveld, P. (1980). Multiple Objective Decision Methods and Regional Planning, North-Holland, Amsterdam.

Osyczka, A. (1984). Multicriterion Optimization in Engineering with FORTRAN Programs, Ellis Horwood Limited, Chichester.

Roy, B. (1985). Méthodologie Multicritère d'Aide à la Décision, Econcmica, Paris.

Sawaragi, Y., H. Nakayama, and T. Tanino (1985). Theory of Multiobjective Optimizations, Academic Press, New York.

Seo, F. and M. Sakawa (1988). Multiple Criteria Decision Analysis in Regional Planning: Concepts, Methods and Applications, D. Reidel Publishing Company, Dordrecht.

Spronk, J. (1981). Interactive Multiple Goal Programming : Applications to Financial Planning, Martinus Nijhoff Publishing, Boston.

Stancu-Minasian, I. W. (1984). Stochastic Programming with multiple objective Functions (translated from Rumanian by V. Giurgiutiu), D. Reidel Publishing Company, Dordrecht.

Steuer, R. E. (1986). Multiple Criteria Optimization: Theory, Computation, and Application, John Wiley & Sons, New York.

Szidarovsky, F., W. E. Gershou, and L. Duckstein (1986). Techniques for Multiobjective Decision Making in Systems Management, Elsevier, Amsterdam.

White, D. J. (1982). Optimality and Efficiency, John Wiley & Sons, New York.

Yu, P. L. (1985). Multiple-Criteria Decision Making: Concepts, Techniques, and Extensions, Plenum Press, New York.

Zimmermann, H.-J. (1985). Fuzzy Set Theory and Its Application, Second Edition (1991), Kluwer Academic Publishers, Dordrecht.

Zimmermann, H.-J. (1987). Fuzzy Set, Decision Making and Expert Systems, Kluwer-Nijhoff Publishing, Dordrecht.

Zeleny, M. (1974). Linear Multiobjective Programming, Springer-Verlag, New York.

Zeleny, M. (1982). Multiple Criteria Decision Making, McGraw-Hill, New York.

INDEX

E

F

G

H

I